SCIENCE AND ITS CONCEPTUAL FOUNDATIONS
David L. Hull, *Editor*

QUANTUM MECHANICS

Historical
Contingency
and the
Copenhagen
Hegemony

JAMES T. CUSHING

THE UNIVERSITY OF CHICAGO PRESS
Chicago & London

James T. Cushing is a professor in the Department of Physics and the Department of Philosophy, University of Notre Dame.

The University of Chicago Press, Chicago 60637
The University of Chicago Press, Ltd., London
© 1994 by The University of Chicago
All rights reserved. Published 1994
Printed in the United States of America
03 02 01 00 99 98 97 96 95 94   1 2 3 4 5
ISBN: 0-226-13202-1 (cloth)
         0-226-13204-8 (paper)

Library of Congress Cataloging-in-Publication Data
Cushing, James T., 1937–
   Quantum mechanics : historical contingency and the Copenhagen hegemony / James T. Cushing.
       p. cm. — (Science and its conceptual foundations)
   Includes bibliographical references and index.
   1. Quantum theory—History.   I. Title. II. Series.
QC173.98.C87   1994   94–8427
530.1′2′09—dc20   CIP

⊗ The paper used in this publication meets the minimum requirements of the American National Standard for Information Sciences—Permanence of Paper for Printed Library Materials, ANSI Z39.48-1984.

FOR TWO SPECIAL PEOPLE
Nimbilasha and Ernan
AND
TO THE MEMORY OF
David Bohm

CONTENTS

Preface	xi
Acknowledgments	xv

ONE
Theory Construction and Selection	1
1.1 A naturalized epistemology	1
1.2 Theory choice—*rational,* but not *unique*	5

TWO
Formalism, Interpretation, and Understanding	9
2.1 The components of a scientific theory	9
2.2 Explanation versus understanding	10
2.3 Examples from physics	12
2.4 Some attempts at understanding quantum phenomena	16
2.5 Contingently necessary conditions for understanding	18
Appendix A derivation of the EPRB correlations	22

THREE
Standard Quantum Theory	24
3.1 Einstein versus Bohr	24
3.2 The formalism	26
3.3 The interpretation	27
3.4 Complementarity/wave-particle duality	32
3.5 Perennial problems	34
Appendix A specific illustration of the measurement problem	40

FOUR
Bohm's Quantum Theory	42
4.1 Bohm's "minimalist" quantum potential theory	42
4.2 New insights	47

viii   Contents

4.3	Is there *complete* observational equivalence?	53
4.4	Various types of nonlocality	56
	Appendix 1 Some technical details of Bohm's program	60
	Appendix 2 Quantum tunneling times	72

FIVE
Alternative Interpretations: An Illustration   76
5.1   The value of the exercise   76
5.2   Neutron interferometry experiments   78
5.3   The "story" told by Bohm   82
      Appendix 1 The quantum mechanics of spin flipping   85
      Appendix 2 A causal interpretation of the Pauli equation   91

SIX
Opposing Commitments, Opposing Schools   96
6.1   Cultural milieux circa 1900–1925   96
6.2   The Forman thesis   97
6.3   Predilections—not uniquely/solely determining   101
6.4   The wave-mechanics route   103
6.5   The matrix-mechanics route   107

SEVEN
Competition and Forging Copenhagen   113
7.1   The challenge of wave mechanics   113
7.2   Copenhagen succeeds   118
      Appendix Pauli's 1927 criticism of de Broglie's theory   122

EIGHT
Early Attempts at Causal Theories: A Stillborn Program   124
8.1   Madelung   125
8.2   De Broglie   126
8.3   Einstein   128
8.4   Kennard, Rosen, Fürth   129
8.5   Von Neumann's "proof"   131
      Appendix 1 Madelung's derivation   135
      Appendix 2 De Broglie's guidance argument   136
      Appendix 3 Einstein's 1927 hidden-variables theory   139
      Appendix 4 Von Neumann's unwarranted assumption   140

NINE
The Fate of Bohm's Program   144
9.1   The initial reception of Bohm's 1952 paper   144

9.2	Continued hostility and general disinterest	152
9.3	Subsequent work	158
9.4	Some recent developments (since 1980)	162
	Appendix 1 Nelson's equations	169
	Appendix 2 Valentini's H-theorem	171

TEN
An Alternative Scenario?   174
- 10.1 What if in 1927 . . . ?   174
- 10.2 A "Bell" theorem   176
- 10.3 Einstein and nonlocality   179
- 10.4 Fertility and growth   186
    - Appendix 1 A simple derivation of Bell's theorem   193
    - Appendix 2 The no-signaling theorem   195
    - Appendix 3 Scalar causal quantum field theory   196

ELEVEN
Lessons   199
- 11.1 The underdetermination thesis   199
- 11.2 Contingency and scientific theories   203
- 11.3 Determinism and indeterminism—equivalent?   207
- 11.4 So, finally . . .   214

Notes   217

References   273

Author Index   301

Subject Index   307

PREFACE

The central theme of this book is that historical contingency plays an essential and ineliminable role in the construction and selection of a successful scientific theory from among its observationally equivalent and unrefuted competitors.[1] I argue that historical contingency, in the sense of the *order* in which events take place, can be an essential factor in determining which of two empirically adequate and fruitful, but observationally equivalent, scientific theories is accepted by the scientific community. This type of actual underdetermination poses questions for scientific realism and for rational reconstruction in theory evaluation. To illustrate this, I examine the possible observational equivalence of two radically different, conceptually incompatible interpretations of quantum mechanics (both based on a common mathematical structure) and contend that an entirely plausible reordering of historical factors could reasonably have resulted in the causal program having been chosen over the Copenhagen one.

My purpose is to foster discussion of the issues raised, not to make pronouncements on them. To pose a series of questions and to discuss a set of possibilities in a fairly focused and concrete form, I do present *a* point of view and I am certain that dissenting responses will follow. I do not claim to *know* the answer to several of these questions (e.g., whether or not there is actual underdetermination between the two theories featured in this book). I have no final word to offer on Bohm's program, since many aspects of it remain to be developed. It is precisely for these reasons that I find the topic interesting. While I do attempt to give some reasonable background to and representation of the standard, or Copenhagen, version of quantum mechanics, I do not devote as much space to it as to Bohm's alternative theory. One reason for this is that standard quantum mechanics has been well represented, both in the technical and in the philosophical literature, as well as in science textbooks, while the causal quantum theory program either is entirely unknown to most scientists and philosophers concerned with foundational problems in quantum mechanics or has been badly presented to them.[2] Parity of space and of analysis of the two hardly seems warranted under the circumstances. It is astounding that there is a formulation of quantum mechanics that has no measurement problem and no difficulty with a classical limit, yet is so little known. One might suspect that it is a *historical* problem to explain its marginal status.[3]

Let me enter a few caveats at the outset. I do not claim that the radical type of underdetermination present in the episode studied here obtains in practice in *all* cases of theory selection.[4] Moreover, I am not concerned with the rather trivial and philosophically uninteresting historical contingency of who did what when. For example, in response to Napoléon's fishing for a favorable comparison between himself and Newton, Lagrange is said to have lamented: "Newton was the greatest genius that ever existed, and the most fortunate, for we cannot find more than once a system of the world to establish."[5] Newton had *discovered* the fundamental laws of mechanics and of gravitation, and there was nothing of comparable significance left for anyone else to accomplish in that area. So, good show for Newton and hard luck for Lagrange. Of course, someone else *might* have discovered these same laws. Then, the laws would bear another's name, but they would be the same laws. This is surely historical contingency, but of a benign and philosophically uninteresting variety. Another type of contingency that is important, especially in the "historical" sciences such as evolutionary biology, concerns the profound, long-term, and really irreversible effects on, say, the future and the very existence of a species that essentially random events produce.[6] Even if we were to accept as unproblematic certain laws or generalizations of science (e.g., a "theory" or concept of evolution via natural selection), contingent factors (such as a chance occurrence that annihilates one species) may remain of overwhelming importance for the subsequent evolution of some species. Again, though, this is not the type of contingency I study in this book. Nor is 'contingency' in the sense of (other) possible worlds having a fundamentally different objective structure of interest to me here. Rather, I discuss the acceptance and rejection of observationally equivalent, alternative, and, indeed, *incompatible* descriptions or theories of our *actual* world. For this purpose, I need not enter into the argument about the existence of *necessary* laws, as opposed to mere descriptions of contingent regularities.[7] After all, there *are* laws and theories (whatever their metalevel epistemic status may be) that are accepted by the scientific community, and I focus on how successful scientific theories come to be accepted.

Specifically, if certain equally plausible conditions, rather than the actually occurring and highly contingent historical ones, had prevailed and the interpretation of quantum mechanics had initially taken a very different route from the Copenhagen one around 1925–1927, would our worldview of fundamental microprocesses *necessarily* have been brought back, by the "internal" logic of science, to our currently accepted picture of an inherently and irreducibly indeterministic nature?[8] Could our present understanding of the behavior of the fundamental laws of nature in terms of an inherently indeterministic physics have been replaced by the apparently diametrically opposed view of absolute determinism? This book argues that the answer to the first question is no and to the second an emphatic yes. This is not to deny that there were already serious conceptual problems for classical physics, even in its own domain of applicability, or to imply that the scientific community did not have good reasons (although not uniquely compelling ones) for making the choices it did during the quantum revolution. I do not charge that scientists acted irrationally in selecting one theory over another. Nor do I believe that an alternative choice would have left us without foundational problems to resolve. The set would be different, though, and a choice between them is up to the reader.

My presentation does not proceed in strict historical sequence. In a sense, I back into my story. I first review the present situation with regard to rival versions of quantum mechanics and show that *today* the standard (or Copenhagen) and the causal (or Bohm) theories of quantum phenomena are both viable, being observationally equivalent (whenever both see a well-formulated question) and logically consistent, even though they are conceptually incompatible with each other. Such considerations are necessary to block any claim that a causal interpretation is incoherent so that the (clever) founders of quantum mechanics would (surely) have spotted that flaw and hence not bothered pursuing such an interpretation. Therefore, the argument will go, the choice in favor of Copenhagen and indeterminism circa 1927 could not have been uniquely demanded by logic and by evidential criteria *alone*. So, how did we arrive at the nearly universally held position that the Copenhagen interpretation—or something fairly close to it—is the *only* acceptable possibility? If one discusses just the present situation as it now stands without asking how we arrived there, a charge of ad hocness is too easily (even if invalidly) raised by an opponent wishing to reject out of hand any interpretation alternative to the accepted, "correct" Copenhagen one. Once I have exposed the essential role of the determinative contingencies that charted our course to the current hegemony, I ask how plausible it is that we might have been brought to a *very* different position had there been a reordering in the temporal sequence of a few key historical events.[9] This book is not intended primarily as an argument in favor of any particular causal interpretation of quantum mechanics over the standard Copenhagen one, but rather one for parity in the consideration of both views. Even if some future development were to tell decisively against, say, Bohm's program, there would still remain the question (with which this book is largely concerned) of just what were the nonevidential grounds on the basis of which causal interpretations were shunted aside in the early days of quantum mechanics.

This is *not* intended to be a technical treatise on David Bohm's theory. Such details can be found elsewhere.[10] Nevertheless, there are several somewhat technical developments presented here since certain claims, such as observational equivalence, do rest on specific mathematical arguments. Enough details are given to indicate how those assertions can be substantiated. I hope that the examples treated at length in some of the appendices will make the book more accessible to students. The reader interested mainly in the historical background and philosophical implications of this case study can simply pass over the equations and the accompanying technical discussions (nearly all of which are confined to appendices and to footnotes) and still see what the overall argument is. The upshot of these mathematical derivations and results is always stated verbally, if often informally, in the text proper. The first two chapters summarize my own views on the epistemic status of our scientific theories and provide, at least for me, a motivation for examining Bohm's alternative to standard quantum mechanics. Many will disagree with my position there and the reader concerned only with the case study itself can simply begin with chapter 3.

Like my previous book on theory construction and selection, this one is intended both for historians and philosophers of science and for scientists interested in foundational issues in quantum mechanics.[11] I have attempted to keep the text accessible to the general reader in any of those areas. Since history, philosophy, and technical physics (but of a fairly elementary variety) are all present, specialists in one or another of these fields will find some of the material quite familiar and elementary. Historians will no doubt judge the history superficial, philosophers the philosophy shallow, and physicists the physics incomplete. My wan hope is that the value of the sum may exceed the sum of the values of the parts. At a minimum, there may be something new and interesting, or at least meaningfully controversial, for all readers.

ACKNOWLEDGMENTS

Preliminary versions of some sections of this book have appeared in previous publications of mine: Cushing, 1990a (section 1.1) (*British Journal for the Philosophy of Science*); Cushing, 1991a (chapter 2) (*Philosophy of Science*); Cushing, 1991b (chapters 6 and 7) and 1992c (chapter 10) (World Scientific Publishing Co.); Cushing, 1992a (section 5.2, chapters 6 and 7) (Plenum Publishing Co.); Cushing, 1992b (section 11.2) (Philosophy of Science Association); Cushing, 1993a (section 11.2) and 1993b (sections 3.4 and 9.4) (Kluwer Academic Publishers). I thank the publishers for allowing me to use this material, often in modified form, in this book.

There are many colleagues whose assistance and criticisms have been of great value to me while I was writing and revising the manuscript of this book. I acknowledge them in notes at appropriate places throughout the text. However, I must recognize in a special way Mara Beller, Arthur Fine, and Don Howard for their exceptional scholarship in areas closely related to this study, for advice, and for encouragement. William Dickson, Sheldon Goldstein, Peter Holland, Don Howard, J. B. Kennedy, Ernan McMullin, John Polkinghorne, Fritz Rohrlich, and Paul Teller generously supplied numerous detailed critical comments on a preliminary draft of this book, as did an anonymous reviewer. Antony Valentini was good enough to send me a copy of his dissertation prior to its publication in book form. None of this, of course, is meant to imply that any of these scholars necessarily agrees with the positions I take here, nor are they to be implicated in any errors in the present work. Simon Capelin provided valuable advice and a useful reference to which I might otherwise not have had access. Gary Bowman and Anthony Lenzo contributed to the final revisions of the manuscript. Finally, I thank Lesley Krueger for typing several drafts of a difficult manuscript and Neal Nash for drawing and lettering the figures in the text.

The National Science Foundation, through its Program in the History and Philosophy of Science, provided partial support for several years while the research for this work was being done (Grants DIR 89 08497 and SBE 91 21476). Any opinions, findings, and conclusions or recommendations expressed in this book are my own and do not necessarily reflect the views of the National Science Foundation.

ONE
Theory Construction and Selection

Before I get into the specifics of a case study, it is only reasonable that I outline my general position on the methodology of science and the status of scientific knowledge. It is within this framework that I subsequently focus on the history of the development of the interpretation of quantum mechanics.

1.1 A naturalized epistemology

It has become common in the philosophy of science to separate scientific practice from the metalevel of the methodological rules used in judging scientific theories and from the metalevel of the goals of science.[1] In a previous book, I argued that any division between scientific practice and a metalevel of the methods and goals of science is largely a false dichotomy.[2] Methodological rules are coupled with and largely derived from scientific practice in such a way that the entire network of practice, methods, and goals evolves together and changes in its essential characteristics.[3] There are normal commonsense demands placed on everyday argument (such as the usual rules of deductive reasoning or some acknowledgment of the importance of facts) and they are important for science as well.[4] Aside from these, though, it is not clear that there are invariant (ahistorical, atemporal) characteristics of science that distinguish it from nonscience.[5] I am not saying that there are no characteristics of scientific practice and reasoning, but only that these need not be universal across all of science for all time unless one makes vague, elastic claims that can be fit to almost anything.[6]

1.1.1 Traditional norms and goals
I consider five requirements that have at times been placed on the philosophy of science and then the difficulties (or impossibilities) that prevent us from satisfying each demand. Starting with the most optimistic, I proceed in a forced retreat to ever less demanding but still unattainable ones. We might hope to establish (1) that science could find empirically adequate,

true theories that we can know to be true and that give us true pictures of the world of a type conducive to scientific realism. This is akin to what has been termed the "Bacon-Descartes ideal."[7] The obvious and today well-known difficulty for this goal is that once-successful scientific theories do change, are overturned, and can be underdetermined by empirical facts. We couldn't know a scientific theory to be true, even if it were. Instrumentalism is one response to this. We might use this Bacon-Descartes goal as one that would be attainable in the best of all possible worlds. We would like this to be an attainable goal, but it simply isn't.

So, adjusting (or downscaling) our demands a bit, we could require (2) that the philosophy of science discover universal characteristics of science, ones good for all times, places, and areas of subject matter. These would include invariant criteria, norms, and methods employed in judging among scientific theories. What is at issue here is not simply a question of an a priori (foundationist or logicist) approach to methodological criteria versus a historicist one. Even if one were to grant that sound, fixed methodological rules cannot be underwritten on the basis of a priori true first principles, he might still hope that such general methodological rules, though ("empirically") discovered from the historical record of science, might remain fixed for all time and apply to all future science. This is certainly a *possibility*. The difficulty for such a position is that case studies show that successful scientific practice, which *does* change, largely defines rationality (i.e., what counts for good reasons or for a good explanation).[8] Closely related to the hope for a fixed set of rules is the requirement (3) that they should be objective and purely rational, in the sense that those rules be independent of social factors. It is again the details of case studies that force us to relinquish such a constraint as unrealistic in actual scientific practice.[9]

Although I have so far concentrated on essential changes in scientific methodology and on sociological factors in the origins of such a methodology, let me mention here two other requirements that have traditionally been assumed to hold for science. We would expect (4) science to provide us with understandable theories as explanations of our world and of the phenomena in it. Quantum mechanics may have done away with this as a possibility for the microworld (see section 2.4). Until very recent times, we have also been accustomed (5) to finding, or being able to construct, theories that are local (obeying the first-signal principle of special relativity or, more loosely, tolerating *no* type of instantaneous action at a distance) or separable (spatially widely distant bodies each having independently specifiable properties) and causal (a specifiable and definite cause or set of causes for every effect). Since we find that we cannot have these, we settle for a theory, standard quantum mechanics, that is "almost" local and "in some sense" causal (i.e., without superluminal transfer of *information*).[10] We make the best compromise we can between our expectations and the constraints of nature. This is just one of the more dramatic examples of a process of accommodation by which we adjust our cannons of rationality to accord with what we can have, not with what we must have.

1.1.2 Attempts at methodological foundations

Just as viable epistemological goals for science have become ever less sweeping as the philosophy of science has paid more attention to the historical details of actual scientific practice, so models or descriptions of that practice have had to become more complex. Let me designate as the "simple model" of science that which represents science as based on induction and the hypothetico-deductive method through the sequence of observation, hypothesis, prediction, and confirmation, leading upward to truth. This model or picture of science has a long tradition. We can see its roots already in Francis Bacon's scheme of cautious ascent to modest generalizations, then descent back to the level of observation, subsequent ascent to higher levels, etc.[11] William Whewell speaks of the epochs of induction, development, verification, application, and extension.[12] There is much correct in this picture, but it is too vague and lacks enough specificity to demarcate science from many other human knowledge-generating enterprises.

To cope with some of the obvious shortcomings both of the simple model (e.g., the change in "true" theories and its failure to accord with actual scientific practice) and of its formalization by the logical positivists, some philosophers of science have suggested as a refinement viewing the scientific enterprise in terms of a vertical three-tiered scheme of practice, methods, and goals, with methods and goals occupying a metalevel above the fray of practice. Even though I have already claimed that such a dichotomy is in many ways false, let me take this model as a place to begin discussion. The level of practice is taken to include both experiments and theories that function to organize, correlate, and explain phenomena, as well as the activity of judging the empirical adequacy of a proposed theory. For many, this would seem to be just about the whole of science. At the metalevel of methods we are concerned with how theories are evaluated. What rules or standards are to be applied in testing and selecting theories? Examples of such criteria are predictive accuracy, simplicity, coherence, and fertility. Also at the metalevel, we find the goals or aims of science, such as true explanations, "mere" calculational and predictive accuracy, control of nature, and the like. Given the obvious corrigibility of our theories, we might hold out for stability or "fixedness" at the metalevel of methods and goals in order to underpin an invariant rationality for science and the knowledge it attains. But how are we to decide the status of these allegedly fixed methods and goals that are to anchor the rationality of science?

A foundationist approach attempts to lay down certain principles that *must* be adhered to. These are based on a priori arguments or upon requirements necessary if one is to avoid the possibility of contradiction. However, this project has not fared well when applied to actual scientific practice. In the real world that practice must often be content with a noncontradiction, arranged through a process of accommodating our initially hoped-for demands with what nature will sustain.

Once a foundationist move to underwrite methodological principles has been abandoned, historical fact remains. General features of practice must be distilled from case studies of episodes in the history of science. The arguments and contingent events that inform the course of development and selection of theories are essential. We must then look to see whether or not characteristics of one area or era of science are truly general across all of science. While such philosophy of science must be based on history (events as they actually occurred), it can be important not to focus exclusively on the form and content of "successful" scientific theories alone. How things might have gone a very different way and why they did not may be as important as the reasons for the "right" choices that science has made. Just as experimental practice and styles of arguing for the "reality" of observed phenomena have varied dramatically in this century, so have types of theorizing.[13] Some of the same case studies cited earlier in relation to the influence of social factors also show changes occurring at all levels. This prevents us from generating a false sense of simplicity or universality for actual science.

1.1.3 *The convergence of scientific opinion*

Now that I have claimed actual change at the metalevel, I must offer some explanation of how scientific opinion manages to converge to a fairly localized position and remains in its neighborhood as that position evolves over time. The door may seem to be open to chaotic change with no real stability. But that does not happen in science. Why? Not just anything can be made to work, because good theories that coherently and accurately cover a broad range of phenomena are difficult to make and only relatively few people have the ability to create them. The pyramid structure of the workforce is part of the answer, but the nature of the phenomena chosen for study is relevant too. If the "facts" to be accounted for were simple enough, then we would in reality face the philosopher's "game" of a set of evidence E and a sequence of theories $T_1, T_2, \ldots, T_n$, all of which are empirically equivalent. There would *in fact* be a large set of theories on hand from which we would be unable to choose *one*. As the phenomena get more complex so that only the best people can, collectively, come with one or two theories, then the choice is between these rather than among many. Convergence is produced by default (or creative ability). Of course, if the data set is too complex (e.g., as in psychology or philosophy), there is no broad convergence of opinion—there are no wholly successful theories.

Even in situations in which there may be observationally equivalent theories (such as the Copenhagen and Bohm versions of quantum mechanics), who gets to the top of the hill first holds the high ground and must be dislodged (if required, not otherwise).[14] At certain critical junctures there is nothing compelling or unique about the direction or particular *course* of evolution of (the history of) scientific theories. Once a successful, stable theory has gained sway, there is a temptation to imbue it with a scientific realist interpretation, both to increase our understanding of its laws and to strengthen its claim to necessity or uniqueness.[15] We have a natural inclination to seek a simple, necessary, and compelling picture of the way the world must be. An invariant set of methodological rules would increase our epistemic security in these final products of science. We may *want* these, but a close attention to the details of scientific practice does not allow at least some of us to harbor these fictions.

1.2 Theory choice—*rational*, but not *unique*

There is an essential corrigibility of the picture of the world given us by science. Some philosophers attempt to delimit methodological rules and heuristic strategies that have proven useful in developing new theories in science. While such guidelines are, as I have already emphasized, typically distilled retrospectively from what has been successful scientific practice, they are still at times elevated to almost prescriptive status for explaining why one theory was chosen over its rivals. We should be cautious about applying such "rules" from the past to new situations (i.e., those that are not already cases used to construct these rules). An example may help here, especially since it involves the development of quantum mechanics.

1.2.1 *Heuristic guidelines*
Heinz Post has discussed heuristic strategies with the aid of which scientists generate a successor (or more general, covering) theory L from a theory S that has been pressed beyond the limits of its validity. One of these

theoretic guidelines to new theories is labeled a *General Correspondence Principle*.[16] This correspondence principle enjoins one to conserve in L, not only the successful empirical consequences of S, but also as many of the explanatorily useful and conceptually desirable features of S as possible. It also requires the new theory L to yield the old one S as a well-defined (mathematical and perhaps even conceptual) limit as some relevant parameter or physically significant quantity is varied. Both Bohr and Bohm did use such heuristic guidelines to generate quantum mechanics from classical mechanics.[17] Post is quite correct that, as a *general* characteristic of theory construction, the procedure of generating a new theory (or of generalizing an old one) "is conservative (as every good scientist is)."[18] Use of general heuristic guidelines by both protagonists does lend support to Post's broad claims. However, as I argue in later sections of this book, the "successful" Copenhagen version of quantum mechanics has far fewer (conceptual *or* formal) characteristics in common with classical mechanics than does the not generally accepted, causal quantum mechanics of David Bohm. In several respects Bohm's quantum mechanics has greater correspondence (in Post's sense) with classical mechanics than does Bohr's. Bohm did call attention to both the past success of (interim) statistical theories finally being underpinned by detailed, classically deterministic successor theories of microentities and to the essentially circular and potentially self-fulfilling nature of Copenhagen's claim to completeness and finality. He employed these observations as guides and motivating factors in his own formulation of a causal, deterministic, realistic interpretation of quantum mechanics.[19] Unfortunately, the competitor that adhered more strongly to this aspect of Post's correspondence principle has lost out—surely an undesirable result for the efficacy of such guidelines as any criterion for judging a theory.

The Copenhagen formulation of quantum mechanics generates a further exception to Post's guideline. The difficulty arises because there is really no suitable limit in which standard quantum mechanics passes over into classical mechanics.[20] He takes this as a sign that a better L theory is needed. I show that Bohm's theory does provide such a limit so that one might attempt to argue that Post's own correspondence principle could be used to select Bohm's quantum mechanics over the Copenhagen one as the proper successor to classical mechanics. Given the pervasive observational equivalence that exists between these two versions of quantum mechanics, a normative argument of that kind would not effectively undermine the current widespread support for Copenhagen. That, at least, is what I hope the reader will be convinced of by the story that unfolds in this book.

Now that I have raised some doubts about the practical value of general heuristic guidelines in evaluating theories, let me enlarge the discussion to other criteria that play a role in theory selection.

1.2.2 A nonevidential element
I begin with a disclaimer or two. I am *not* centrally concerned with the question of the *refutation* of scientific theories. That is, in terms of the title of a well-known collection of essays on the underdetermination thesis, *Can Theories Be Refuted?*, I would be willing to give away a yes answer in the following sense.[21] Even if one grants that there are many theories that can reasonably be rejected on evidential grounds—those are the *easy* ones—there still remain other cases, I claim, in which viable, fertile theories have been rejected. Practical underdetermination may not *always* exist, but this book contends that there is at least *one* important instance in which genuine underdetermination could exist. This case may not be concerned with the mere compatibility of two essentially different theories with the presently *available* data, but may involve a much deeper-seated indistinguishability. While a choice can be, and has been, made on the basis of nonevidential criteria, the question must then be faced of the basis for such criteria and of the role historically contingent factors have played in fashioning them.

What is in doubt here is a belief in the (at least effective) uniqueness of a correct scientific theory, with the selection process being "objective" and not involving in any ineliminable fashion "subjective" criteria such as coherence, beauty, simplicity, or minimum mutilation.[22] By coherence I most specifically do *not* mean just lack of logical contradiction, since both versions of quantum mechanics treated in this book are logically consistent and neither is pejoratively ad hoc in nature. Scientists typically take for granted the practical uniqueness of successful scientific theories (in any given era). Einstein, in an address delivered before the Berlin Physical Society in 1918 on the occasion of Max Planck's sixtieth birthday, tells us that

> the supreme task of the physicist is to arrive at these universal elementary laws from which the cosmos can be built up by pure deduction. There is no logical path to these laws; only intuition, resting on sympathetic understanding of experience, can reach them. In this methodological uncertainty, one might suppose that there were any number of possible systems of theoretical physics all equally justified; and this opinion is no doubt correct, theoretically. But the development of physics has shown that at any given moment, out of all conceivable constructions, a single one has always proved itself decidedly superior to all the rest. Nobody

who has really gone deeply into the matter will deny that in practice the world of phenomena uniquely determines the theoretical system, in spite of the fact that there is no logical bridge between phenomena and their theoretical principles.[23]

Notice that, curiously enough, Einstein allows the *theoretical* (i.e., logical) possibility of more than one empirically adequate theory, but then goes on to make the (rather startling) declaration of faith that at any given time the "world of phenomena" (which sounds pretty objective) uniquely determines one theory as superior to all others.[24] Now, Einstein to the contrary notwithstanding, there *are* people who have looked carefully at the development of certain major episodes in the history of physics and who have concluded that factors other than just "the world of phenomena" have been essential for specific theory choice and that, but for contingency, the final choice might have turned out other than it did.[25] This is *not* to claim that just *any* theory can be made to work, but rather to emphasize that logic and physical phenomena *alone* are not enough to rule out or reject some theories as viable candidates. Nor does this deny Einstein's claim that in practice one theory *is* finally chosen as "decidedly superior to all the rest." The question, then, is *how*, in fact, is such a choice made?

Heisenberg, in his retrospective reconstruction of what was going on in Copenhagen in 1926, stated a similar view about there being just *one* interpretation possible for quantum mechanics.

> I wanted to start from the fact that quantum mechanics as we then knew it already imposed a unique physical interpretation of some magnitude occurring in it . . . so that it looked very much as if we no longer had any freedom with respect to that interpretation. Instead, we would have to try to derive the correct general interpretation by strict logic from the ready-to-hand, more special interpretation.[26]

It is just such (even practical or effective, as opposed to merely conventional) uniqueness that the specific case developed in this book calls into question.

TWO
Formalism, Interpretation, and Understanding

I now discuss the structure of and some demands made on scientific theories. With that in mind, it is helpful to distinguish among empirical adequacy, formal explanation, and understanding as goals of science. While no a priori criterion for understanding should be laid down, there may be inherent limitations on the way we are able to understand explanations of physical phenomena. I examine several recent contributions to the exercise of fashioning an explanatory discourse to shape to our modes of understanding the formal explanation provided by quantum mechanics. The question is whether we are capable of truly comprehending quantum phenomena, as opposed to simply accepting the mathematical formalism and certain irreducible quantum correlations. The reader can decide whether the understandability of a successful, contending scientific theory is an epistemic virtue worthy of consideration in selecting one theory over another.

2.1 The components of a scientific theory

A scientific theory can be seen as having two distinct components: its formalism and its interpretation. These are conceptually separable, even if they are often entangled in practice. To simplify matters, my remarks will be restricted to theories in modern physics. Here a formalism means a set of equations and a set of calculational rules for making predictions that can be compared with experiment.[1] We shall see in chapters 3 and 4 that both standard quantum mechanics and Bohm's version use the same set of rules for predicting the values of observables. The physical interpretation refers to what the theory tells us about the underlying structure of these phenomena (i.e., the corresponding story about the furniture of the world—an ontology). Hence, *one* formalism with *two* different interpretations counts as *two* different theories. Such a use of the terms 'formalism' and 'interpretation' is similar to and consistent with, even if a bit technically less explicit than, what is typically done in historical/philosophical analyses of quantum theory.[2] An interpretation is formulated after an only partial examination of a formalism, since one never exhausts *all* of the implications of a (mathematical) formalism.

2.2 Explanation versus understanding

Consider three levels on which scientific theories function: empirical adequacy, formal explanation, and understanding.[3] These are three distinct goals of a scientific theory, even though it may not always be possible to draw a sharp dividing line between one of these hierarchical levels and the next. The distinctions and relations I discuss may apply only to the physical sciences, the area of science from which illustrative examples will be chosen. For pure mathematics there may be no meaningful difference between my 'explanation' and my 'understanding,' since then abstract relations, established by logical deduction or implication, would seem to be all that is available to us.

The nature of explanation in science is a topic much discussed by philosophers of science and I certainly cannot hope to resolve that issue here.[4] My interest is in the explanation and understanding of laws, as opposed to single events.[5] Since I am concerned with the difference between an explanation and an understanding of physical phenomena, let me first distinguish between the terms 'explain' and 'understand' as I use them. The *Oxford English Dictionary* tells us that to 'explain' is "to make plain or intelligible," while to 'understand' is "to grasp the idea of [or] to apprehend clearly the character or nature of," with to 'apprehend' meaning "to lay hold of with the intellect." *Webster's International Dictionary* defines to 'understand' as "to achieve a mental grasp of the nature, significance, or causal explanation of something." Here 'understanding' signifies something more than formal 'explanation'. Basically, I see explanation as a means to achieve the goal of understanding.[6]

Empirical adequacy consists essentially in getting the numbers right, in the sense of having a formula or an algorithm that is capable of reproducing observed data. An example is a phenomenological or semiempirical calculational scheme, such as some parts of the old quantum theory or, once upon a time, Bode's law for planetary orbits.

An *explanation* is provided by a successful formalism with a set of equations and rules for its application, such as nonrelativistic quantum mechanics, which is usually taken (incorrectly, I argue later) to include the Copenhagen interpretation. This is basically an explanation in terms of entailment and is really equivalent to the concept of explanation in the D-N (deductive-nomological) or covering-law model. Such an explanation, while quite precisely definable and objective, does not in itself give us understanding of the phenomena subsumed under the law in question.

Here the emphasis is on unification via derivability from a more general framework. While unification and reduction are important aspects of an explanation that can produce understanding, these alone are not enough. It is precisely with quantum effects, such as the Einstein-Podolsky-Rosen-Bohm (EPRB) correlations, that this becomes especially clear, as I argue below. At this level, the theory is most suitably (although not *necessarily*) given an instrumentalist interpretation.[7]

Finally, *understanding* is possible once we have an interpretation of the formalism that allows us to grasp the character of and the relations among the phenomena. This is typically associated with an interpretation that can plausibly be defended as a realistic one (in the sense of scientific realism).[8] There are *pragmatic* aspects to understanding and these can include contextual factors, like psychological ones common to our species.[9] Regulative principles, say ideals of natural order, are also relevant, although they may be era dependent.[10] These features go beyond the purely epistemic. Understanding would be a pragmatic "bonus" that some theories possess and that may be relevant for our acceptance of them. In part, I go a "historical"/"psychological" route by considering what features have been common in producing a sense of understanding in episodes from the history of science.[11] As a paradigm of an explanation that *can* (or may) produce understanding for physical processes, I take a causal explanation, consisting either of direct cause-effect between phenomena and events or of a common cause located in the past of the collection of phenomena under consideration.[12] The ultimate goal is to construct a framework that is empirically adequate, that explains the outcomes of our observations, and that finally produces in us a sense of understanding how the world could possibly be the way it is. It remains an open question whether all three of these goals are simultaneously attainable.

My argument here really begins from the intuition, based on experience and on some history of physics, that *understanding* of physical processes involves a story that can, in principle, be told on an event-by-event basis. This exercise often makes use of picturable physical mechanisms and processes. I believe that William Whewell, in his *History of the Inductive Sciences*, adumbrates such a position when he distinguishes between a *law* and a *cause* or between the *formal* and the *physical* stages in the development of a theory.[13] Einstein spoke of principle theories (such as thermodynamics or special relativity) and constructive theories (such as kinetic theory or the Lorentz ether-based theory of the electron). In his opinion, the virtues of the former were epistemic security (of the deduced consequences once the empirically discovered principles had been accepted) and generality of applicability, while the latter had the advantage of providing clarity of understanding. Einstein characterized these two types of theories as follows:

> We can distinguish various kinds of theories in physics. Most of them are constructive. They attempt to build up a picture of the more complex phenomena out of the materials of a relatively simple formal scheme from which they start out. . . . When we say we have succeeded in understanding a group of natural processes, we invariably mean that a constructive theory has been found which covers the processes in question.
>
> Along with this most important class of theories there exists a second, which I will call "principle-theories". These employ the analytic, not the synthetic, method. . . .
>
> The advantages of the constructive theory are completeness, adaptability, and clearness, those of the principle theory are logical perfection and security of the foundations.[14]

This is, for me, very much the difference between formal explanation and understanding.

It is possible that the only genuine difference between 'explanation' and 'understanding' is mere psychological acclimation. Any further distinction may be just a leftover (historically conditioned) prop from our classical (physics) worldview, one that must simply be exorcised. My doubt, though, is that probabilistic 'explanations' produce no really satisfying understanding when a visualizable causal story is in principle blocked.

2.3 Examples from physics

I list in table 2.1 some examples from physics to illustrate the differences among empirical adequacy, explanation, and understanding.

Table 2.1

Empirical adequacy	Explanation	Understanding
Kepler's first and second laws	Newton's laws of motion and gravitation	"ether" (general relativity)
Boyle's law	formalism of statistical mechanics	kinetic theory (model) of gases
EPRB correlations	formalism of quantum mechanics	Bohm interpretation

I discuss the first two examples rather briefly and then the third one at greater length.

2.3.1 Newtonian gravitational theory

Suppose that (once we have learned, observed, or been told that the orbit of a planet is a *plane* curve) we are given (not necessarily on the basis of a derivation from any theory) Kepler's first law of planetary motion in the form[15]

$$r(\theta) = \frac{b^2/a}{1 - \varepsilon \cos \theta} \tag{2.1}$$

(with $\varepsilon = \sqrt{1 - (b/a)^2}$), as well as his second law of equal areas swept out in equal times. This gives the distance r (measured from the focus at which the sun is located) of the planet from the sun in terms of the angular location (θ) of the planet. The formula is an empirically adequate representation of the observational data, but (at this level) we have not the slightest idea *why* this particular form of relation should obtain. It is a straightforward (but not wholly trivial) exercise to prove that Newton's second law of motion,

$$F = m\frac{d^2 r(t)}{dt^2}, \tag{2.2}$$

plus his law of universal gravitation,

$$F_{grav} = -G\frac{Mm}{r^2}\hat{r}, \tag{2.3}$$

together can be solved to yield eq. (2.1) and the equal-areas law. Such a deductive argument certainly provides a formal explanation of eq. (2.1), but it gives us no understanding of what physical process causes the planet to follow an elliptical orbit. One attempt at a causal explanation would be to invoke the notion of (instantaneous) action at a distance.[16] It seems implausible that anyone fully *understood* such action at a distance (for over two hundred years). In fact, William Whewell claimed that one of the appeals of the Cartesian doctrine of vortices was its "picturability," essentially in terms of contact forces.[17] In a sense, Einstein's general theory of relativity provided an understandable (picturable) causal explanation in terms of a curved space-time background (whose specific structure is determined by the distribution of masses) through which gravitons (or gravitational waves) propagate to transmit physical influences of one mass upon another.[18] This space-time plays the role of an "ether" in acting as a background through which disturbances can propagate at a finite velocity.

2.3.2 The kinetic theory of gases

Another elementary example from classical physics is provided by Boyle's law,

$$PV = const. \qquad (2.4)$$

(with T a constant), which relates the pressure P and the volume V of an "ideal" gas (or a real gas at low enough pressure and high enough absolute temperature T). At this level, eq. (2.4) is just a phenomenological law summarizing a class of empirical data. We can obtain this result from the formalism of statistical mechanics.[19] Using the postulate of equal a priori probabilities for accessible states satisfying the equilibrium conditions of the system and the microcanonical ensemble appropriate for the discussion of an isolated system in which the total energy E is a constant, one calculates the entropy S in terms of the volume of the hypersphere enclosed by the energy surface in the phase space of N (noninteracting) particles. This yields the internal energy of the gas as a function of the entropy and the volume of the gas. That, plus the thermodynamic relations for the absolute temperature T and the pressure P in terms of appropriate partial derivatives, implies that

$$PV = NkT, \qquad (2.5)$$

which (for constant T) is Boyle's law.[20]

This derivation (or deductive argument) gives us no understanding of the physical mechanism that has led to the result of eq. (2.4). A causal picture story can be generated by considering a simple model of this ideal gas as consisting of "point" particles that suffer perfectly elastic collisions with the rigid walls of a cubical container of edge length ℓ. Elementary applications of energy and momentum conservation then lead to[21]

$$PV = \frac{2}{3}E = const. \qquad (2.6)$$

This gives us a type of understanding of eq. (2.4) that is much different from any we have of eq. (2.1) in classical gravitational theory.

2.3.3 Quantum mechanics and the EPRB experiment

Now consider the situation for quantum phenomena as illustrated by the well-known Bohm version of the Einstein-Podolsky-Rosen (EPR) thought experiment shown schematically in figure 2.1.[22] There two observers, in

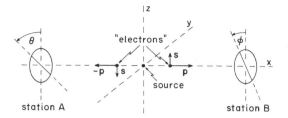

Figure 2.1 The EPRB thought experiment

spatially separated regions (A and B), can each make one of two choices of directions (instrument setting θ_1, θ_2 and ϕ_1, ϕ_2, respectively) along which to measure the transverse spin projections of each emitted "electron," and for each setting there are two possible outcomes for these spin projections at stations A and B.[23] The distinctions I have made among empirical adequacy, explanation, and understanding can be illustrated by starting with the joint probabilities $P(++ | \theta, \phi)$ for Bohm the EPR-Bohm (EPRB) experiment. Here $++$ refer to the outcomes of the spin measurements (both "up" in this case) at the two stations, and θ, ϕ refer to the directions (perpendicular to the line of flight) along which the spins are observed. There is also a $P(+ - | \theta, \phi)$ for the probability of getting a spin up (+) at station A, while a spin down (−) obtains at B. Given the observed data for a series of such experiments, a modern-day Ptolemy— or perhaps better, Kepler—without use of quantum mechanics or any other theory might find, within experimental errors, the following purely phenomenological representation of the data:

$$P(++ | \theta, \phi) = P(-- | \theta, \phi) = \frac{1}{2}\sin^2\left(\frac{\theta - \phi}{2}\right) \quad (2.7)$$

$$P(+ - | \theta, \phi) = P(- + | \theta, \phi) = \frac{1}{2}\cos^2\left(\frac{\theta - \phi}{2}\right). \quad (2.8)$$

This representation of these joint probabilities is surely empirically adequate. In fact, it is little more than equivalent to the data itself. At this stage we have neither an explanation of these results nor any understanding of how they come about. If someone gives us the formalism of quantum mechanics, then we can formally explain the results of eqs. (2.7) and (2.8) by performing a calculation of the type given in the appendix to this chapter.[24] However, this formal explanation of these joint distributions gives us no understanding of how these outcomes have been produced in nature. In other words, an intelligible *interpretation* is missing. We have no understanding of what the formalism means, in terms of underlying physical processes.

Bell's original work showed that no determinate, local theory can account for the outcome of the EPRB experiment.[25] Much subsequent analysis has sharpened our understanding of precisely what must be assumed to obtain Bell-type inequalities.[26] For this EPRB situation, Bell inequalities are the necessary and sufficient conditions on the joint distributions for a common-cause explanation of the actually observed outcome of the experiments.[27] Since these inequalities are violated, then, at least for this *one* experiment (and there are others), no common-cause explanation is possible. A reasonable list of exhaustive explanations for the observed results is (i) mere chance or coincidence, (ii) a direct causal link between the two spatially separated stations, and (iii) a common-cause explanation located in the light cone of the common past of the system and apparatus. Explanation (iii) is blocked by the violation of the Bell inequalities and (ii) (apparently) by the first-signal principle of special relativity interpreted as prohibiting any type of instantaneous action at a distance.[28] I put aside (i) for the moment as no explanation at all. This situation is often summarized by saying that there is *in principle* no possibility of accounting for the joint distributions of eqs. (2.7) and (2.8) with what is usually termed a local, realistic theory.[29] Is it possible, in such circumstances, to produce any explanation that allows us to understand or comprehend how the observed phenomena come about?

2.4 Some attempts at understanding quantum phenomena

I have indicated some moves—specifically, traditional direct causal links and common-cause explanations—that are blocked in any attempt to produce an understanding of quantum phenomena.[30] That is a negative message. There does exist a positive explanatory framework (quantum mechanics) for quantum events. To provide some understanding of this explanatory scheme is a very desirable goal, one pursued mainly by the scientific realist. It seems to be a part of our human nature (at the very least in more recent Western tradition) to want this something extra that goes beyond mere formal explanation. There have been many, often realist, attempts at interpreting standard quantum mechanics in a broadly Copenhagen spirit. Let us see whether some of the more recent ones succeed in providing us with any understanding of the phenomena.

A quantum realist can take seriously an 'ontic blurring' of many of the variables that can be observed in a quantum system.[31] The notion of "quantum particle" is introduced for an object that exhibits wave-particle duality to distinguish it from the traditional "classical particle." The nonseparability characteristic of quantum systems is taken to indicate the "holistic character" of such systems. In a sense, what has been done is to take some of the unique aspects of the quantum formalism and then assign names and ontic status to them. This in itself does not produce any sense of understanding of the physical phenomena. If we say that quantum realism is that realism required by quantum mechanics, we have not thereby helped anyone to comprehend just what quantum realism *is* as a representation of the world.

One can return to an earlier suggestion by Heisenberg that we introduce a new class of physical entity, *potentia*, into our theory (and into our ontology).[32] Such a move promises to help our understanding of how definite results appear or actualize in one localized region of space-time. This is an attempt to cope with an aspect of what is commonly referred to as the reduction of the wave packet in quantum mechanics (to be discussed more fully in section 3.5.1) which is *the* central problem for the standard version of quantum mechanics.[33] A spontaneous reduction mechanism, which is to some an appealing modification of the usual quantum formalism, has also been proposed to cope with the measurement problem.[34] However, either of these possible resolutions of the measurement problem would leave us, at the level of the actual physical phenomena, with the same physics puzzle about *how* correlations of the EPRB type are produced in spatially separated systems. And this, of course, is *the* problem for this paradigm quantum riddle.[35]

A more radical move is to claim that particularism (basically, the separability and independence of spatially distinct systems) is a fallacy we fall prey to because of our classical heritage.[36] In place of this possibly refuted concept one can posit that relational properties of physical objects may not simply supervene wholly upon nonrelational properties of localizable individuals, but that a type of *relational holism* in which the objects have inherent relations among themselves is essential.[37] Even aside from the remaining serious problem of the passage from the microrealm, where relational holism supposedly holds, to the macrorealm, where it is not obviously prevalent, does the assertion of the phenomenon of relational holism provide us with a concept we can *understand* for physical systems?[38]

Another possibility is that, since traditional direct causal links and common causes are impossible for quantum correlations, these quantum correlations themselves ought to be taken as the irreducible brute facts or the primitive givens upon which we build any further explanations.[39] We are urged to buy wholeheartedly into an indeterministic worldview and to abandon any hope of understanding how these correlations are produced. Can we do better than wholesale surrender?

2.5 Contingently necessary conditions for understanding

The point is not that any explanation must stop somewhere so that there must always exist irreducible primitives in an explanation. Rather, not all sets of primitives are equally acceptable for producing understanding.[40] It is a matter of fashioning an understandable explanatory discourse versus giving mere redefinitions of terms to paper over our ignorance.

2.5.1 Local realism

In section 2.3 I used some examples of explanatory discourses from the history of physics in an attempt to clarify this distinction. Let me return to one of those examples. Newton's extremely successful formal law of universal gravitation was soon interpreted in terms of instantaneous action at a distance. Newton himself neither defended nor claimed to understand this concept.[41] Those before him and many of his contemporaries found this incomprehensible as a physical mechanism. However, the formalism worked so well in 'explaining' the data that action at a distance was essentially bracketed as a problem for two hundred years or so. The phrase "action at a distance" continued to be used, but it is unclear that anyone really understood it as a physical process. After Einstein's general theory of relativity, it would be considered ridiculous to defend action at a distance because one now has a causal story to replace it. Action at a distance had originally been a conceptual problem, was then largely ignored (as a worry) for two centuries and has finally been rejected as physically meaningless.[42] This was a failed attempt at an intelligible explanatory discourse. General relativity provided a successful one in terms of *local* interactions.

Why has the need for a local explanation had such a hold on us? It could simply be a culturally conditioned requirement. In various guises we can locate the roots of what we today term the locality/nonlocality debate as far back as the origins of our Western philosophical tradition. It has been suggested that this debate may be a continuation of attempts to grapple with a very old philosophical problem—that of the one and the many.[43] Parmenides, Empedocles, and Zeno of Elea held that being is one and that there could be no plurality or change.[44] These influences, of course, predated Plato and Aristotle and had a reincarnation in the third century A.D., in a neo-Platonic interpretation as a divine *one* out of which all reality emanates. By contrast, change and becoming were essential concepts in Aristotle's philosophy. Newton conceived of an active and omnipresent God, whose sensorium was absolute space. This space acted upon matter, but was not in turn influenced by matter. It also served, as Immanuel Kant suggested in his transcendental aesthetic applied to our conception of space, to separate objects from one another.[45] I believe that it is not unreasonable to see Émile Meyerson's concept of cause as an atemporal logical identity as being related to this ancient Eleatic doctrine.[46] In quite recent times, Louis de Broglie, in a famous 1948 issue of *Dialectica,* attempted "to apply the concept of complementarity to the relations which exist between the constituents of a system and the system as a whole."[47] He saw, even in classical mechanics, a loss of the individuality of the parts of a system as a result of their mutual interaction via the potential energy.

In the long debate about action at a distance versus contact action we also find a concern with the effective individuation of particles or subsystems and with distant action.[48] Belief in contact action goes back to Aristotle and Eudoxus, based on the self-evident regulative principle that a thing cannot act where it is not. For them, causal explanations were essential for true science. For Descartes, too, there could be no vacuum and what may appear to be empty is actually filled with an ether. After the *Principia,* there were basically two camps on the "gravity dilemma."[49] The one, among whose members were the Cartesians, Leibniz, and Huygens, maintained that an ether was required to allow any intelligible explanation of gravitational phenomena. The other, notably represented by Newton's ally, the cleric Samuel Clarke, took gravity to be evidence for God's action. Even though something cannot act where it is not, God is everywhere and, hence, "instantaneous" action between spatially separated bodies is no mystery. The basic motivating factor in demanding contact action was that of *intelligibility.*[50] Maxwell put the matter quite succinctly when he argued for "a force of the old school—a case of *vis a tergo*—a shove from behind."[51]

Even in so simple and visualizable process as a "perfectly elastic" collision between two billiard balls, we can push just so far our explanation and understanding of precisely what goes on during the instant of collision. Finally, there is something we do not understand. However, we can, as it were, isolate our ignorance into a little "sphere" around each ball and leave that unknown alone. This allows us to follow *locally* the time evolution of each part of our causal picture story. Contact forces (*broadly* construed—as I have done for general relativity above) are a paradigm of *local* interactions—hence their importance or privileged position in explanatory discourses. John Bell stressed the great desirability, perhaps even the indispensability, of locality in a physical theory (hence his "local *beables*").[52] With long-range quantum correlations (as with instantaneous classical gravitational action at a distance) our problem is *compounded* by the nonlocal nature of the interaction. The "trouble" leaks out, so to speak, over all of space. In the realm of quantum phenomena, especially correlations like those of the EPRB experiments, over and above some ultimate, irreducible mystery that always attends *any* explanation (in terms of some set of givens), there is the apparently nonlocal nature of the effects.[53]

My concern has been that, for physical processes, a satisfying understanding may be possible for us only when we are able to tell a local, causal picture story of those processes. If this is so, then we are, by definition, doomed, according to the standard (Copenhagen) interpretation of quantum mechanics, to be unable to understand quantum phenomena (and quantum mechanics). Now the world *may* just be that way. In our human intellectual history, we have been forced to acknowledge that we are unable to fulfill certain perceived intellectual (psychological?) needs. Not all of the desiderata that we may want (or need for understanding) in a physical theory according with the phenomena of our actually existing world can be mutually compatible. The problem then becomes one of finding (if possible) an interpretation that is compatible with those regulative principles we most value.

2.5.2 Visualizable causal models and understanding

While locality is a constraint we have traditionally imposed on our theories, we have at times (as with classical gravitational theory) been willing to yield on it. But even then we have required a theory that allowed a *causal* story to be sustained in terms of interactions among parts of a system. This has enabled us to picture, in some sense, what is going on. Quantum-field theory is at least as puzzling and incomprehensible as ordinary nonrelativistic quantum theory, yet some versions of it allow a greater degree of visualizability than others. Richard Feynman found visualization important in his formulation of quantum electrodynamics.[54] Feynman's formulation (in terms of his diagrams) is much more in ("everyday") use by practicing theoretical physicists than are the much more abstract Schwinger-Tomonaga ones.[55]

A desire for picturable models or explanations could be rooted in our patterns of thought.[56] This is a theme that goes back to Kant, who sought to underpin physical inference by the way we think about objects. A line of argument that modes of thought are permanent, or at least endure for centuries, was used by Émile Meyerson.[57] His *principle of lawfulness*, related to our need to know *how* phenomena are related, can be rooted in the advantage (or survival value) of purposeful action or prevision.[58] His *principle of causality*, related to our desire to know *why* certain observed regularities obtain, is a manifestation of a fundamental and universal characteristic of human thought and modes of understanding. For Meyerson, they have the status of contingent but universal principles of the human mind. His was a search for such principles that the human mind uses in framing theories. He anchored these, especially causality, in our demand for something permanent and self-identical (e.g., conservation laws).[59] Although Meyerson accounted only for lawfulness, but not for causality, on the basis that it simplifies our process of comprehending the phenomena of the world (its survival value), it would seem that causality can also confer the advantage of our being able simply and efficiently to reduce (and hence, to deduce thereafter) our laws to the understanding of basic causal mechanisms. In his *Identity and Reality*, he tells us that "science also wishes to make us *understand* nature" as a reflection of man's causal instinct.[60] His belief in these innate urges or needs of the intellect is again reflected in "atomism has its roots in the depth of our spirit," as *prior* to experience.[61] Can we ever hope to purge mankind of this need for a causal (perhaps, even, a picturable) explanation of physical phenomena? Harald Høffding, a philosophical mentor to the young Niels Bohr, regarded anything that is not subject to a causal explanation as beyond the limits of intelligibility.[62]

In a similar vein, Gaston Bachelard in his *The New Scientific Spirit* emphasized the dialectic of science, stressing that psychology and epistemology are equally important components in a representation of the functioning scientific enterprise.[63] His pragmatic attitude toward foundational philosophical problems was nicely encapsulated in the marvelous slogan: "Science in effect creates philosophy."[64] While citing Meyerson on the stability of modes of thought that endure for aeons, Bachelard disagreed that there are any permanent forms of rationality and did consider the possibility of changes in these patterns or modes, but only on a time scale characteristic of or limited by the rate of evolution of the human brain.[65]

I suggest that we attempt to generate theories that make the world comprehensible in terms of our inherent patterns of thought. Understandability in a successful scientific theory is an important pragmatic virtue, both for visualizing fundamental phenomena for new research and for allowing a wider audience to comprehend that theory. Even an only partial attainment of this goal by a theory (e.g., Bohm's program) should not be totally discounted or, even worse, taken as a negative mark against it, when its competitor (Copenhagen) does even less well on this score.[66] 'Causality' and 'locality' are logically distinct concepts, with causality being the more central in my scheme of understanding. The actual nonlocality demanded by nature turns out to be of a fairly benign variety: we cannot signal with it and it does not so entangle the world as to prevent us from doing science as we have traditionally known it. We are able to construct a less incomprehensible, more nearly picturable, representation of the physical universe with Bohm than with Copenhagen. We do have the option of giving up locality while maintaining a visualizable causality. The choice is ours to make on purely pragmatic grounds.

The origins of the uneasiness about nonlocality may be more psychological than logical.[67] It is fitting that I should give David Bohm's view on this question:

> For several centuries, there has been a strong feeling that non-local theories are not acceptable in physics. It is well known, for example, that Newton felt uneasy about action-at-a-distance and that Einstein regarded this action as *"spooky"*. One can understand this feeling, but if one reflects deeply and seriously on this subject one can see nothing basically irrational about such an idea. Rather it seems to be most reasonable to keep an open mind on the subject and therefore to allow oneself to explore this possibility. *If the price of avoiding non-locality is to make an intuitive explanation impossible, one has to ask whether the cost is not too great.*[68]

My evident sympathies with views such as those on contact action expressed by Maxwell earlier clearly put me into the camp of what has been termed the *mechanistic* view of physical processes, which can be rather bluntly identified as "an obsolete hangover from an unenlightened past [and one that] reaffirms the convictions of a number of famous nineteenth-century scientists (Maxwell, Kirchhoff) who saw the aim of all science in the discovery of models which allow an understanding of phenomena by their interactions in time and space."[69] That is a not unfair characterization of the position I have advocated here.

Appendix   A derivation of the EPRB correlations

The basic geometry and the meaning of the relevant variables are shown in figure 2.1. The initial state of the system (say, the two electrons just after they have been emitted) is represented as

$$\Psi_0 = \frac{1}{\sqrt{2}}\left[\begin{pmatrix}1\\0\end{pmatrix}\otimes\begin{pmatrix}0\\1\end{pmatrix} - \begin{pmatrix}0\\1\end{pmatrix}\otimes\begin{pmatrix}1\\0\end{pmatrix}\right] \quad (2.9)$$

and the final state as

$$\psi_\pm^A(\theta)\otimes\psi_\pm^B(\phi). \quad (2.10)$$

Here $\psi_\pm^A(\theta)$ are the eigenvectors of the projection of the spin operator σ onto $\hat{n}$ where

$$\boldsymbol{\sigma} = \sigma_x \hat{i} + \sigma_y \hat{j} + \sigma_z \hat{k}, \qquad \hat{n} = (0, -\sin\theta, \cos\theta). \tag{2.11}$$

The angle θ is shown in figure 2.1. Specifically, we have

$$\boldsymbol{\sigma}_A \cdot \hat{n}_A \, \psi_{\pm}^A(\theta) = \pm \psi_{\pm}^A(\theta). \tag{2.12}$$

There is, of course, a similar relation for the $\psi_{\pm}^B(\phi)$. An explicit representation of these $\psi_{\pm}(\theta)$ is

$$\psi_+(\theta) = \begin{pmatrix} \cos\left(\dfrac{\theta}{2}\right) \\ -i \sin\left(\dfrac{\theta}{2}\right) \end{pmatrix}, \tag{2.13}$$

$$\psi_-(\theta) = \begin{pmatrix} -i \sin\left(\dfrac{\theta}{2}\right) \\ \cos\left(\dfrac{\theta}{2}\right) \end{pmatrix}. \tag{2.14}$$

The transition probabilities are computed as

$$P(\pm\pm \mid \theta, \phi) = |\langle \Psi_0 | \psi_{\pm}^A(\theta) \otimes \psi_{\pm}^B(\phi) \rangle|^2 \tag{2.15}$$

to obtain eqs. (2.7) and (2.8).[70]

For reference later in chapter 10, the correlation for the spin measurements at stations A and B of figure 2.1 can be found as

$$\begin{aligned} & P(++|\theta, \phi)(+1)(+1) + P(--|\theta, \phi)(-1)(-1) \\ & + P(+-|\theta, \phi)(+1)(-1) + P(-+|\theta, \phi)(-1)(+1) \\ & = \sin^2\left(\frac{\theta-\phi}{2}\right) - \cos^2\left(\frac{\theta-\phi}{2}\right) = -\cos(\theta-\phi). \end{aligned} \tag{2.16}$$

The same result can also be obtained by a direct calculation of the expectation value

$$\langle \Psi_0 | \boldsymbol{\sigma}_A \cdot \hat{n}_A \otimes \boldsymbol{\sigma}_B \cdot \hat{n}_B | \Psi_0 \rangle, \tag{2.17}$$

where Ψ_0 is the state given in eq. (2.9).

THREE
Standard Quantum Theory

The standard view of quantum mechanics, accepted almost universally by practicing physicists and often by philosophers of science concerned with such issues, is what may be (somewhat elusively) termed the Copenhagen interpretation. This interpretation requires complementarity (e.g., wave-particle duality), inherent indeterminism at the most fundamental level of quantum phenomena, and the impossibility of an event-by-event causal representation in a continuous space-time background. In this chapter I discuss some of the counterintuitive and problematic aspects of this theory. Even among mainstream physicists, there can be discomfort about the present status of quantum mechanics. Murray Gell-Mann has observed:

> *Quantum mechanics* [is] that mysterious, confusing discipline, which none of us really understands but which we know how to use. It works perfectly, as far as we can tell, in describing physical reality, but it is a 'counter-intuitive discipline', as the social scientists would say. Quantum mechanics is not a theory, but rather a framework within which we believe any correct theory must fit.[1]

3.1 Einstein versus Bohr

The two figures who dominated the physics of the first half of the twentieth century were Albert Einstein and Niels Bohr. Relativity and quantum mechanics will surely remain the major watersheds of this century in our conceptions of the physical world. It also appears reasonably unproblematic that Einstein will continue to be seen not only as the almost sole founder of relativity as a program but also to have set, in theoretical physics, a style of thought that has endured and proven to be extraordinarily fruitful for the developments in modern physics. Quantum mechanics is even more profound in its fundamental physical and philosophical implications than is relativity. Bohr is remembered for his seminal 1913 paper on a semiclassical model for the hydrogen atom (and for his quantization rule for the angular momentum).[2] It is much less clear that his fundamental insights into quantum mechanics will be of lasting value. While the formalism of quantum mechanics has proven to be correct (i.e., predictively accurate), Bohr's "insights" into its interpretation may have obfuscated many of the fundamental issues. These could turn out to have been, eventually, counterproductive for genuine progress on the interpretative problems of quantum mechanics. The conventional wisdom has been that Einstein was simply wrong about some of his central tenets concerning the physical world—the existence of an objective, observer-independent reality, the necessity for causal (essentially deterministic) explanations for physical processes, and the locality/separability of the physical world. Bohr is usually depicted as the victor in the "Bohr-Einstein" debates, even though it is not always made clear just what issues were at stake there or in precisely what sense Bohr actually clarified the points being contended.

Their differences (after the 1935 Einstein-Podolsky-Rosen paper) can be cast in terms of Einstein's ontological commitment to separability as a necessary requirement for the individuation of physical systems versus Bohr's demand, on the basis of "experimental" results, that discontinuous changes in the state of a microsystem preclude objectively associating, throughout an interaction, any classical state with such a microsystem.[3] So strong a commitment to separability by Einstein was, as we shall see, *not* necessary for doing science as we have traditionally known it.[4] Bohr's slip from epistemology (based on observability) to ontology (as a necessary discontinuity and as the impossibility of "classical" trajectories throughout an interaction) was, as I show below, not only logically unjustified but also not demanded, either by experiment or by the formalism of quantum mechanics.[5] Some claim that Bohr "accorded much higher value to *understanding the phenomena,* than merely getting the numbers right."[6] If that is so, then he might have come closer to that goal with a less counterintuitive interpretation than his own. Bohr's concept of complementarity was never fleshed out with a coherent ontology, while Einstein's comprehensible ontology did not appear consistent with quantum theory. It will become clear that Bohm's theory recovers all of the statistical predictions of standard quantum theory and has a largely classical ontology.

Central to Bohr's vision was his "quantum postulate"—the *discontinuous* transition of an atomic system from one stationary state to another during an interaction. We are told he held "that the uncertainty principle implied that the classical ideal of causality, strict determinism, would have to be replaced by a new goal for an adequate description of the behavior of a system, that expressed by statistical determinism."[7] Like Bohr, Pauli also believed in statistical causality as an essential feature of the world demanded by quantum mechanics.[8] It remains unclear *precisely* what "statistical causality" means, aside from being a term generated to hide our ignorance of an alleged feature of the world. Bohr is sometimes represented as a realist (in the sense of not denying the existence of an objective physical reality).[9] Does the following gloss on Bohr's doctrine really help us to understand such a reality?

> Complementarity suggests developing an ontological conception of an independent reality . . . not describable by the terms of experience. Beyond such a generalization, little else can be said regarding the positive nature of such an ontology.[10]

One *can* consistently buy into the picture of complementary descriptions in which a space-time representation is eschewed, but this Bohrian point of view is not unequivocally *demanded* by empirical evidence. Although conventional wisdom about quantum mechanics holds that it is *in principle* impossible that position (and, hence, a trajectory in space and time) is a possessed property of a microsystem, Bohm's theory accomplishes just that. One ought not to accept, as logically required, constraints that are more restricting than nature actually dictates.

Einstein, unlike Bohr, felt that a coherent description of the macrorealm and of the microrealm should be able to be given in the *same* terminology.[11] He believed that the terms of a theoretical description should not be made directly dependent upon operational criteria.[12] Einstein took Bohr's arguments to be largely circular or question begging since the conclusion (about the impossibility of certain goals for or demands upon a theory) follows only if one already accepts a positivistic criterion of what constitutes a reasonable goal for a theory.

3.2 The formalism

Entire books have been written on the formalism of (nonrelativistic) quantum mechanics and I intend here only to outline in the briefest (if somewhat vague) form, in terms of a few simple rules, the types of postulates that are usually employed in making quantum-mechanical calculations.[13] The reader should appreciate that such an ahistorical, "cold-facts" approach is the way that quantum mechanics (and science, generally) is presented to students. This will often be the mind-set that a physicist brings to a first consideration of the possibility of an alternative interpretation of quantum mechanics.

My representation of the formalism of quantum mechanics is the following:[14] (i) a state vector ψ—a vector, in a Hilbert space $\mathcal{H}$, representing the state of the physical system; (ii) a dynamical equation, the Schrödinger equation

$$H\psi = i\hbar \frac{\partial \psi}{\partial t} \qquad (3.1)$$

(where $\hbar$ is Planck's constant divided by 2π) giving the time evolution of the state vector ψ under the influence of the Hamiltonian H for the physical system; (iii) a correspondence between (hermitian) operators A in $\mathcal{H}$ and physical observables a. These physical observables a can take on only the eigenvalues a_j where

$$A\psi_j = a_j \psi_j ; \qquad (3.2)$$

(iv) the expectation value for a series of observations (of the quantity a) given as $\langle \psi | A | \psi \rangle$; and (v) a projection postulate (either explicitly or effectively assumed) upon measurement

$$\psi = \sum_k c_k \psi_k \to \psi_j , \qquad (3.3)$$

where (for ψ normalized to unity) $|c_j|^2$ is the probability of obtaining the result a_j (i.e., the chance of ending up in the eigenstate ψ_j).

3.3 The interpretation

Even if we accept these as a sufficiently adequate representation of the formalism of quantum mechanics, we still cannot give a simple and concise statement of *the* Copenhagen version of quantum mechanics because the term 'Copenhagen' has been used to refer to so many variations, all allegedly based on Bohr's writing.[15]

3.3.1 *Individual systems or ensembles only?*

The quantum theorist Henry Stapp attempted to isolate (or actually, perhaps, to *define*) the logical essence of what he takes to be the Copenhagen interpretation of quantum mechanics.[16] There Stapp's version of my rules (i) through (v) above is given a somewhat more pragmatic or operationalist gloss in terms of rules for encoding preparation and measurement procedures in wave functions, a unitary time-displacement operator to effect the translation of the initial state forward in time, and, finally, a set of rules for computing the probability (in the sense of the relative frequency of occurrence) of obtaining specified results upon a prescribed type of measurement or observation (for a *given* initial state, of course). He eschews what he terms the "absolute-ψ approach," in which the wave function is taken as referring to the *actual* state of a physical (micro- or macro-) system.[17] In another analysis of various "standard" interpretations of quantum mechanics, Leslie Ballentine distinguishes what he terms the Bohr-Copenhagen view, in which the formalism is taken as referring to *individual systems* (akin to Stapp's absolute-ψ approach), from Einstein's statistical interpretation, according to which the formalism applies to *ensembles* only.[18] Ballentine observes that "the statistical interpretation, which regards quantum states as being descriptive of ensembles of similarly prepared systems, is completely open with respect to hidden variables."[19] Stapp's explication of the Copenhagen interpretation seems fairly close to Ballentine's statistical interpretation. My use of 'Copenhagen interpretation' accords more closely with Ballentine's use of that term (i.e., Stapp's absolute-ψ approach). One commitment of a Copenhagen interpretation would seem to be to Bohr's concept of complementarity (of mutually exclusive descriptions of a system).

These two representations of the interpretation of quantum mechanics are relatively recent constructions. Let us go back to the views of some of the founders of quantum mechanics.

3.3.2 Bohr on interpretation

As is typical for his writing on broader philosophical issues, Bohr's pronouncements on the interpretation of quantum mechanics are often difficult to understand and at times just plain opaque. In a 1925 article, written after Heisenberg's matrix-mechanics formulation had been published, Bohr gave his own view on the new mechanics. It was formulated in terms of directly observable quantities only, required a renunciation of mechanical models in space and time, and indicated an inherent limitation on our means of visualization.[20] The difficulties to which Bohr referred here were the failures to give a classical-mechanical account of the interactions between atoms and light. As a consequence, Bohr suggested that these fundamental revisions in our way of representing the world *may* be necessary, but he did not yet foreclose alternatives.

In the version of his Como lecture published in *Nature* in 1927, Bohr also made clear his positivist (perhaps operationalist) stance when he wrote that the quantum postulate "implies a renunciation as regards the causal space-time co-ordination of atomic processes."[21] A space-time representation and causality, which were characteristic of classical theories, became for him complementary but mutually exclusive means of description.[22] By 1927 Bohr was citing the Heisenberg uncertainty relations as further evidence of the *impossibility* of simultaneously having a causal description (conservation laws or sharp values of energy and momentum) and a space-time description (sharp locations in space-time) of atomic processes. These are his own (perhaps somewhat odd) definitions of 'causal' and 'space-time'. One could certainly put another gloss on those terms and still hope for a causal description in a space-time background. In a 1929 lecture recounting the development of quantum mechanics, Bohr told his audience that "only by a conscious resignation of our usual demands for visualization and causality" is it possible to advance.[23] Not only was this what *did* happen, but it had now become the only *possible* way to progress.

Recalling the debates at the 1927 Solvay congress, Bohr pointed out differences in some details of the interpretation of quantum mechanics. For Dirac the appearance of actual observed phenomena was produced by a choice on the part of "nature," while for Heisenberg the choice was that of the "observer." Bohr wanted no part of volition for nature or determinative influence for the observer. While Dirac and Heisenberg would have reality become definite only upon measurement or observation, Bohr, apparently, would see an individual phenomenon (in whatever sense), different aspects of which we can choose to observe.[24] One of his clearer statements about the doctrine of the impossibility of a deterministic description of atomic phenomena is contained in a 1955 lecture.[25] What Bohr gave there is really an argument against being able to make certain types of determinations simultaneously, rather than against the possibility of a description of the type he rejected. His was an essentially positivist approach, one very different from Einstein's. It is clear that, at a minimum, Bohr was committed to complementary descriptions and to the impossibility of a deterministic quantum theory.

3.3.3 *Heisenberg on interpretation*

I have already discussed the Copenhagen interpretation of quantum mechanics as represented by Henry Stapp in 1972 after he had had extensive correspondence with Heisenberg.[26] But what were Heisenberg's *own* views on this interpretation and how did they differ in essentials from those, say, of Bohr and of Born? In his *Physics and Philosophy*, Heisenberg laid out quite specifically the finality (or completeness) of that interpretation. He claimed that the concepts and language of classical physics, while limited by the uncertainty relations, were essential to describe experimental results and could not be improved upon.[27] This was a prohibition against any change in the structure of the interpretation. How did he argue for it? He claimed that probabilities play an ineliminable role due to the uncertainty relations, but that these probabilities were of a fundamentally different type from those encountered in classical physics. Quantum-mechanical probabilities represented not the course of events in time, but rather the *tendency* for events. A connection with reality could be made only when "a new measurement is made to determine a certain property of the system."[28] Here Heisenberg introduced the notion of *potentia* for one actual event or another, but there were to be *no* actual events until measurement. Using the familiar gedanken γ-ray microscope, he then described the result produced in attempting to observe an electron in its atomic orbit and, of course, the electron was disturbed in the process. He concluded that "there is no orbit in the ordinary sense" and that use of such a concept "in quantum theory . . . would be a misuse of the language which . . . cannot be justified."[29] As he pointed out immediately thereafter, the central question, which he left open for the moment, was whether this prohibition against orbits was about epistemology or ontology.[30] In explaining the time evolution of the probability function, Heisenberg used the change of this function upon observation (i.e., the collapse of the wave packet) to conclude *quite generally* that "the space-time description of the atomic events is complementary to their deterministic description."[31] Such a claim has force only if one already grants that a probabilistic description is, in principle, the best one can have. In attempting to block any possible *description* of a system between measurements or observations—"why one would get into hopeless difficulties if one tries to describe what happens between two consecutive observations"[32]—Heisenberg rehearsed the double-slit experiment and showed that one must necessarily be frustrated in any attempt to determine through which slit a single light quantum went. From this he concluded that the concept of the quantum-mechanical "probability function does not allow a description of what happens between two observations" and that "therefore, the transition from the 'possible' to the 'actual' takes place during the act of observation."[33]

In summary, then, the essential elements of Heisenberg's Copenhagen interpretation are complementary descriptions (among which are mutually exclusive deterministic and space-time ones), a reality that becomes actual or definite only upon observation, and a commitment to the completeness of such a description. Nowhere is it *proven* that alternatives are not possible (i.e., that the indeterminism of quantum mechanics is actually ontological, rather than simply epistemological). A *consistent* interpretation has been presented, but not necessarily the *only possible* one compatible with observations. The extent to which these arguments could constitute a "proof" would depend, as I have stated earlier, upon the acceptance of a positivistic view. That Heisenberg did accept just such a view he made explicit when he discussed the "impossibility of interpretations truly alternative to the Copenhagen one.[34]

3.3.4 *Born on interpretation*

Finally, I turn to Max Born's characterization of the Copenhagen interpretation. In his introductory popular book *The Restless Universe* (1936), Born summarized the lessons of wave mechanics in terms of his own statistical interpretation. There he stated that "in the quantum theory it is the *principle of causality*, or more accurately that of *determinism*, which must be dropped and replaced by something else."[35] The only way he saw out of "the dilemma [the dual entity of wave and particle] ... [is] the *statistical interpretation* of wave mechanics [according to which] *the waves are waves of probability.*"[36] For Born, these waves, apart from their objective reality, had to have something to do with the subjective act of observation. With this new form of causality, only the *probability* of subsequent events was to be governed by exact laws.[37]

Born took the only physically meaningful questions to be those that could be answered operationally. Hence, for him, the trajectory of an atomic particle during interaction or scattering was not a proper topic of discussion for quantum mechanics. In the 1936 edition of his *Atomic Physics*, Born explicitly said that we are not justified in speaking of the existence of a "particle" unless we can determine both its position and its momentum simultaneously and exactly.[38] He has here laid down a criterion for what is and what is not justified, but there was no independent argument that one *must* (logically, or on pain of contradiction) accept this stricture. Bohr's principle of complementarity is of central importance for Born and only by use of all complementary experimental arrangements, Born claimed, can we obtain the maximum information possible for an atomic system.[39]

This brief overview certainly is not an exhaustive cataloging of various Copenhagen interpretations. However, one can see here a certain set of common commitments: complementarity, completeness of the description (in terms of the state vector or probability amplitude), a prohibition against any possible alternative causal description in a space-time background, and a positivistic attitude.[40] Henceforth, I use the expression 'Copenhagen interpretation' in this sense.

Applications of the formalism of quantum mechanics to (idealized) position-momentum measurements, double-slit arrangements, and the like have often been taken to lead to a picture, or interpretation, in which definite space-time trajectories cannot be maintained, specific possessed values of observables (such as position) are not possible at *all* times, event-by-event causality must be abandoned, the process of measurement (i.e., the projection postulate or collapse of the wave function) assumes a central and highly problematic role in nature, and the passage to a classical limit (in terms of an underlying physical ontology) defies any coherent description. An examination of the formalism in specific EPRB-type experiments shows the nonseparable nature of the theory, which gives rise to correlations that may imply the existence of nonlocal influences between spatially separated regions (really, at spacelike separations).[41] On the Copenhagen interpretation of quantum mechanics, physical processes are arguably, at the most fundamental level, both inherently indeterministic and nonlocal. The ontology of classical physics is dead.[42] This will become especially clear in the measurement process as described by standard quantum theory.

3.4 Complementarity/wave-particle duality

The principle of complementarity is usually conceded as being Bohr's most profound contribution to the foundations of quantum mechanics. It is also the doctrine of his that produces the most dispute as to its precise meaning. Anyone who has actually attempted to read much of Bohr's writing on causality and complementarity is likely to find it, at best, obscure. It is one thing to understand *how*, historically, Bohr came to his position and quite another to *assess* that position.[43] Mara Beller has reconstructed the historical setting of Bohr's 1927 Como lecture to decipher the *original* meaning of his complementarity principle.[44] She contends that Bohr in 1927 saw the resolution of the paradoxes of atomic physics to lie in the union of Schrödinger's continuous wave mechanics with the quantum postulate of discrete energy states.[45] Bohr argued for a complementarity between space-time and causal descriptions. It was Heisenberg who later (in his indeterminacy paper) reinstated particles in space-time.[46] A problematic aspect of Bohr's method of arriving at such overarching views is that he typically did so not via universal arguments or proofs claiming broad epistemic conclusions, but rather via detailed considerations of highly specific thought experiments. One must always be concerned about sweeping generalizations based on an examination of highly specific cases. Bohr felt that he was *forced* to adopt the complementarity viewpoint by empirical results.[47] This is a good example of an unwarranted slide from consistency to necessity. Toward the end of his life, Bohr remarked: "I think that it would be reasonable to say that no man who is called a philosopher really understands what is meant by complementary descriptions."[48] One obvious explanation for this situation is that the doctrine itself is flawed or poorly stated. More recently, Bohr's complementarity has been characterized as a general framework for the description of nature, rather than as some sharp, well-defined principle.[49]

Any formal doctrine of a symmetrical wave-particle duality and the holism of object-apparatus were *later* developments of Bohr's thoughts on complementarity.[50] According to duality, a physical system (e.g., an electron or a photon) behaves *either* as a wave *or* as a particle, depending upon the context or environment. Pauli, in 1933, gave the first consistent discussion of wave-particle duality.[51] Only in response to the challenge of the 1935 EPR paper did Bohr emphasize the key role played by such contextuality.[52]

Let me make a somewhat lengthy parenthetical comment here lest some purists be disturbed by what I take to be a rather minor issue in the present context. In this book I often use wave-particle duality as an example of Bohr's concept of complementarity.[53] Is this fair? I do not know of any "smoking-gun" citation in which Bohr states explicitly that such duality *is* a case of complementarity. Heisenberg, though, did say that "Bohr considered the two pictures—particle picture and wave picture—as two complementary descriptions of the same reality."[54] According to Pauli, if "the use of a classical concept excludes that of *another,* we call both concepts . . . *complementary* (to each other), following Bohr."[55] Is that a reasonable gloss on Bohr? Such a reading is often given, but that does not address the basic question of accuracy. Let me look at just a few of Bohr's own explanations of complementarity. In his 1927 Como lecture, where he first enunciated such a concept, Bohr spoke of "a complementarity . . . [of] the properties of light and material particles."[56] Nearly two decades later he stated that "evidence obtained under different experimental conditions cannot be comprehended within a single picture, but must be regarded as *complementary* in the sense that only the totality of the phenomena exhausts the possible information about the objects."[57] This essential element of ambiguity constitutes the dilemma of the corpuscular and wave properties of electrons and photons, where we employ contrasting pictures.[58] He believed that we must "make a choice between different complementary types of phenomena we want to study."[59] Bohr stated explicitly that we must "adopt a new mode of description designated as *complementary* in the sense that any given application of classical concepts precludes the simultaneous use of other classical concepts which in a different connection are equally necessary for the elucidation of the phenomena."[60] Does it really do violence to this mysterious concept of complementarity to offer wave-particle duality as an illustration of Bohr's view of representing physical reality in terms of classical concepts? I think not.[61] And, even if one takes exception with me on this point, I do not believe that anything essential in my overall argument in this book turns on the use of such an illustration.[62]

By whatever historical route, Bohr did arrive at a doctrine of mutually exclusive, incompatible, but necessary classical pictures in which any given application emphasizing one class of concepts *must* exclude the other.⁶³ This is actually a *hypothesis* appended to the formalism of quantum mechanics.⁶⁴ In the traditional, interference-type experiments the formalism of quantum mechanics (i.e., the Heisenberg uncertainty relation for position and momentum) *does* guarantee that the interference pattern vanishes whenever "which-path" information is obtained.⁶⁵ An experimental arrangement has recently been suggested in which *tunneling* (a wavelike characteristic) occurs with *anticoincidence* (which supplies "which-path" information).⁶⁶ It has been claimed that this specific proposal "reveals that the formalism of quantum mechanics *does* allow a situation where the Bohrian notion of 'mutual exclusiveness' of classical pictures ceases to be applicable."⁶⁷ As in the case of the neutron-interferometry experiments to be discussed in section 5.2, one may find a causal account *appealing* in such a situation, but one is not *forced* to subscribe to it.⁶⁸ What I want to emphasize here is that there is a viable alternative to Bohr's broadly construed concept of complementarity. But neither logic nor empirical considerations demand either choice. In Bohrian complementarity we have another example of his general philosophical prohibitions overstepping the necessary logical implications of the mathematical formalism of quantum mechanics (i.e., in categorically forbidding certain possibilities).

3.5 Perennial problems

Ever since its inception, the Copenhagen interpretation of quantum mechanics has been plagued with certain counterintuitive, and arguably even inconsistent, implications of its view of the measurement process and of the classical limit. Although I give an elementary discussion of this measurement process here, I hold that these examples do exhibit the essence of the conundrum that saddles the standard version of quantum theory. Even though the measurement problem has been around for nearly seventy years now, it has resolutely resisted all attempts at a coherent solution.⁶⁹

3.5.1 *Measurement*
Let me illustrate the measurement process as described by the Copenhagen interpretation of quantum mechanics by means of an idealized thought experiment to measure the spin of an electron.⁷⁰ (I gloss over a difficulty Bohr long ago pointed out: that the uncertainty relations preclude one from using an apparatus to measure the magnetic moment of a *free* electron.⁷¹ There are practical modifications necessary for an actually realizable experiment, but I do not discuss those here.) Figure 3.1 shows

Standard Quantum Theory 35

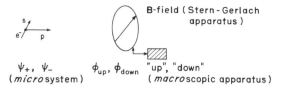

Figure 3.1 An idealized measurement process

a microsystem (an "electron") whose spin projection we wish to determine by use of a magnetic field. I denote by ψ_+, ψ_- the eigenvectors for the spin-up and spin-down states, respectively, of this microsystem. Similarly, ϕ_{up} and ϕ_{down} are the eigenvectors for the macroscopic apparatus corresponding to an output (or "meter reading") for the microsystem having been in the spin-up or -down state. The basic idea is that initially the two systems are far apart and do not interact. Then, the electron enters a region of interaction with the measuring apparatus, and the state of the macrosystem becomes correlated with that of the microsystem. In an ideal situation, we would have the final state of the apparatus uniquely and perfectly correlated with what the state of the microsystem was *before* the interaction and leave undisturbed the initial state of the microsystem (a "faithful measurement"). The two systems then separate, the interaction ceases and we read off the output ("up" or "down") of the apparatus to learn what the state of the microsystem had been *before* the measurement. In the appendix to this chapter, I outline the formal description that quantum mechanics gives of this process. Here I give an informal verbal summary of this description.

If we knew (of course, there would then be no need to do a measurement to ascertain this fact) that the electron was in the state ψ_+ and that the apparatus was in the ("neutral" or "idling") state ϕ_0 (corresponding to no reading), then prior to any interaction between these two subsystems, we would represent the state of the entire system by $\Psi(t=0) = \psi_+ \phi_0$. It is relatively straightforward to exhibit a specific interaction between the electron and the measuring apparatus that will, via the Schrödinger equation, carry this initial state into the final (or "out") state $\Psi_{out}(t) = \psi_+ \phi_{up}$, which is just what we have required of a faithful measurement process that does not disturb the microsystem.[72] Similarly, *if* we knew that the electron was in the state ψ_- and that the apparatus was again in the state ϕ_0 (corresponding to no reading), then prior to any interaction between these two subsystems, we would represent the state of the entire system by $\Psi(t=0) = \psi_- \phi_0$. The interaction between the two subsystems would lead to the result $\Psi_{out}(t) = \psi_- \phi_{down}$. So far, there does not seem to be any measurement problem at all.

However, we do not know, *prior to a measurement*, that the state of the electron is either ψ_+ or ψ_-, but only (at best) that it is in the superposition given by[73]

$$\psi_0 = \alpha\psi_+ + \beta\psi_-, \qquad |\alpha|^2 + |\beta|^2 = 1. \tag{3.4}$$

It is now endgame because the linearity of the Schrödinger equation (cf. appendix to this chapter) *requires* that, after the same interaction as previously, the final state must be[74]

$$\Psi_0 \equiv (\alpha\psi_+ + \beta\psi_-)\phi_0 \xrightarrow[t\to+\infty]{} \Psi_{out} = \alpha\psi_+\phi_{up} + \beta\psi_-\phi_{down}. \tag{3.5}$$

Unfortunately, the state on the far right side of this equation does not correspond to a definite state for a *macroscopic* apparatus. Taken literally, this result would say that the macroscopic apparatus is in a superposition of reading (or printing out) both a "plus" and a "minus," or a "yes" and a "no." We *never* observe such macroscopic superpositions. A strict falsificationist would have to conclude that quantum mechanics has been refuted. This is the measurement problem, since the theory predicts results that are in clear conflict with observation. It is at this point that the standard program invokes the "reduction" of wave packet upon "observation" to resolve this problem (or do away with it by fiat):

$$\alpha\psi_+\phi_{up} + \beta\psi_-\phi_{down} \xrightarrow[\text{discontinuously}]{} \begin{cases} \psi_+\phi_{up} \text{ with probability } |\alpha|^2 \\ \psi_-\phi_{down} \text{ with probability } |\beta|^2. \end{cases} \tag{3.6}$$

Various attempts to account for this "miracle" are at the heart of the measurement problem.

Let me now consider briefly what might constitute a resolution of this problem. If we hope that quantum mechanics should apply to individual systems (i.e., to the individuals that make up an ensemble of similarly prepared systems), then we might want the formalism itself to account for the collapse (or "projection") indicated in eq. (3.6). The superposition on the left side of eq. (3.6) would then become just *one* of the terms on the right side of that equation (i.e., the other term would disappear, as it were). Many arguments and proofs exist that this is mathematically impossible as long as the time evolution of the state vector is governed by the (linear) Schrödinger equation. There is no physically acceptable Hamiltonian H that, when inserted into the Schrödinger equation (eq. [3.1]), could produce a time evolution like that required by eq. (3.6).[75] There is a long history of (problematic) attempts to modify the Schrödinger equation to accomplish the reduction of eq. (3.6), but these have their own difficulties and I do not discuss them here.[76]

One can, of course, abandon any pretense that quantum mechanics applies to *individual* systems and retreat to the statistical, or ensemble, level. If we then abbreviate the state on the left side of eq. (3.6) as

$$|\psi\rangle = \alpha|\psi_1\rangle + \beta|\psi_2\rangle \qquad (3.7)$$

and maintain the probability interpretation of $|\alpha|^2$ and $|\beta|^2$, we would demand that the (ensemble) average value for some observable represented by the hermitian operator A should be

$$\langle A \rangle = |\alpha|^2 \langle A \rangle_1 + |\beta|^2 \langle A \rangle_2 , \qquad (3.8)$$

where $\langle A \rangle_1 \equiv \langle \psi_1|A|\psi_1 \rangle$, the value of $\langle A \rangle$ when the system is in state $|\psi_1\rangle$, and similarly for $\langle A \rangle_2$. However, according to rule (iv) of section 3.2, $\langle A \rangle$ should be given by $\langle \psi|A|\psi \rangle$. In this case, a direct calculation yields

$$\langle A \rangle \equiv \langle \psi|A|\psi \rangle = |\alpha|^2 \langle \psi_1|A_1|\psi_1 \rangle + |\beta|^2 \langle \psi_2|A|\psi_2 \rangle \qquad (3.9)$$
$$+ \alpha^* \beta \langle \psi_1|A|\psi_2 \rangle + \alpha \beta^* \langle \psi_2|A|\psi_1 \rangle ,$$

which has additional interference, or cross, terms of the form $\langle \psi_1|A|\psi_2 \rangle$. The first gambit for the optimist is that these interference terms can be kept arbitrarily small for *all* observables A and for *all* states $|\psi\rangle$, for then the empirical adequacy of (an instrumentalist reading of) the formalism would be (effectively) salvaged.[77] But this is not readily done either.[78]

If the state of the system is such that eq. (3.8) obtains, then the system is said to be in a *mixed* state and an ignorance interpretation of which definite, "classical" value actually exists is tenable, whereas if eq. (3.9) (or, equivalently, eq. [3.7]) holds, then the system is in a *pure* state. (Obviously, a mixed state cannot be represented by a state vector.) It is often claimed that the measurement problem consists in arranging to have a pure state reduce to a mixture. One should appreciate, though, that while this would make the calculational scheme empirically adequate at the level of ensembles, it still would not account for the actual production of *one* (and *only* one) *specific* result for each member of the ensemble. Even this limited objective can be achieved only with considerable contrivance. There are (arguably artificial) mathematical models, involving *strictly infinite* limits in time or in the number of particles, for which the cross terms can be made to vanish.[79] However, as long as the time or number of particles is finite (no matter *how* large), the interference terms in eq. (3.9) need not be negligibly small (i.e., there are, in principle, observational consequences of their presence). Bell has raised objections to such approaches.[80] He put the matter very nicely:

> The continuing dispute about quantum measurement theory is not between people who disagree on the results of simple mathematical manipulations. Nor is it between people with dif-

ferent ideas about the actual practicality of measuring arbitrarily complicated observables. It is between people who view with different degrees of concern or complacency the following fact: so long as the wave packet reduction is an essential component, and so long as we do not know exactly when and how it takes over from the Schrödinger equation, we do not have an exact and unambiguous formulation of our most fundamental physical theory.[81]

3.5.2 Schrödinger's cat

Related to the measurement problem is a paradox generated by Schrödinger in 1935.[82] Schrödinger suggested coupling a microsystem (a "uranium" nucleus or atom) and a macrosystem (a live cat). Things are so arranged that, if the nucleus decays (with a characteristic lifetime τ_0), it triggers a device that kills the cat, as indicated schematically in figure 3.2.

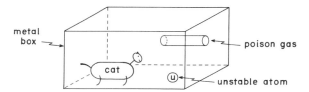

Figure 3.2 Schrödinger's cat

The point of the exercise now is to give a quantum-mechanical description of the time evolution of this coupled system. Let $\Psi(t)$ denote the wave function for the combined system, ϕ that for the cat, and ψ that for the atom. Then the initial state of the system is just

$$\Psi_0 = \Psi(t=0) = \phi_{\text{live}} \psi_{\text{atom}} . \qquad (3.10)$$

This initial state evolves into

$$\Psi(t) = \alpha(t) \phi_{\text{live}} \psi_{\text{atom}} + \beta(t) \phi_{\text{dead}} \psi_{\text{decay}} . \qquad (3.11)$$

The probabilities of interest are

$$P_{\text{live}}(t) = |\langle \phi_{\text{live}} \psi_{\text{atom}} | \Psi(t) \rangle|^2 = |\alpha(t)|^2 \sim e^{-t/\tau_0} , \qquad (3.12)$$

$$P_{\text{dead}}(t) = |\langle \phi_{\text{dead}} \psi_{\text{decay}} | \Psi(t) \rangle|^2 = |\beta(t)|^2 \sim 1 - e^{-t/\tau_0} . \qquad (3.13)$$

As time goes on, things look less and less good for the cat's survival. *Before* we look at the system, $\Psi(t)$ represents a *superposition* of a live and a dead cat. *After* we look, the state function is reduced (live *or* dead).

Schrödinger in effect posed the central question: What does $\Psi(t)$ represent? Possible answers are: (a) *our state* of knowledge (quantum mechanics is incomplete) and (b) the *actual state* of the system (there is a sudden change upon observation). If we choose (a) (which is what Schrödinger feels we *must* do intuitively), then quantum mechanics is *incomplete* (i.e., there are physically meaningful questions about the system that it cannot answer—*surely* the cat was *either* alive *or* dead *before* we looked). On the other hand, choice (b) saddles us with the measurement problem (and with a vengeance). The collapse of the wave function becomes an actual *physical* process that must be explained.

3.5.3 The classical limit

To complete my brief rehearsal of some of the common woes of the standard interpretation of quantum mechanics, I sketch the problem of a classical limit. A theory that supersedes a previous one whose domain of validity has been established must reduce to the old one in a suitable limit. Thus, in special relativity there is a ubiquitous parameter $\beta = v/c$ such that, when $\beta << 1$, the equations of special relativity approach those of classical mechanics. In general relativity, the limit of weak gravitational fields (or small space-time curvature) leads to Newtonian gravitational theory. Similarly, there is a geometrical-optics limit of the wave theory of light that obtains when diffraction effects are negligible (roughly, when the typical dimensions of objects encountered by the light are large compared to the wavelength λ of the light). There is a related and well-known limit in which the equations of wave optics can be replaced by the equations of classical particle mechanics (the so-called eikonal approximation).[83]

If quantum mechanics is to be a candidate for a fundamental physical theory that replaces classical mechanics, then we would expect that there is a suitable limit in which the equations of quantum mechanics approach those of classical mechanics.[84] It is often claimed that the desired limit is $\hbar \rightarrow 0$. But $\hbar$ is not a dimensionless constant and it is *not* possible for us to set it equal to zero.[85] A more formal attempt at a classical limit is Ehrenfest's theorem, according to which expectation values satisfy Newton's second law as[86]

$$\langle F \rangle = m \frac{d^2 \langle r \rangle}{dt^2}. \tag{3.14}$$

This really only implies something to the effect that the centroid of the packet follows the classical trajectory.[87] But wave packets spread and eq. (3.14) is just *not* the same as $F = m\,a$. A similar formal attempt is the

WKB (Wentzel-Kramers-Brillouin) or eikonal approximation which is often advertised as a classical limit of the Schrödinger equation.[88] Again there is not a well-defined limit (in terms of a dimensionless parameter) in which one obtains exactly the equations of classical mechanics for *all* future times.

What is in question here is the consistency of quantum mechanics. If we cannot obtain a classical description of macroscopic objects in a suitable limit, then we do not have a theory that is applicable to both the microdomain and the macrodomain. This remains a serious unresolved problem for traditional quantum mechanics.

Appendix A specific illustration of the measurement problem

Here I sketch the formal account that standard quantum mechanics gives of the measurement process indicated in figure 3.1. Before the systems interact, the total wave function Ψ is a simple product of the wave functions for the two subsystems

$$\Psi = \psi \cdot \phi . \qquad (3.15)$$

Suppose we *knew* that the initial configuration was

$$\psi_0 = \psi^+, \; \phi_0 = \phi_0 \text{ (no reading)}, \qquad (3.16)$$

so that

$$\Psi(t=0) = \psi_+ \phi_0 . \qquad (3.17)$$

The time evolution of Ψ is governed by the Schrödinger equation

$$H \, \Psi(t) = i \hbar \, \frac{\partial \Psi(t)}{\partial t}, \qquad (3.18)$$

where the Hamiltonian

$$H = H_0 + H_I \qquad (3.19)$$

consists of a "free" term H_0 (for each of the two separate, noninteracting systems) and an interaction piece H_I (that turns on and then back off). We can write the formal solution to eq. (3.18) as[89]

$$\Psi(t) = e^{-iHt/\hbar} \, \Psi(t=0). \qquad (3.20)$$

Under the action of H, the *continuous* time evolution of $\Psi(t)$ is[90]

$$\Psi(t) \xrightarrow[t \to +\infty]{} \Psi_{\text{out}}(t) = \psi_+ \phi_{\text{up}} \qquad (3.21)$$

if a decent measuring device is at hand. Similarly, we would have

$$\psi_- \phi_0 \xrightarrow[t \to +\infty]{} \psi_- \phi_{\text{down}} . \qquad (3.22)$$

However, we typically do *not* know in advance whether the state of the microsystem is ψ_+ or ψ_-. Rather, the microsystem is represented by a superposition

$$\psi_0 = \alpha \psi_+ + \beta \psi_-, \qquad |\alpha|^2 + |\beta|^2 = 1, \tag{3.23}$$

and the macrosystem is still in its "ready" (prereading) state

$$\phi_0 = \phi_0, \tag{3.24}$$

so that

$$\Psi(t=0) = (\alpha \psi_+ + \beta \psi_-) \phi_0. \tag{3.25}$$

Since the Schrödinger equation is *linear*

$$H(\Psi + \Phi) = H\Psi + H\Phi, \tag{3.26}$$

it follows from eqs. (3.21) and (3.22) that the Ψ_0 of eq. (3.25) *must* evolve as

$$\Psi_0 \equiv (\alpha \psi_+ + \beta \psi_-)\phi_0 \xrightarrow[t\to+\infty]{} \Psi_{out} = \alpha \psi_+ \phi_{up} + \beta \psi_- \phi_{down}. \tag{3.27}$$

This is just the result stated in eq. (3.5) of the text.

FOUR
Bohm's Quantum Theory

Since the *formalism* of quantum mechanics is not identical with, and need not include, the Copenhagen *interpretation*, there exists the possibility of an alternative, observationally equivalent theory based on that formalism. In this chapter I present such an alternative that preserves particle trajectories and event-by-event causality in a space-time background.[1] In a sense, Bohm's 1952 work can be seen as an exercise in logic—proving that Copenhagen dogma was not the only logical possibility compatible with the facts. I stay quite close to Bohm's original style and notation so that the reader can get some sense of that seminal paper.[2] That will be important for appreciating the historical development of later chapters.

4.1 Bohm's "minimalist" quantum potential theory

In essence, Bohm accepted the formalism of quantum mechanics and showed that more microstructure is consistent with it than had previously been appreciated.[3] The calculational rules given in section 3.2 still give average values for observed quantities, even though there will turn out to be actual particles and trajectories for Bohm. I take special care in elucidating the status of an *effective* projection postulate (rule [v] of section 3.2) as part of the description of the measurement process in his theory.

4.1.1 The formalism
Bohm began with the nonrelativistic Schrödinger equation (eq. [3.1]) and, by means of a mathematical transformation of the type discussed in appendix 1.1 to this chapter, rewrote the basic dynamics of quantum mechanics in a "Newtonian" form. More specifically, he first expressed the wave function $\psi(x, t)$ in the form

$$\psi = R \exp(iS/\hbar), \tag{4.1}$$

where $R(x, t)$ and $S(x, t)$ are *real* functions. (This can *always* be done for any complex function ψ.) In terms of the phase $S(x, t)$ of ψ, a velocity field $v(x, t)$ is defined by the "guidance" condition

$$p = mv = \nabla S. \qquad (4.2)$$

The dynamical equation for the motion of this particle of mass m (and momentum $p = m\,v$) then becomes

$$\frac{dp}{dt} = -\nabla(V + U). \qquad (4.3)$$

Here V is the usual classical potential energy and U is the so-called quantum potential that is given in terms of the wave function ψ as

$$U \equiv -\frac{\hbar^2}{2m}\frac{\nabla^2 R}{R}. \qquad (4.4)$$

The total energy E must now include *all* forms of energy, most particularly the quantum potential U.[4]

The result of eq. (4.3) is *exact* and *no* approximations have been made in passing from the Schrödinger equation to it. This quantum potential U produces highly nonclassical effects on the motion of the particle. Equation (4.3) differs from the basic dynamical equation of standard classical mechanics precisely by the quantum-potential term and it is U that is responsible for the quantum "properties" (or behavior) of the particle. This U has the classically unexpected feature that its value depends sensitively on the *shape*, but not necessarily strongly on the *magnitude*, of $R = |\psi|$ so that U need not fall off with distance as V does.[5] Although eq. (4.3) has a strong formal resemblance to the familiar $F = m\,a$, its content and meaning are *very* different. For instance, when the classical potential V vanishes, the quantum potential U need not. This means that a particle can be accelerated even when there is no classical force acting on it (or, equivalently, that straight-line motion need not obtain when the classical force vanishes). This is important in accounting for interference patterns, as in the double-slit arrangement.

By eq. (4.2), the initial particle momentum p_0 must be restricted to

$$p_0 = \nabla S(x_0, t_0), \qquad (4.5)$$

where x_0 is the initial ("hidden") position of the particle.[6] Thereafter, the particle will stay on this flow line (or trajectory) of the field defined by eq. (4.2).[7] This constraint between x_0 and p_0 is necessary for consistency. Once the initial position x_0 has been specified, the trajectory $x(t)$ is uniquely specified. On the causal interpretation, the future behavior of a particle is determined at the outset by its initial position and by the state of the apparatus (represented by ψ).

All of the mathematical details aside, what Bohm did was to take the Schrödinger equation, which has the form of a wave equation and hence naturally invites a wave (or, perhaps, a wave-particle) interpretation, and reexpressed it in a form similar to Newton's second law of motion, which naturally invites a particle interpretation in terms of trajectories. Because of the influence of the quantum potential, these trajectories are very sensitive to the initial positions of the particles.[8]

As in standard quantum theory, we can consider

$$P(x, t) = |\psi(x, t)|^2 = R^2 \qquad (4.6)$$

as the probability density of our ensemble of particles moving in the velocity field defined by ∇S. The value of the distribution of momenta given by eq. (4.2) averaged over the P of eq. (4.6) is just that given by the usual quantum-mechanical prescription of the expectation value of the momentum operator.[9] In other words, eq. (4.3) with the quantum potential U and eq. (4.6) for the probability distribution produce the same results for average values as the usual quantum formulation in terms of expectation values of the operators x_{op} and p_{op}, where $[x_{op}, p_{op}] = i\hbar$. We can see that a sharply peaked initial distribution $|\psi(x_0)|^2$ will produce large (quantum) forces on a particle (since the U of eq. [4.4] essentially depends upon the second derivative of R) that will vary greatly with the exact position x_0 of the particle in the distribution. It is this great sensitivity to initial conditions that would subsequently produce a large spread in the momenta for the particles in such a packet. The specific shape of the initial distribution $|\psi(x_0)|^2$ depends on the apparatus that prepares the beam and it can be determined only by gathering statistical information about that apparatus.

Bohm observed that:

> This probability density is numerically equal to the probability density of particles obtained in the usual interpretation. In the usual interpretation, however, the need for a probability description is regarded as inherent in the very structure of matter, whereas in our interpretation, it arises, as we shall see . . . , because from one measurement to the next, we cannot in practice predict or control the precise location of a particle, as a result of corresponding unpredictable and uncontrollable disturbances introduced by the measuring apparatus. Thus, in our interpretation, the use of a statistical ensemble is (as in the case of classical mechanics) only a practical necessity, and not a reflection of an inherent limitation on the precision with which it is correct for us to conceive of the variables defining the state of the system.[10]

All of these formal developments can be extended to a many-body system, and I consider that at length in section 9.4.

4.1.2 The conceptual framework[11]

The usual statistical predictions of the standard (Copenhagen) theory are recovered from Bohm's theory provided *all* three of the following mutually consistent assumptions are made: (i) that the ψ-field satisfies Schrödinger's equation; (ii) that the particle momentum is restricted to $p = \nabla S(x)$; and (iii) that we do not predict or control the precise location of the particle, but have, in practice, a statistical ensemble with probability density $P(x) = |\psi(x)|^2$ (the use of statistics is, however, not inherent in the conceptual structure, but merely a consequence of our ignorance of the precise initial conditions of the particle).[12] Notice that ψ plays two conceptually very different roles here: (1) as determining the influence of the environment on the microsystem and (2) as determining the probability density P.[13] It is not a *logical* or an *a priori* necessity that the same function need play both of these roles. (Complete predictive equivalence with standard quantum mechanics follows *only* if this is the case.) In fact, the *primary* conceptual role for ψ in Bohm's theory is (1) (i.e., [i] and [ii] above). That is, (i) and (ii) *alone* would constitute a perfectly complete and coherent system of mechanics and, in a sense, *are* the essence of Bohm's theory. (It would, however, have the slight drawback that it would *disagree* with observations in our actual world.) This would be a thoroughly deterministic scheme and no concept of probability would be needed, although there is certainly room for such a concept in this framework. Bohm later gave an argument to show that, even if initially $P \neq |\psi|^2$, still P would be driven, through random interactions, to the equilibrium distribution $P = |\psi|^2$ (where it would remain by virtue of the continuity equation).[14]

For Bohm, all observations ultimately reduce to position measurements.[15] This is not unreasonable since there *is* something special about *coordinate* space—we exist there (not, say, in momentum space).[16] All instrument outputs are ultimately readings in position space.[17] It might be objected that this favored status of spatial coordinates in Bohm's theory is a shortcoming since a general representational symmetry is an essential feature of and a *calculational* convenience for the standard version of quantum mechanics. However, this *formal* convenience of being able to do calculations in any basis is available for Bohm too.

A not uncommon objection to Bohm's theory is that there is no mutual causality between the guiding wave and the particle—the former affects the latter, but not vice versa. This is held to be an embarrassment. For a proper understanding of action-reaction in Bohm's (1952) theory, the correct approach (cf. section 9.4) is to begin with the many-body case (really, the universe) as governed by Bohmian mechanics and then deduce the dynamics for subsystems, including one-particle systems guided by a "wave" representing the influence of the environment on this microsystem. (The equations *do* turn out to be just the ones Bohm began with for the one-particle system.) Then Ψ is the wave function for the entire system and it time-evolves according to the Schrödinger equation (which includes the mutual interactions of the particles as represented by the classical potential V). This wave function in turn determines the motion of the particles in the system. The point is that the various parts of the system *do* influence each other (through V via the Schrödinger equation). The wave function encodes the holistic state of the system and the interactions among its parts. Who says that, in this quantum regime, our intuitions about action-reaction from the domain of classical physics need be reliable?

Bohm provided a dramatic illustration of the fact that different observations or measurements correspond to different environments (e.g., a particular slit being opened or closed in a double-slit arrangement)—and hence to different wave functions ψ—so that different outcomes (or "trajectories") result. He considered a "free" particle confined to the region between two perfectly reflecting walls.[18] In this case, the boundary conditions imposed on the wave function make the phase of ψ independent of the spatial coordinate.[19] Then by eq. (4.2), the speed of the particle must be *zero*. This, as Einstein pointed out, is very counterintuitive since, whenever we observe such a particle, it is always moving (either to the right or to the left) with a momentum whose value corresponds precisely to the quantized energy level of this well.[20] That energy is *never* zero, nor is the *observed* velocity. As Bohm had already explained in his 1952 paper, while the particle is in the well, its kinetic energy is zero and all of its energy is (quantum) potential. In order to observe a *free* particle, we must, say, remove the confining walls suddenly.[21] This dramatically changes the wave function from one localized between the two walls to a superposition of oppositely moving wave packets now no longer confined to remain within the well (which no longer exists). Such a change in the wave function changes the velocity (or equivalently, the quantum potential). This property of zero velocity also occurs in the ground state of the hydrogen atom where the Coulomb force of attraction by the nucleus is exactly balanced by the repulsive force generated by the quantum potential.[22]

The lesson from this is that the measured value of an observable need not (and in general cannot) be the value that existed before the measurement process.[23] This truly reflects Bohr's concept of the wholeness of quantum phenomena and the spirit of his principle of complementarity. How a microsystem behaves depends upon its environment—an observed value is *contextual*. Bohm's theory is further committed to genuine nonlocality in nature, but this is a not directly observable or controllable action at a distance.

At this level of the causal interpretation, or "theory," we still have no understanding of the physical origin of the highly nonlocal quantum potential U that is responsible for those nonseparable features that are the hallmark of specifically quantum phenomena. Nevertheless, we are arguably better off with regard to understanding than with the Copenhagen interpretation. I have already suggested a reasonable analogy with classical Newtonian gravitational theory with its (instantaneous) action at a distance. That property remained a mystery, even though a causal story could still be told about, say, planetary motion. This new quantum theory does provide us with an ontology of actual particles moving along continuous (even if at times irregular) trajectories in space-time. Such an ontology is not nearly as radical a departure from that of classical physics as is that associated with the Copenhagen version. While the nonlocality of Bohm's theory may appear unpalatable to some, the Copenhagen interpretation has the same nonseparable (mathematical) structure and other bizarre features as well.[24] One of the central lessons we may draw from Bell's theorem and from the analysis resulting from it is that such nonlocality appears to be a feature of our *world*, not just of this or that *theory* of physical phenomena.[25] That being the case, nonlocality itself gives us little reason to choose Copenhagen over Bohm. One can then turn to other criteria, such as intelligibility, simplicity, fertility, and the like.

4.2   New insights

I can now begin to compare and contrast Bohm's quantum theory with the standard one. Some of the most perplexing problems for the Copenhagen interpretation are simply absent from Bohm's theory.

4.2.1   No measurement problem

One of the most beautiful aspects of Bohm's paper is his treatment of the measurement problem (which becomes a nonproblem).[26] In this theory, measurement is a dynamical and essentially a many-body process. There is, as we shall see, *no* collapse of the wave function and hence no measurement problem. The basic idea is that a particle always has a definite position between measurements. There is no superposition of properties, and "measurement" (or observation) is an attempt to discover this position.

Before turning to the example he used in his 1952 paper, I first illustrate the basic argument with the familiar case of the determination of the angular momentum of an atom.[27] Figure 4.1 shows a Stern-Gerlach

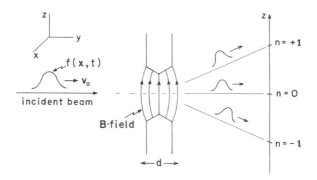

Figure 4.1 A Stern-Gerlach apparatus

device in which an inhomogeneous magnetic field is used to produce a spatial separation among the various angular momentum components of an incident beam of atoms.[28] The situation diagrammed in the figure corresponds to an angular-momentum-one state (i.e., $\ell = \hbar$, $\ell_z = 0, \pm \hbar$). The incident wave packet $f(x)$ moves with a velocity v_0 along the y-axis. This function $f(x)$ (e.g., a Gaussian) is fairly sharply peaked about $x = 0$. The initial state of the atom is a superposition of angular momentum eigenstates $\psi_n(\xi)$ of the atom.[29] Therefore, the initial wave function for the system before the atom has entered the region of the magnetic field is

$$\Psi_0(x, \xi, t) = f(x - v_0 t) \sum_{n=-1}^{1} c_n \psi_n(\xi) . \qquad (4.7)$$

Beginning at time $t = 0$, the packet encounters the magnetic field (which is confined to a region of width d) and takes a time $\Delta t = d/v_0$ to pass through the field.[30] The interaction between the inhomogeneous B-field and the magnetic moment ($\mu = \mu \ell$) of the atom produces on the atom a net force in the z-direction: up for $\ell_z = +\hbar$, none for $\ell_z = 0$, and down for $\ell_z = -\hbar$. This deflects the atom accordingly. Once the packet emerges from the B-field, the three components of the packet diverge along separate paths. After sufficient time has elapsed, the three emergent packets no longer overlap (i.e., they have essentially disjoint supports).[31] The wave function has by then evolved into

$$\Psi(x, \xi, t) = \sum_{n=-1}^{1} c_n f_n[x - x_n(t)] e^{i\varphi_n} \psi_n(\xi) . \qquad (4.8)$$

Here $x_n(t)$ are the particle trajectories indicated on the right of figure 4.1, and the φ_n are simply constant phases.

The description so far has been the standard quantum-mechanical account of a measurement. The next step would be to apply the projection postulate once an atom has been observed (say in the $n = +1$ packet). One would simply "erase" the other two packets. For Bohm, however, the situation is different. The probability of finding *the* atom at some particular position on the screen at the far right of figure 4.1 is

$$P(x, \xi, t) = \sum_{n=-1}^{1} |c_n|^2 f[x - x_n(t)]|^2 |\psi_n(\xi)|^2 . \qquad (4.9)$$

There are no interference (or "cross") terms here because the various f_n no longer overlap (i.e., they have disjoint supports). After *the* particle has been found in *one* packet, it cannot be in either of the others and has negligible probability of crossing to another one (because P effectively vanishes between the packets). It is also clear that once there is no overlap between the packets, the guidance conditions (eq. [4.2]) and the quantum potential (eq. [4.4]), *when evaluated at the location of the particle*, involve only that packet in which the particle actually exists. The remaining two packets become irrelevant for further calculations.[32] In this example, we have learned something about the internal state of the microsystem (i.e., the ξ or the ℓ_z) by determining the location (x) of its center of mass. This type of correlation is at the heart of measurement processes. It also follows from eq. (4.9) that $|c_n|^2$ gives the probability of finding the particle in the nth packet.

There still remains, though, the in-principle possibility of recombining these "empty" waves with the one containing the particle. However, once the particle has coupled to a macroscopic apparatus (e.g., a measuring instrument from which we get a reading or output), this becomes a practical impossibility.[33] The basic reason is that this complex system has a wave function that is defined on a multidimensional configuration space, and this wave function becomes (once a measurement has been accomplished) a sum of disjoint pieces on that space.[34] In my example, the configuration space is six-dimensional (x and ξ).

Bohm's own illustration of a measurement process was the inelastic scattering of a particle by a force center. He developed the quantum-mechanical solution for the scattering of an incident wave packet by a hydrogen atom initially in its ground state $\psi_0(x)$ with energy $E_0 = (\hbar^2/2m)k_0^2$, as shown schematically in figure 4.2. I can summarize the results of the mathematical calculations (cf. appendix 1.2) for this scattering process as follows. The initial wave function for the system (incoming electron in the beam and the hydrogen atom with its electron in the

50   Chapter Four

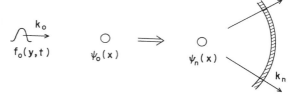

Figure 4.2   An inelastic scattering process

ground state) is simply the product $\Psi_i = f_0 \psi_0$ prior to any interaction between the two. Long after the (physically and mathematically very complicated) scattering process has taken place and the scattered electron is far separated from the hydrogen atom (which here serves as the scattering center), the wave function becomes a sum of terms (i.e., a superposition) corresponding to the various possible outcomes of the scattering. For example, the incident electron could be elastically scattered, leaving the hydrogen atom in its ground state. Or the incident electron could be inelastically scattered, leaving the hydrogen atom in one of its excited states. This is just the standard quantum-mechanical scattering formalism and is nothing peculiar to Bohm's theory. Each of the terms in this (asymptotic) superposition consists of a product of the outgoing spherical wave (or "shell") for the scattered electron and the hydrogen atom wave function corresponding to the energy level that the atom is in after the interaction. Crucial for Bohm's interpretation of the measurement process is that each of these outgoing scattered spherical wave shells is separated from its nearest neighbor by *macroscopic* distances. There is (essentially) *no* overlap between these shells. This means that, after the scattering process has been completed, the total wave function for the entire system reduces to a sum of terms that are *disjoint* (as functions of the coordinate-space variables). Once the scattered electron is found in a given shell, we know *with certainty* that the hydrogen atom is in the energy level corresponding to the energy of the scattered electron. (It is precisely because the speed of propagation of these spherical shells depends upon the energy a scattered electron would have if it were in that shell that these shells separate in space.) The rest of the "pieces" of the wave function subsequently play no role in determining either the probability P or in generating the quantum potential U.[35] The scattered electron is now *known* to be in just *one* of these shells and the shells no longer overlap—implying zero probability of the electron ever being in the region between these shells—thus forbidding the electron's moving over to another shell.

For either of these examples, *if* we had known x_0 (i.e., the hidden variable that I denote as λ_0) for the particle *exactly*, we could have (in principle) predicted in which of the nth outgoing packets it would be located.[36] Due to the chaotic nature of the motion in the scattering region, even small uncertainties in these λ_0 produce large variations in the final location of the scattered particle.[37] Because the spatial distribution of these systems is given by $P = |\psi(x, t)|^2$ (a point to be discussed more fully in section 9.4.1), we are reduced to statistical or probabilistic predictions. Since the quantum potential U acts only on the particle at its *actual* location (which we can "discover" by an observation), only that particular outgoing wave packet containing the particle can have any future influence on the particle. Therefore, we can effectively just drop from the sum all of the other wave packets.[38] This produces the same result for calculations (and for the *future* time evolution of Ψ) as the "reduction of the wave packet" in the Copenhagen interpretation, but there is no actual (physical or mysterious) collapse over space. Because the quantum potential depends more on the *form* than on the *amplitude* of R (cf. eq. [4.4]), we can "renormalize" this one packet to keep the new total probability equal to unity (i.e., adjust the overall normalization constant for the packet so that its integral over all space is unity). Since the new wave function is effectively just a *single* term from the sum, it factors into a simple product, and the quantum potential reduces to a sum of two terms, one referring only to the particle and the other only to the hydrogen atom. That is, after the scattering, these two systems are once again *independent*.

Bohm discussed the necessity of having a microsystem interact, effectively irreversibly, with a macroscopic measuring device that has many degrees of freedom to make it practically impossible (i.e., overwhelmingly improbable) for these "lost" wave packets to interfere once again with the one actually containing the particle.[39] The process of measurement is a two-step one in which (i) the quantum states of the microsystem are separated (by an in-principle reversible interaction) into nonoverlapping parts and (ii) a further (irreversible) interaction with a macroscopic apparatus registers the actual result.[40] By the term 'irreversible' here I mean 'practically impossible' in the sense that, even if the pieces of the scattered waves for the microsystem being "observed" could be made to overlap again in the future, the wave functions for all of the microscopic coordinates of the macroscopic recording apparatus could not.[41]

> In principle, it is of course possible for such a reversal to take place, but the probability is so small that this would not be likely, even over the age of the universe (e.g., as unlikely as it is for water placed on ice to boil). If such an event were to happen in a quantum measurement, it would simply not be recognized as a valid

result in physics, because it could not be reproduced (e.g., it could not be distinguished from a mistaken observation). If we accept such an explanation in thermodynamics, it seems that we should also do so here.[42]

4.2.2 The classical limit
One would expect no particular problems in obtaining the classical equations of motion from those of Bohm's theory since the basic *exact* dynamical equation (eq. [4.3]) already has the *form* of Newton's second law. This intuition is essentially correct since, loosely speaking, when $U/V <<$ 1 (a *dimensionless* parameter), eq. (4.3) becomes just the classical equation of motion. The requirement truly is $U \to 0$ (in the sense of $U/V \to 0$), rather than anything like $\hbar \to 0$. There will be some states (certain ψ) that have no classical limit. I do not go into a detailed discussion of the technicalities of the classical limit, since those can be found elsewhere.[43]

There is no mismatch between Bohm's ontology and the classical one regarding the existence of trajectories and the objective existence of actual particles. Conceptually, a continuous passage from the microdomain to the macroworld is possible.[44] The nonlocality of the theory is manifest through the quantum potential U and, when U vanishes, so do the nonlocal connections. On the other hand, when ∇U is large (as happens, for example, in regions where $|\psi| = R$ is small and rapidly changing), the motion can be "very irregular and complicated, resembling Brownian motion more closely than it resembles the smooth track of a planet around the sun."[45]

4.2.3 The uncertainty relations
The status of the Heisenberg uncertainty relations is much different in Bohm's theory from their status in standard quantum theory. They are practical limitations on the accuracy with which we can simultaneously *know* or *measure* the values of noncommuting observables, but not on the *existence* of such quantities. Because of the initial uncertainty in the position of a microsystem (say, an "electron"), any attempt to predict the value of some observable q associated with the electron can result only in the *probability* of obtaining some given value for q. Averaging over these hidden variables leads to the statistical spread in a series of repetitions. Bohm also showed that, once we have ascertained the value of q (by measurement), we can again predict only the probability of the outcome of a measurement of a noncommuting observable p.[46] From Bohm's perspective, the indefiniteness in the correlation between the state of the system and of the apparatus *is* the uncertainty.[47]

Whatever the interpretation, from the standard formalism of eqs. (3.1) through (3.3), which Bohm accepts, follows the Heisenberg uncertainty, or indeterminacy, relation. The usual Hilbert space–based arguments are still available to obtain the relation

$$\Delta x \, \Delta p \geq \frac{\hbar}{2} \qquad (4.10)$$

for operators x and p that satisfy $[x, p] = i\hbar$. These are *formal* relations for the dispersions Δx and Δp and do not *necessarily* have a direct physical interpretation in terms of a *spread* about fairly sharply peaked values of x or of p (since Δx and Δp depend upon the specific state or distribution used to compute them).[48] Bohm's discussion shows how the act of measurement, through the influence of the quantum potential, can disturb the microsystem and thus produce an uncertainty in the outcome of a measurement.[49] That is, the *formal* derivation of the Heisenberg uncertainty relations goes through as before, but now we have some understanding of how the spreads come about physically. It also becomes clear that the value of the "observable" (except for the *position*) obtained upon measurement need not (and usually does not) correspond to any one value necessarily possessed by the microsystem before the measurement.[50] We usually cannot infer from the result of a measurement what the value of a variable was prior to the measurement. Bohm made just this point: "However, in our suggested new interpretation of the theory, the so-called 'observables' are, as we have seen . . . , not properties belonging to the observed system alone, but instead potentialities whose precise development depends just as much on the observing apparatus as on the observed system."[51]

4.3   Is there *complete* observational equivalence?[52]

Since these two versions of quantum mechanics have the same set of equations available for calculation, it might seem as though they should be observationally completely equivalent.[53] If the calculation of any observable quantity is well posed in both theories, then both theories will calculate, or produce, the same answer when the (common) formalism is applied. This appears to have been Bohm's own view since, when asked directly in 1986 whether there were any new predictions from his model, Bohm responded, "Not the way it's done. There are no new predictions because it is a new interpretation given to the same theory."[54] John Bell made a similar point, but a bit more circumspectly: "It [the de Broglie–Bohm version of nonrelativistic quantum mechanics] is experimentally equivalent to the usual version insofar as the latter is unambiguous."[55]

Could it be, though, that a certain class of phenomena might correspond to a well-posed problem in one theory but not in the other? Or might the additional micro-ontology (i.e., particles and definite trajectories) of Bohm's theory issue in a prediction of an observable where Copenhagen would just have no definite prediction to make? These two theories would not each have made definite predictions that disagreed. Rather, one would have made a definite prediction and the other would simply remain agnostic on the question. I suggest that there *may* be a difference for an observable that cannot be represented by a hermitian operator and I assess the likely observational consequences of this.

In standard quantum mechanics the time t is simply a *parameter*, not an *operator*, and there is, in general, no well-defined meaning to the transit time of an individual particle between two points (because, of course, the particle aspect of an object need not be the appropriate description under all circumstances). To find the average value of an observable one usually computes the expectation value of the corresponding hermitian operator (i.e., rule [iv] of section 3.2). This cannot be done to find an average value for transit times since time cannot be represented by a hermitian operator.[56] For Bohm's theory, by contrast, the transit time of a particle between any two points is conceptually well defined.

As an illustration of the type of "experiment" one might do to probe these conceptual differences, consider a beam of particles incident upon a potential barrier as shown in figure 4.3.[57] Some of the particles in the

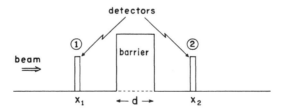

Figure 4.3 Quantum tunneling times

beam will be transmitted through the barrier and some will be reflected. As an idealized type of thought experiment, suppose that, particle by particle (i.e., event by event), we knew the time $t_j^{(1)}$ at which the jth particle passes x_1 (i.e., the time at which detector 1, assumed perfectly efficient, fires) and then the time $t_j^{(2)}$ at which it passes detector 2 (if it is transmitted), or $t_j^{(3)}$ at which it (again) passes detector 1 (if it is reflected). With these as data, one can then compute the time $\tau_j = t_j^{(2)} - t_j^{(1)}$ (or $t_j^{(3)} - t_j^{(1)}$) each particle spent in the region (x_1, x_2) and then the *dwell time* τ_D, which is the average time spent by the particles ($j = 1, 2, 3, \ldots, N$) between points x_1 and x_2. Similarly, one can define a *reflection time* τ_R, which is the average time spent in (x_1, x_2) by the *reflected* particles, and a *transmission time* τ_T. If we grant for discussion's sake at the moment that these τs are accessible experimentally (i.e., that the various t_js can be inferred), then we can ask a quantum theory to calculate these quantities. Much thought and effort have gone into how one might coherently calculate these τs in standard quantum mechanics, but there appear to be *no* candidates that are totally consistent.[58] The upshot of this work is that it "leaves open the question of the length of time a transmitted [reflected] particle spends in the barrier region. It is not clear that a generally valid answer to this question exists."[59] This is not totally surprising since, for standard quantum mechanics, there is no particle concept available in the barrier region.

Bohm's theory can be, and has been, used to calculate these various τs.[60] There are difficult (both theoretical and experimental) technical questions that are still unresolved about these tunneling times.[61] It remains unclear whether or not a convincing test is possible in this area. If it should turn out that the predictions of Bohmian mechanics for the τs can be compared with experiment and are found to be in error, then that would count as a refuting instance, a failure for Bohm's program.[62] Copenhagen would be neither supported nor refuted there, since it is unable to make any unambiguous predictions. Bohm's program is clearly at more risk (i.e., falsifiable) here. Or it could happen that Bohm's predictions cannot be compared with experiment (perhaps for in-principle or technical reasons) and a case of evidential underdetermination would still remain.

But what if the Bohm predictions for the τs *agreed* with experiment while Copenhagen had to remain on the sidelines, no prediction in hand? Is this success additional evidential support of Bohm over Copenhagen? While it may appeal philosophically to some to take such additional empirical success of one theory, even in the absence of any *incorrect* prediction by the other theory, as decisive between the two, it is more likely that the other school (here, Copenhagen) would simply appropriate whatever *formula* had been generated and attribute it to a *heuristic* argument that produced a useful result.[63] So much is at stake—two *radically* different ontologies or worldviews—that the dominant school would more likely claim to have "learned" how to calculate (in this case) than to grant reality to the microstructure that still remains not directly observable.

Because these two versions of quantum mechanics have so far proven to be observationally equivalent, I shall continue to refer to them as such. That is, I simply bracket the concern about possible *future* observational difference until I discuss in section 11.2.2 some methodological issues relevant to underdetermination.

4.4 Various types of nonlocality

Throughout this book I use interchangeably the terms 'nonlocal' and 'nonseparable' and I now discuss my reasons for not distinguishing between the concepts of nonlocality and of nonseparability.[64] *Logical* distinctions do exist between them and I begin by laying out some basic terminology common in the literature on Bell's theorem and its implications.[65] The context for these comments is an EPRB-type correlation experiment previously illustrated in figure 2.1. At spatially separated stations A and B, experimenters can choose instrument settings (labeled i and j for each station respectively) and the possible outcomes of the corresponding measurements are denoted by x and y.[66] In this notation, $P(x, y \mid i, j)$ represents the joint probability of obtaining a result (or outcome) x at station A when the setting (or parameter) i has been chosen by the experimenter at A, while the result y obtains at station B when setting j has been chosen there.[67]

4.4.1 Bell's nonlocality

In John Bell's celebrated theorem, a set of parameters λ (the hidden variables) is assumed to give a complete state specification for the system.[68] At the fine-grained level, the appropriate joint probabilities are denoted (in my notation) as $P_\lambda(x, y \mid i, j)$. The λ are distributed according to a weighting function $\rho(\lambda)$ (normalized to unity) so that the observed probabilities are given as

$$P(x, y \mid i, j) = \int P_\lambda(x, y \mid i, j)\, \rho(\lambda)\, d\lambda \, . \tag{4.11}$$

The marginal $P_\lambda(x \mid i, j)$ is defined as

$$P_\lambda(x \mid i, j) \equiv \sum_y P_\lambda(x, y \mid i, j). \tag{4.12}$$

A similar expression holds for $P_\lambda(y \mid i, j)$. From eq. (4.11) it follows that eq. (4.12) holds as well for the $P(x \mid i, j)$ and $P(y \mid i, j)$ in terms of the $P(x, y \mid i, j)$. The conditional probability $P_\lambda(x \mid i, j, y)$ is *defined* by the relation

$$P_\lambda(x, y \mid i, j) = P_\lambda(x \mid i, j, y)\, P_\lambda(y \mid i, j) \, , \tag{4.13}$$

with a similar definition for the $P(y \mid i, j, x)$ in terms of $P(x, y \mid i, j)$ and $P(x \mid i, j)$. Here we condition on *possible* outcomes y (i.e., not on outcomes that can never occur).

Bell assumed that there could not be *any* instantaneous influence from one region on another that is spatially separated from it. Arthur Fine has characterized as *Bell locality* the more restricted constraint that the outcome at one station cannot instantaneously be influenced by the kinds of measurements made at another station.[69] On the basis of his own version of locality, Bell assumed that the P_λ could be factored as[70]

$$P_\lambda(x, y \mid i, j) = P_\lambda(x \mid i)\, P_\lambda(y \mid j)\,, \quad (4.14)$$

where $P_\lambda(x \mid i)$ and $P_\lambda(y \mid j)$ are the probabilities for outcomes at one station irrespective of what happens or is done at the other station. The condition of equation (4.14) is sometimes termed *factorizability*.[71] To give away the end of my story early on, let me say that when I use the term 'nonlocality' I am referring to a violation of Bell's own version of locality. That is, instantaneous influences (of whatever type) between (spacelike) separated regions constitute my nonlocality. As Fine points out, a *restricted* Bell locality *alone* is not sufficient to underwrite the factorizability of eq. (4.14). If factorizability fails, then such locality need not be taken as refuted.[72] If Bell locality fails (as it does in Bohm's theory), so will factorizability. The real concern is whether or not a violation of locality allows superluminal signaling (or communication—an observable effect), as opposed to (mere) superluminal influences (which need not be accessible to observation).

4.4.2 *Locality versus separability*

Having defined my 'nonlocality', I should now explain why I do not make use of several distinctions that are possible. Jon Jarrett has shown that factorizability is implied by the conjunction of two other properties: Jarrett locality and Jarrett completeness. *Jarrett locality* (also termed 'parameter independence' by Abner Shimony and simply 'locality' by Don Howard) means that the single distribution of outcome x at station A (given the state specification λ) is independent of the choice j made at station B;[73] i.e.,

$$P_\lambda(x \mid i, j) = P_\lambda(x \mid i)\,. \quad (4.15)$$

[A similar condition holds for $P_\lambda(y \mid i, j)$.] If this is violated (and it is *not* in quantum mechanics), then there *could* be a conflict with special relativity, in the following sense.[74] Jarrett argued that, *provided* we could control the parameters λ, then superluminal communication would be possible with the marginals (hence, Shimony's *controllable* nonlocality). Quite simply, if eq. (4.15) were violated, the statistics gathered at station A for outcome x would depend not only upon the choice i made there, but also upon the choice j made at the distant station B. By prior arrangement (of what choice the experimenter at B would make if a given occurrence did or did not take place at station B), the two spatially separated experimenters could communicate via this correlation and do so instantaneously—a message would be sent "faster than light."[75] The caveat about being able to control the parameters λ is crucial, as Jarrett himself has admitted.[76]

However, in Bohm's theory the hidden variables $\{x_0\}$ are *not* under our control. Quantum correlations in Bohm's theory can be used to signal *only* when $P \neq |\psi|^2$. As I discuss further in section 9.4.1, it has been shown that, *in Bohm's theory,* "instantaneous signaling is possible *if and only if* $P_0 \neq |\psi_0|^2$."[77] Such signaling does require an *entangled* state. Recall that for Bohm *all* measurements are ultimately position measurements so that the only probabilities relevant to this discussion are the $P(x_1, x_2, \ldots, x_n)$ (i.e., probabilities of finding various particles at certain *positions*). Our inability to "beat" the Heisenberg uncertainty relations prevents us from controlling the $\{x_0\}$ (*the* hidden variables for Bohm) well enough to signal. It has further been established "that the uncertainty principle holds if $P = |\psi|^2$, but is generally violated otherwise."[78] Quantum equilibrium (i.e., $P = |\psi|^2$) is the key to making (Bohmian) nonlocality uncontrollable.[79] While there are uncontrollable superluminal influences, these cannot, in Bohm's theory, be used to send a signal (i.e., to communicate). Does this mean the special theory of relativity (STR) is violated? Well, if the first-*signal* principle means no superluminal *communication* (or signaling), then no; but if it means no superluminal *influences,* then yes. I return to this question in chapter 10. Bohm's theory is nonlocal in the sense of violating eq. (4.15), but that need not generate any *empirical* conflict with special relativity.

Jarrett completeness (Shimony's outcome independence and Howard's separability) means that the conditional probability for outcome x at station A (given the state specification λ) is independent of the measurement outcome y at station B:

$$P_\lambda(x \mid i, j, y) = P_\lambda(x \mid i, j) . \quad (4.16)$$

That is, this conditional probability becomes a marginal.[80] If this condition fails (as it does in quantum mechanics) while eq. (4.15) holds, then common-cause explanations are not possible.[81] However, a violation of eq. (4.16) (which *is* the case for quantum mechanics) does not allow superluminal signaling.[82] This is the content of the no-signaling theorems.[83] Shimony has labeled a violation of eq. (4.16) *uncontrollable nonlocality.* It should be clear that Bohm's theory does satisfy eq. (4.16) (i.e., it is a completely deterministic theory). If specification of (λ, i, j) completely determines (in principle) the outcomes (x, y), then additional conditioning on y is superfluous—y has *only* one allowed value—and cannot affect the probability (which can be only 0 or 1 in a deterministic theory). The point of Jarrett's analysis is just this. From the very meaning of joint and conditional probabilities in eq. (4.13), we see that if Jarrett locality (eq. [4.15]) and Jarrett completeness (eq. [4.16]) hold, then eq. (4.14) follows at once. So (Howard's) locality and separability together imply factorizability.

Let me give a short list of correspondences among these various types of locality, etc.:[84]

(i) (Jarrett or Howard) locality = (Shimony) parameter independence = eq. (4.15) respected (outcomes at one station are independent of the choices made [*i.e.*, parameter settings] at the other station);
(ii) (Shimony) controllable nonlocality = eq. (4.15) violated (a *possibility* of superluminal signaling);
(iii) (Jarrett) completeness = (Howard) separability = (Shimony) outcome independence = eq. (4.16) respected (outcomes at one station are independent of the outcomes at the other station); and
(iv) (Shimony) uncontrollable nonlocality = eq. (4.16) violated (*no* possibility of superluminal signaling).

4.4.3 *Nonlocality in Bohm's theory*

In Bohm's nonlocal, deterministic theory, eq. (4.15) is violated and eq. (4.16) respected *at the level of the hidden variables*. However, these conditions need not hold for the actually accessible (or integrated) probabilities (cf. eq. [4.11]). For Bohm, the "experimental" (or integrated) probabilities are just those of standard quantum mechanics (i.e., once the subscript λ has been dropped, they satisfy eq. [4.15] and violate eq. [4.16]).[85] So it is not clear that Howard's locality and separability concepts are particularly useful *in this case*.[86] If one were to *insist* upon such terminology here, then the P_λs would be nonlocal and separable, while the (integrated) Ps would become local and nonseparable.[87] The no-signaling proofs of quantum mechanics *are* relevant for the nonlocality of Bohm's theory just because only the *observable* correlations are accessible to us. From the perspective of Bohmian mechanics, quantum equilibrium ($P = |\psi|^2$), rather than a locality/separability distinction, is the relevant condition for no signaling.

In Bohm's theory the (complete) specification of the state of a system is (Q, Ψ), where Q stands for the configuration-space coordinates of the particles. The state vector Ψ will generally be entangled (or nonseparable).[88] It would seem not unreasonable to consider the theory nonseparable. There can be nonlocality for a *single* particle (via the quantum potential), but there is further entanglement when the state vector is not separable. What should this be termed: nonlocality or nonseparability? Why not just (Bell) nonlocality? That is basically what I do in this book.

I take 'separability'/'nonseparability' to be a technical or mathematical property of a state vector Ψ. If Ψ is a *single* product of the state vectors of the subsystem, then there is no entanglement and the *state vector* is separable.

On this question of the proper concept of and terminology for nonlocality, it is useful to return to Bell's own initial paper on the famous inequality. There he worked with expectation values, not with probabilities. The basic quantities in his proof were the actual outcomes (i.e., the observed values) $A_\lambda(i, j)$ and $B_\lambda(i, j)$.[89] Notice that, as one might expect in a *deterministic* theory, A_λ does not depend upon B_λ or vice versa. On the basis of (Bell) locality, he then required that

$$A_\lambda(i, j) = A_\lambda(i), \quad (4.17)$$

$$B_\lambda(i, j) = B_\lambda(j), \quad (4.18)$$

so that the correlation $\langle AB(i, j)\rangle$ becomes

$$\langle AB(i, j)\rangle \equiv \int A_\lambda(i, j)\, B_\lambda(i, j)\, \rho(\lambda)\, d\lambda = \int A_\lambda(i)\, B_\lambda(j)\, \rho(\lambda)\, d\lambda, \quad (4.19)$$

which then leads to the famous inequality.[90] What would separability (in Howard's sense) even *mean* for the $A_\lambda(i, j)$ and $B_\lambda(i, j)$? Equations (4.17) and (4.18) are a locality condition in the sense of eq. (4.15). But separability, according to eq. (4.16), is to have something to do with independence of the *outcome* at the *other* station. However, the A_λs do not depend upon the B_λs (and vice versa). In terms of the physical observables (A_λ and B_λ), locality seems to be *the* relevant concept.[91] In the case of Bohm's *deterministic* theory, it is not helpful to frame the question of (Bell) locality in terms of a conjunction of Howard's locality and separability. Hence, I use simply the term 'locality' (i.e., loosely speaking, Bell locality) and 'nonlocality' (perhaps "Bell nonlocality"?) and do not treat Howard's two concepts as meaningfully distinct for the physics of Bohm's theory.

Once what I claim to be a fairly benign type of nonlocality has been accepted in Bohm's theory, several long-standing difficulties, such as the measurement problem and the existence of a classical limit, simply evaporate.

Appendix 1 Some technical details of Bohm's program

In this appendix I give a few of the mathematical calculations that are necessary to demonstrate some of the claims made in the text about

Bohm's theory. Here I stay, for the most part, very close to Bohm's original presentation in his 1952 paper.[92] For further reference the reader is encouraged to consult Peter Holland's recent book (1993a) on Bohm's theory, David Bohm and Basil Hiley's (1993) recapitulation, and Antony Valentini's monograph (1994) on the subject.

1.1 The transformation of Schrödinger dynamics

Bohm's basic idea is the following. Beginning with the (nonrelativistic) Schrödinger equation (which is *accepted, not derived*, here)

$$i\hbar \frac{\partial \psi}{\partial t} = -\frac{\hbar^2}{2m} \nabla^2 \psi + V\psi, \qquad (4.20)$$

one defines two *real* functions R and S as

$$\psi = R \exp(iS/\hbar). \qquad (4.21)$$

Substitution of eq. (4.21) into eq. (4.20) and separation of the real and imaginary parts of the resulting expression yields

$$\frac{\partial R}{\partial t} = -\frac{1}{2m}[R\nabla^2 S + 2\nabla R \cdot \nabla S], \qquad (4.22)$$

$$\frac{\partial S}{\partial t} = -\left[\frac{(\nabla S)^2}{2m} + V - \frac{\hbar^2}{2m}\frac{\nabla^2 R}{R}\right]. \qquad (4.23)$$

The *quantum potential U* is defined as

$$U \equiv -\frac{\hbar^2}{2m}\frac{\nabla^2 R}{R}. \qquad (4.24)$$

With the identification

$$P = R^2 = |\psi|^2, \qquad (4.25)$$

eq. (4.22) can be rewritten as

$$\frac{\partial P}{\partial t} + \nabla \cdot \left(P \frac{\nabla S}{m}\right) = 0. \qquad (4.26)$$

If U were identically zero, then eqs. (4.23) and (4.26) together would represent a continuous "fluid" of particles of momentum

$$p = \nabla S \qquad (4.27)$$

following well-defined classical trajectories.[93] With this assignment for $p = mv$ the $P = |\psi|^2$ of eq. (4.26) can be given the interpretation of a probability density for the distribution of particles, since eq. (4.26) then becomes the standard continuity equation. Notice that this is all very similar to what one normally does in quantum mechanics, where a (probability) current density is defined as

$$j = \frac{i\hbar}{2m}(\psi\nabla\psi^* - \psi^*\nabla\psi) = \frac{R^2}{m}\nabla S \equiv P v, \quad (4.28)$$

from which it follows that

$$v = \frac{1}{m}\nabla S. \quad (4.29)$$

However, even when $U \neq 0$, we can use eq. (4.27) to write[94]

$$\frac{dp}{dt} = v_j \frac{\partial}{\partial x_j}(\nabla S) + \frac{\partial}{\partial t}(\nabla S) = \frac{1}{m}p_j\frac{\partial}{\partial x_j}(\nabla S) + \frac{\partial}{\partial t}(\nabla S)$$
$$= \frac{1}{m}\nabla S \cdot \nabla(\nabla S) + \frac{\partial}{\partial t}(\nabla S) = \nabla \cdot \left[\frac{1}{2m}(\nabla S)^2 + \frac{\partial S}{\partial t}\right]. \quad (4.30)$$

The last form of this expression, plus eq. (4.23), imply that

$$\frac{dp}{dt} = -\nabla(V + U), \quad (4.31)$$

which can be rewritten as

$$\frac{dp}{dt} = F, \quad (4.32)$$

where F is the negative of the gradient of the potential energy, $V + U$. The potential energy now includes the familiar "classical" potential energy V as well as the "quantum" potential energy U.

This entire scheme is readily extended to the many-body problem. For two particles (each taken to have mass m for simplicity here), the quantum potential is

$$U(x_1, x_2) = -\frac{\hbar^2}{2m}\frac{1}{R}(\nabla_1^2 R + \nabla_2^2 R). \quad (4.33)$$

Since $R = R(x_1, x_2)$, this many-body quantum potential "entangles" the motion of the various particles. This is a reflection of the fact that the overall wave function $\psi(x_1, x_2, t)$ of the combined system does not (usually) separate (as a *single* product). For the special case of *independent* systems, the wave function factors into a single product as

$$\psi(\mathbf{x}_1, \mathbf{x}_2, t) \equiv R(\mathbf{x}_1, \mathbf{x}_2, t) \exp[iS(\mathbf{x}_1, \mathbf{x}_2, t)/\hbar] = \psi_1(\mathbf{x}_1, t)\psi_2(\mathbf{x}_2, t)$$
$$= R_1(\mathbf{x}_1, t) R_2(\mathbf{x}_2, t) \exp\left\{\frac{i}{\hbar}[S_1(\mathbf{x}_1, t) + S_2(\mathbf{x}_2, t)]\right\}. \quad (4.34)$$

In that case, the quantum potential becomes a simple sum

$$U = U(\mathbf{x}_1) + U(\mathbf{x}_2) \quad (4.35)$$

and we have (of course) two independent systems.

1.2 Measurement as "discovery" for Bohm

For the experimental arrangement shown in figure 4.1, we can represent the incoming packet $f(\mathbf{x})$ as

$$f(\mathbf{x}) = \int g(\mathbf{k}) e^{i\mathbf{k}\cdot\mathbf{x}} d\mathbf{k}, \quad (4.36)$$

where $g(\mathbf{k})$ is a function fairly sharply peaked about $\mathbf{k} = \mathbf{k}_0 = (0, k_0, 0)$.[95] A wave packet constructed as

$$f(\mathbf{x}, t) = \int g(\mathbf{k}) e^{i(\mathbf{k}\cdot\mathbf{x} - \hbar k^2 t/2m)} d\mathbf{k} \quad (4.37)$$

represents a sharply peaked packet of the shape of f, propagating to the right with velocity $\mathbf{v}_0$ and satisfying the free Schrödinger equation

$$i\hbar \frac{\partial f(\mathbf{x}, t)}{\partial t} = -\frac{\hbar^2}{2m} \nabla_x^2 f(\mathbf{x}, t). \quad (4.38)$$

It is for this reason that I often write $f(\mathbf{x}, t)$ simply as $f(\mathbf{x} - \mathbf{v}_0 t)$ in what follows. The initial wave function $\Psi_0(\mathbf{x}, \xi, t)$ is given in eq. (4.7).[96] We must now find the solution to

$$i\hbar \frac{\partial \Psi(\mathbf{x}, \xi, t)}{\partial t} = (H_0 + H_1) \Psi(\mathbf{x}, \xi, t). \quad (4.39)$$

Here H_0 is the Hamiltonian for the incoming packet (cf. the right-hand side of eq. [4.38]) and also for the internal degrees of freedom of the atom.[97] Since the only relevant degrees of freedom for the atom are the angular variables (i.e., the angular momentum), I display explicitly only those variables as in the interaction Hamiltonian

$$H_1 = -\mu \mathbf{B} \cdot \frac{\boldsymbol{\ell}}{\hbar} = -\mu B(z) \frac{\ell_z}{\hbar} \approx -\mu(B_0 + B'_0 z)\frac{\ell_z}{\hbar}, \quad (4.40)$$

where μ is the magnetic moment of the atom and where an expansion has been made of $B(z)$ about $z = 0$. During the (short) time Δt that the atom is in the region of the magnetic field, the appropriate Schrödinger equation is just eq. (4.39) with $H_0 = 0$.[98] If the solution to that equation is written as

$$\Psi(x, \xi, t) = \sum_{n=-1}^{1} f_n(x, t) \psi_n(\xi) \qquad (4.41)$$

and substituted into the Schrödinger equation and the orthogonality of the ψ_n is used, the result is

$$i\hbar \frac{\partial f_n}{\partial t} = -\mu n(B_0 + B_0'z)f_n, \; 0 \leq t \leq \Delta t = d/v_0 . \qquad (4.42)$$

Since $f_n(x, t=0) = c_n f(x)$, this equation is easy to integrate to obtain

$$f_n(x, t) = c_n f(x) \, e^{i\mu n(B_0 + B_0'z)t/\hbar}, \; 0 \leq t \leq \Delta t. \qquad (4.43)$$

Equation (4.36) can then be used to write down f_n at $t = \Delta t$. The free solution for $t > \Delta t$ is then seen to be (cf. eq. [4.37])

$$\Psi(x, \xi, t) = \sum_n c_n \psi_n(\xi) \int dk \, g(k) \, e^{i\mu n B_0 \Delta t/\hbar}$$

$$\times \exp \left\langle i \left\{ k_x x + k_y y + \left(k_z + \frac{\mu n B_0' \Delta t}{\hbar} \right) z \right. \right.$$

$$\left. \left. - \frac{\hbar}{2m} \left[k_x^2 + k_y^2 + \left(k_z + \frac{\mu n B_0' \Delta t}{\hbar} \right)^2 \right] t \right\} \right\rangle$$

$$= \sum_n c_n f[x - x_n(t)] \, e^{i\varphi_n} \psi_n(\xi). \qquad (4.44)$$

Here the emergent trajectories are given as $x_n(t) = (0, v_0 t, \mu(n/m)B_0'\Delta t\, t)$. This is just eq. (4.8).

Next, I consider some of the calculational details for the scattering process illustrated in figure 4.2. If $f_0(y, t)$ represents the incident wave packet as

$$f_0(y, t) = \int \tilde{f}_0(k - k_0) \exp\left(i\, k \cdot y - \frac{i\hbar k^2 t}{2m} \right) dk, \qquad (4.45)$$

where $\tilde{f}_0(k - k_0)$ is a function sharply peaked about $k = k_0$, then the initial wave function Ψ for these two independent systems (i.e., the wave packet and the hydrogen atom) prior to any interaction between them is the product state

$$\Psi_i(x, y, t) = \psi_0(x) \exp(-iE_0 t/\hbar) f_0(y, t) . \qquad (4.46)$$

The final-state wave function after the scattering process has been completed is the superposition

$$\Psi(x, y, t) = \Psi_i + \sum_n \psi_n(x) \exp(-iE_n t/\hbar) f_n(y, t), \quad (4.47)$$

where $\psi_n(x)$ is the hydrogen-atom wave function for the energy level E_n and the $f_n(y, t)$ are outgoing (asymptotically spherical) wave packets (i.e., they are the expansion coefficients of Ψ in terms of the complete set ψ_n). That is, in the distant past (i.e., $t \to -\infty$) the initial asymptotic state is

$$\Psi \xrightarrow[t \to -\infty]{} \Psi_i \quad (4.48)$$

and one must then compute the "out" state that Ψ approaches as $t \to +\infty$. One can use standard Green's-function techniques to develop the asymptotic scattering solution to the time-dependent Schrödinger equation.[99] The result is that

$$\Psi(x, y, t) \xrightarrow[t \to +\infty]{} \Psi_i(x, y, t)$$
$$+ \sum_n \psi_n(x) \exp\left(-\frac{iE_n t}{\hbar}\right) \int \tilde{f}_0(k - k_0) \frac{1}{r} \exp\left[ik_n r - i\left(\frac{\hbar k_n^2}{2m}\right)t\right] g_n(\theta, \phi; k) \, dk, \quad (4.49)$$

with $y = (r, \theta, \phi)$, $k_n = (k_n, \theta, \phi)$ and

$$\frac{\hbar^2}{2m} k_n^2 = \frac{\hbar^2}{2m} k_0^2 + E_0 - E_n. \quad (4.50)$$

Equation (4.50) states the conservation of energy.[100] Here k_0 is the wave vector of the incident packet (i.e., $v_0 = \hbar k_0/m$), k_n that for the outgoing scattered (spherical) wave corresponding to the scattering center (here, the hydrogen atom) having been left in the (excited) state of energy E_n. The function $g_n(\theta, \phi, k)$ is a known (but complicated) scattering amplitude whose precise form need not concern us here.[101] Since the wave packet function $\tilde{f}_0(k - k_0)$ is sharply peaked about $k = k_0$, the overwhelmingly major contribution to the integral in eq. (4.49) arises from that region in which the phase is stable in the exponential of the integrand. Therefore, the nth outgoing spherical packet is centered about

$$r_n = \frac{\hbar k_n}{m} t. \quad (4.51)$$

That is—and this is the crucial observation for blunting the criticism raised by Pauli in 1927 (cf. section 7.2.2)—these outgoing spherical shells (or packets) are *far* separated in space and (effectively) do *not* overlap (because the speeds of propagation $v_n = \hbar k_n/m$ are different).[102]

The physical import of all of this mathematics is just the following. Since the initial state Ψ_i of eq. (4.46) is a simple product, the quantum potential U (eq. [4.33]) reduces to a sum of two terms (as in eq. [4.35]) and the incident particle and the scattering center behave as *independent* systems. Once the particle (in the wave packet f_0) enters the interaction region and Ψ takes the form of the superposition of eq. (4.47), the guidance conditions $p_j = \nabla_j S$ (where $j = 1$ refers to the particle and $j = 2$ to the electron in the hydrogen atom) produce very complicated and "entangled" motions for each part of the system. Eventually, the scattered wave packets separate and no longer overlap. Since $P = |\Psi|^2$ is the probability of finding a particle at a given location and since (asymptotically) P vanishes outside of these separated packets, the particle must (on the Bohm interpretation) be in *one* of these scattered packets. The probability of finding the outgoing particle in the nth outgoing spherical packet is then equal to this P integrated over the spatial extent of this volume. The result is proportional to

$$|\tilde{f}_0(0)|^2 \int |g_n(\theta, \phi; k_0)|^2 \, d\Omega \, . \tag{4.52}$$

This is the same result obtained from the Copenhagen interpretation.[103]

Let me recast these results into a form that may make their relation to the measurement problem more evident. The wave function for this problem (cf. the sum in eq. [4.49]) has the form[104]

$$\Psi(x, y) \to \sum_n \psi_n(x) \, \Phi_n(y) \, , \tag{4.53}$$

where the $\{\Phi_n(y)\}$ have macroscopically disjoint supports (i.e., the region in which any one of these Φ_n is nonzero [its support] is separated by a macroscopic distance from the regions in which the other Φ_ns do not vanish).

In Bohmian mechanics the state of a system is specified by (Q, Ψ) where Q stands for the configuration space values of the coordinates of the particles and Ψ is the usual quantum-mechanical wave function. It will prove useful here and later to distinguish between the "free" variables q and the actual coordinates of the particles Q. Let x, X denote the variables of the subsystem of interest and y, Y those for the rest of the "universe," so that

$$q = (x, y), \qquad Q = (X, Y) \, . \tag{4.54}$$

If, in an observation or "measurement," the scattered particle is "seen" in the nth outgoing packet (of eq. [4.49]), then Y must be in the support of $\Phi_n(y)$. In that case, write the wave function $\Psi(x, y)$ for the "universe" (here, just our two-particle system) of eq. (4.53) as

$$\Psi(x, y) = \psi_n(x) \, \Phi_n(y) + \Psi^\perp(x, y) \, , \tag{4.55}$$

where $\Phi_n(y)$ and $\Psi^\perp(x, y)$ have macroscopically disjoint supports in y (i.e., no overlap). It is, in fact, quite a general feature that a wave function for a compound system has the form of eq. (4.55) after a measurement, for then the *subsystem* just measured has a wave function, as will now be shown explicitly. Unless such special conditions obtain, the wave function for a subsystem of a larger system has no meaning (i.e., the overall state is an *entangled* one).

Consider the probability distribution $P(X, Y)$ for the particles in the system and the guidance condition for a coordinate X_1 (say) in the measured system. Since

$$P(X, Y) = |\Psi|^2 \qquad (4.56)$$

and we know that $Y \in supp\ \Phi_n(Y)$, it follows that $X \in supp\ \psi_n(X)$ (recall the *disjoint* structure of the y supports in eq. [4.55]). Therefore, we obtain

$$P(X, Y) \sim |\psi_n(X)\Phi_n(Y)|^2 \qquad (4.57)$$

and, if we were now to integrate out the Y coordinates, the probability distribution governing the subsystem X would be

$$P(X) \sim |\psi_n(X)|^2 . \qquad (4.58)$$

Because the *actual* particles are now known (i.e., have been "discovered") to be in these two regions, the probability distribution can be governed *only* by ψ_n and Φ_n (*not* by $\Psi^\perp$). Since Bohmian mechanics is left invariant when the wave function is multiplied by any constant, we can "renormalize" ψ_n to write eq. (4.58) as[105]

$$P(X) = |\psi_n(X)|^2 . \qquad (4.59)$$

The probability distribution for the measured subsystem is determined *only* by the effective wave function $\psi_n(x)$. What about the motion of the particles of this subsystem, though?

Consider the guidance condition of eq. (4.29) for, say, the coordinate X_1.

$$\dot{X}_1 = \frac{1}{m}\nabla_1 S\Big|_{\substack{x=X\\y=Y}} = \frac{\hbar}{m}Im\left(\frac{\nabla_1\Psi}{\Psi}\right)\Big|_{\substack{x=X\\y=Y}} = \frac{\hbar}{m}Im\left(\frac{\nabla_1\psi_n}{\psi_n}\right)\Big|_{x=X} . \qquad (4.60)$$

The last equality above again follows from the disjoint structure of the supports in y. Before we evaluate $y = Y \in supp\ \Phi_n$, there are *many* terms in $(\nabla_1 \Psi/\Psi)$. All but *one* vanish when we restrict y to $supp\ \Phi_n$ (i.e., the region, or "shell," in which the scattered particle is discovered in the example of eq. [4.49]). Similarly, if one computes the quantum potential U of eq. (4.33), he finds (in the case of disjoint y supports)

$$U(X,Y) = -\frac{\hbar}{2m} \frac{(\nabla_1^2 + \nabla_2^2)R}{R}\bigg|_{\substack{x=X\\y=Y}} = -\frac{\hbar^2}{2m}\left(\frac{\nabla_1^2|\psi_n|}{|\psi_n|} + \frac{\nabla_2^2|\Phi_n|}{|\Phi_n|}\right)\bigg|_{\substack{x=X\\y=Y}}$$
$$= U_1(X) + U_2(Y) \qquad (4.61)$$

With regard to the subsystem whose coordinates are X, it is the wave function $\psi_n(x)$ *alone* that is relevant for the probability distribution and for the subsequent motion of the particles in that subsystem.

We see, then, that the arguments presented here in some detail for the example illustrated in figure 4.2 can be extended (via eqs. [4.54] and [4.55]) to the general case. In Bohm's picture, no actual collapse of the wave function takes place upon measurement, but one does effectively obtain just the "reduced" wave function (ψ_n in our example). I refer to this example again in section 9.4 when I discuss Bohmian mechanics applied to a large system (the "universe") from which one extracts a subsystem for measurement. Another simple illustration of the time evolution of a wave function into disjoint pieces is provided by the "measurement" example of the following section.

1.3 Measurement and uncertainty

The analysis given above of a scattering process as a paradigm of a measurement can readily be extended to see how the uncertainty relations come about and how, in general, the act of measurement disturbs the microsystem. As an illustration of the type of argument used, let us consider a microscopic system, with spatial coordinates x, having an observable q (corresponding to the hermitian operator Q) whose value we wish to determine by measurement.[106] The macrosystem ("measuring device") to which an interaction couples the microsystem has a coordinate y (think of the proverbial "pointer reading") that we observe and its conjugate momentum is P_y. As in the standard theory of measurement, we can take for a suitable interaction Hamiltonian

$$H_I = aQP_y, \qquad (4.62)$$

where a is a constant determining the strength of this interaction.[107] The Schrödinger equation that must be solved is then

$$i\hbar \frac{\partial \Psi}{\partial t} = H_I \Psi . \qquad (4.63)$$

For simplicity in the following, I take the spectrum of q to be discrete. Let $\psi_q(x)$ be the orthonormal eigenvectors of Q

$$Q\psi_q(x) = q\,\psi_q(x) . \qquad (4.64)$$

Since $P_y \to -i\hbar\, \partial/\partial y$, eq. (4.63) becomes

$$\frac{\partial \Psi}{\partial t} = -a Q \frac{\partial \Psi}{\partial y}. \tag{4.65}$$

Because the $\psi_q(x)$ form a complete set of vectors, the $\Psi(x, y, t)$ can be expanded as

$$\Psi(x, y, t) = \sum_q \psi_q(x) f_q(y, t). \tag{4.66}$$

Substitution of eq. (4.66) into eq. (4.65) and use of the completeness of the orthonormal $\psi_q(x)$ yield

$$\frac{\partial f_q(y, t)}{\partial_t} = -aq \frac{\partial f_q(y, t)}{\partial y}. \tag{4.67}$$

A simple change of variables shows that the solution of eq. (4.67) is[108]

$$f_q(y, t) = f_q^0(y - aqt), \tag{4.68}$$

where

$$f_q^0(y) \equiv f_q(y, t = 0). \tag{4.69}$$

Therefore, the required solution is

$$\Psi(x, y, t) = \sum_q \psi_q(x) f_q^0(y - aqt) \tag{4.70}$$

Suppose that initially (i.e., at $t = 0$ here) the macrosystem is represented by a wave packet $g_0(y)$, centered at $y = 0$ and of width Δy and that the microsystem is in the state

$$\psi_0(x) = \sum_q c_q \psi_q(x) \tag{4.71}$$

so that

$$\Psi_0(x, y) = \psi_0(x) g_0(y) = g_0(y) \sum_q c_q \psi_q(x). \tag{4.72}$$

Comparison of eqs. (4.72) and (4.70) (for $t = 0$) implies that

$$f_q^0(y) = c_q g_0(y). \tag{4.73}$$

The wave function for the interacting micro-macrosystem is then just

$$\Psi(x, y, t) = \sum_q c_q \psi_q(x) g_0(y - aqt). \tag{4.74}$$

Let the discrete values of q be separated by δq.[109] Since the centroid of the qth packet is at $y = aqt$, the distance δy between neighboring packets is

70   Chapter Four

$$\delta y = at\, \delta q. \tag{4.75}$$

For large enough values of t, these packets will be far enough separated in y-space so that

$$\delta y \gg \Delta y. \tag{4.76}$$

Here Δy is the accuracy to which we observe the apparatus coordinate. By the same type of argument given in the discussion of the scattering process in the previous section, the macrosystem will (definitely) be in *one* of these spatially separated packets, corresponding to *one* particular value q of the microvariable. The probability of obtaining a given value q upon observation is just $|c_q|^2$, as in the usual interpretation. Once a specific value of q has been revealed upon observation, the effective wave function "reduces" (upon renormalization) to

$$\Psi(x, y, t) \rightarrow \psi_q(x)\, g_0(y - aqt). \tag{4.77}$$

The act of measurement effectively "reduces" the wave function of the measured system. This is all very much a repetition of the type of story told for the two examples of appendix 1.2. We now see that the insights and results based on those simple cases have turned out to be quite general for any measurement process.

A *subsequent* measurement (through an interaction $H_I = bRP_z$) of a variable r represented by an operator R that does not commute with Q would lead, by a repetition of these same manipulations, to

$$\Psi(x, z, t) = \sum_r a_{qr}\, \phi_r(x)\, g_0(z - brt) \rightarrow a_{qr}\, \phi_r(x)\, g_0(z - brt) \tag{4.78}$$

since the initial state of the microsystem (at the *new* $t = 0$) is

$$\psi_q(x) = \sum_r a_{qr}\, \phi_r(x). \tag{4.79}$$

Equation (4.79) is just the expansion of the eigenstate $\psi_q(x)$ in terms of the eigenstates $\phi_r(x)$ of the operator R. The probability of a given outcome r, for a fixed value q before the measurement, is

$$|a_{qr}|^2 = |\langle \psi_q | \phi_r \rangle|^2. \tag{4.80}$$

This is a familiar result from standard quantum mechanics. Since noncommuting measurements necessarily disturb the wave function, it is not possible to predict the specific outcomes that will be produced, but only the probability of possible outcomes.

1.4   Bohm's original mixing argument for $P \rightarrow |\Psi|^2$

Shortly after he had introduced his new theory, Bohm gave an argument[110] to show that, even if initially $P \neq |\psi|^2$, still P would be driven, through random interactions, to the equilibrium distribution $P = |\psi|^2$.[111] Once $P = |\psi|^2$, the continuity equation will maintain this equality. It is well known that, since ψ satisfies the Schrödinger equation of eq. (4.20), then $|\psi|^2$ also satisfies the continuity equation.[112] For *both* of these continuity equations, the velocity v is given by eq. (4.29). However, the continuity equation is a first-order, linear partial differential equation and its solution $P(x, t)$ is uniquely determined everywhere and for all time once its initial value $P(x, t_0)$ is given over all of space.[113]

The largely heuristic illustration employed by Bohm consisted of an ensemble of identical systems (e.g., hydrogen atoms) each in a doubly degenerate level of energy E_0. The system has axial symmetry (about the z-axis) so that the wave function has the form of a superposition (represented by the coefficients c_1 and c_2)

$$\psi = g(\rho, z) (c_1 \cos \phi + c_2 \sin \phi) \exp(-iE_0 t/\hbar), \quad (4.81)$$

where ρ is the radial polar coordinate and ϕ the azimuthal angle. For a statistical ensemble of these systems, all with the same wave function ψ but with a statistical distribution of positions for the electron in each atom, suppose P_0 is such that $P_0 \neq |\psi|^2$. The guidance condition of eq. (4.29) shows that the only motion of the electron is in the ϕ-direction (i.e., ρ remains fixed) and the subsequent ϕ-motion turns out to be periodic. Since only ϕ varies during the motion of the electron in its circular orbit and since ϕ is periodic in time, P_0 will also be periodic in time. Hence, there can be *no* way that this *isolated* system can have P_0 approach $|\psi|^2$.

Therefore, Bohm allowed each member of this ensemble to undergo a series of random interactions (but ones gentle enough to leave ψ always in a superposition of the form of eq. [4.81]) that will change the coefficients c_i ($i = 1, 2$). Bohm showed that no matter what the initial distribution of the $\{c_i\}$ over the ensemble, the final (i.e., asymptotic in time) distribution of the $\{c_i\}$ is *uniform* (i.e., any allowed set of values is as likely as any other). In particular, he considered all of the different *initial* sets of $\{c'_i\}$ that could produce (via these random interactions) a *given final* set c_i (i.e., a specific wave function, say ψ_f). Corresponding to each of the initial c'_i (in the distant past) there would have been a particular set of initial coordinates (say x') for the electron. Since the time evolution of these coordinates is governed by the guidance condition (eq. [4.29]) and since that condition depends upon the c'_i, it is no surprise that, as Bohm proved, the distribution of these coordinates (i.e., particle positions) becomes randomized also. This affects the distribution P for the ensemble, and Bohm showed that $P \to |\psi|^2$. Another way to understand the sequence of influences is to appreciate that the time evolution of P is governed by the continuity equation

$$\frac{\partial P}{\partial t} + \nabla \cdot (P\boldsymbol{v}) = 0 \qquad (4.82)$$

and that (in a Bohmian picture) this v is given by the guidance condition of eq. (4.29). But it is the wave function ψ that enters into the guidance condition to determine v. This means that ψ (here, the c_j) determines the time evolution of P. Hence, variations in the c_j (as in Bohm's example here) will produce variations in the time evolution of P. For the particular case he considered, Bohm was able to demonstrate that this chain led to $P \to |\psi|^2$.

As nice as this model calculation is, it is highly specific and need not have any implications for a *general* proof that P *necessarily* approaches $|\psi|^2$. In Section 9.4.1 I return to this question and show that a mixing argument, of the type familiar from classical statistical mechanics, can be fashioned to establish the desired result more generally.

Appendix 2   Quantum tunneling times

For the thought experiment of figure 4.3 discussed in section 4.3, we can use the values of τ_j to find the *dwell time* τ_D, which is defined as the average time spent by the particles ($j = 1, 2, 3, \ldots, N$) between points x_1 and x_2, as

$$\tau_D = \frac{1}{N} \sum_{j=1}^{N} \tau_j. \qquad (4.83)$$

Similarly, if N_R particles are reflected and N_T transmitted (where $N_R + N_T = N$), then one can calculate a *reflection time* τ_R, which is the average time spent by the *reflected* particles in (x_1, x_2), as

$$\tau_R = \frac{1}{N_R} \sum_{\{N_R\}} \tau_j, \qquad (4.84)$$

and also a transmission time as

$$\tau_T = \frac{1}{N_T} \sum_{\{N_T\}} \tau_j, \qquad (4.85)$$

Since the reflection and transmission coefficients are defined, respectively, as

$$R = \frac{N_R}{N} \tag{4.86}$$

$$T = \frac{N_T}{N}, \tag{4.87}$$

it follows as an identity that

$$\tau_D = R\tau_R + T\tau_T. \tag{4.88}$$

At first glance, this quantum-mechanical tunneling problem might not even appear worthy of much serious thought, since it is essentially a standard homework problem in introductory quantum mechanics courses.[114] Let V_0 be the (rectangular) barrier height in figure 4.3 and d the width. For a plane wave incident from the left, an elementary application of the Schrödinger equation leads to the wave function[115]

$$\psi(x,k) = \begin{cases} e^{ikx} + \sqrt{R}\, e^{i\beta - ikx}, & x < 0 \\ \chi(x,k), & 0 < x < d \\ \sqrt{T}\, e^{i\alpha + ikx} & x > d \end{cases} \tag{4.89}$$

with

$$T(k) = \frac{4k^2\kappa^2}{D}, \qquad R(k) \equiv 1 - T(k), \tag{4.90a}$$

$$D = 4k^2\kappa^2 + k_0^2 \sinh^2(\kappa d), \tag{4.90b}$$

$$E = \frac{\hbar^2 k^2}{2m} < V_0, \qquad V_0 - E = \frac{\hbar^2 (k_0^2 - k^2)}{2m} = \frac{\hbar^2 \kappa^2}{2m}, \tag{4.90c}$$

$$\chi(x, k) = A\, e^{\kappa x} + B\, e^{-\kappa x}, \tag{4.90d}$$

$$\alpha(k) + kd = \beta(k) + \frac{\pi}{2} =$$
$$-\tan^{-1}\left[\frac{\kappa^2 - k^2}{2\kappa k} \tanh(\kappa d)\right].^{116} \tag{4.90e}$$

From these solutions we can construct a wave packet as

$$\psi(x, t) = \int f(k)\psi(x, k)\, e^{-iE(k)t/\hbar}\, dk, \tag{4.91}$$

where $f(k)$ is sharply peaked about some k_0. A stationary-phase approximation to this integral yields (for $x > d$) the well-known result

74   Chapter Four

$$\frac{d\alpha}{dk} + x - \frac{1}{\hbar}\frac{dE}{dk}t = 0 \equiv \frac{d\alpha}{dk} + x - v(k)t. \qquad (4.92)$$

The packet "falls behind" by an amount $\delta x = d\alpha/dk$ compared to where it would have been had the barrier not been present. A similar calculation can be made for $x < 0$. This leads to the classic definition of the transmission and reflection phase times as[117]

$$\tau_T^\varphi = \frac{d + d\alpha/dk}{v(k)}, \qquad \tau_R^\varphi = \frac{d\beta/dk}{v(k)}. \qquad (4.93)$$

However, these are *asymptotic* (phase) times and do not represent (just) true barrier-crossing times because self-interference effects are included in, for example, τ_R^φ.[118]

A popular attempt to derive an expression for the dwell time τ_D in standard quantum mechanics begins with the observation that the probability of finding the "particle" in the barrier region is

$$P(t, d) = \int_0^d dx |\psi(x, t)|^2. \qquad (4.94)$$

If we next assume that the average time spent in the barrier region is given as[119]

$$\tau_D = \int_0^\infty dt\, P(t, d), \qquad (4.95)$$

then formal mathematical manipulations lead to the result[120]

$$\tau_D(k) = \frac{1}{v(k)} \int_0^d dx\, |\psi(x, k)|^2. \qquad (4.96)$$

This is a heuristic argument only, not a *proof*. The expression of eq. (4.96) and the asymptotic phase times of eqs. (4.93) do not satisfy the constraint of eq. (4.88), which is a necessary condition for consistency.[121] In spite of many attempts, there appear to be *no* consistent candidates for τ_R and τ_T.[122]

However, for Bohm's theory the motion of a particle is completely deterministic so that once its initial position x_0 has been given [corresponding to $p_0 = \partial S(x)/\partial x|_{x_0}$], the time τ spent between x_1 and x_2 is determined [i.e., $\tau = \tau(x_0)$]. These initial positions are distributed according to $|\psi(x_0)|^2$. The various τs can (in principle) be computed as[123]

$$\tau_D = \int \tau(x_0) \, |\psi(x_0)|^2 \, dx_0 \,, \tag{4.97}$$

$$\tau_R = \frac{1}{R} \int_{\{R\}} \tau(x_0) \, |\psi(x_0)|^2 \, dx_0 \,, \tag{4.98}$$

$$\tau_T = \frac{1}{T} \int_{\{T\}} \tau(x_0) \, |\psi(x_0)|^2 \, dx_0 \,, \tag{4.99}$$

while R and T are simply

$$R = \int_{\{R\}} |\psi(x_0)|^2 \, dx_0 \,, \tag{4.100}$$

$$T = \int_{\{T\}} |\psi(x_0)|^2 \, dx_0 \,. \tag{4.101}$$

Here $\{R\}$ and $\{T\}$ are, respectively, those subsets of particles that will be reflected and those that will be transmitted. These definitions satisfy eq. (4.88) identically. The scattering of a particle by a barrier has been studied numerically in Bohmian mechanics and these $\tau(x_0)$ can be found (at least numerically).[124] There are several technical questions that remain open—both experimental and theoretical ones—about the actual possibility of a comparison of eqs. (4.97) through (4.99) with experimental observations. For instance, it is unclear whether or not the motion of the particle along its trajectory may be so sensitive to the details of the *shape* of the initial wave packet that no clean quantitative estimate can be made that could meaningfully be compared with experiment.[125]

FIVE
Alternative Interpretations: An Illustration

Now that I have presented Bohm's version of quantum mechanics and compared it with the standard one, it may be helpful for the reader if I apply both theories to an actual experimental situation. Modern technology has advanced to the stage that what had previously been mere gedankenexperiments are now actually performed in the laboratory. This allows clean and direct tests of a theory, free from the messy approximations so often necessary to make contact between theory and experiment. There is also considerable pedagogical value because a fairly straightforward application of the basic formalism of quantum mechanics (rules [i] through [v] of section 3.2) yields results immediately relevant to real observations. Such an experiment is discussed in this chapter and the corresponding calculation, independent of any commitment to Bohm or to Copenhagen, is carried through in appendix 1 to this chapter. I ask the reader to think about the results of these calculations, which agree with experiment, in terms of the physical stories that go with the two interpretations of the previous chapters.

First, though, I consider what possible value such an alternative interpretation can have, given they both are observationally equivalent.[1] This will lead me back to the issue of understanding that was discussed in chapter 2.

5.1 The value of the exercise

When David Bohm's new quantum theory entered the scene in the early 1950s, the prevailing view was a strong one on the completeness and finality of the standard interpretation of quantum mechanics. These sentiments are well illustrated in an article written by the ever-loyal Leon Rosenfeld and directed at the generation that had grown up since the great struggles that gave birth to quantum mechanics. He told these young physicists unequivocally that raising doubts about the correctness of the basic ideas of standard quantum theory was futile because the laws of quantum mechanics are not something we have invented, but an aspect of nature we have discovered.[2] He went on to discuss complementarity and the demise of determinism and then claimed that, since we cannot speak of all physical quantities with unlimited precision, we must replace the older concept of causality with that of statistical causality. He wanted to avoid a loss of rationality and any charge of subjectivism (due to the essential role played by the observer) once determinism and rigid, exact laws of nature (which are formulated without reference to any observer) have yielded to this statistical causality. To do so, Rosenfeld essentially redefined 'objectivity' as "simply the possibility of guaranteeing that the account of the phenomena will convey equivalent information to all observers, that it will consist of statements intelligible to all human beings."[3] For him, quantum theory did this with its transformations that allowed one to pass from the point of view of one observer to that of another. He assured his audience that they were indeed fortunate to have been given so fine an instrument by those older and wiser than they. Hence, we can only bow down before the inescapable lessons forced on us by nature.

Given such an attitude, one might feel that there is little motivation for or value in studying an interpretation alternative to the standard one. But even if Copenhagen were accepted as empirically adequate and consistent, the question of *understandability* remains. Does the Copenhagen interpretation give us a description of the world that we can understand in any meaningful sense of that term? That question is certainly open to debate. The quest for a more (nearly) understandable worldview can be a motivating factor in seeking another interpretation of a quantum formalism.[4] This was concisely and elegantly stated by David Bohm in his classic 1952 paper:

> The usual interpretation of the quantum theory is self-consistent, but it involves an assumption that cannot be tested experimentally, *viz.*, that the most complete possible specification of an individual system is in terms of a wave function that determines only probable results of actual measurement processes. The only way of investigating the truth of this assumption is by trying to find some other interpretation of the quantum theory in terms of at present "hidden" variables, which in principle determine the precise behavior of an individual system, but which are in practice averaged over in measurements of the types that can now be carried out. In this paper and in a subsequent paper, an interpretation of the quantum theory in terms of just such "hidden" variables is suggested. It is shown that as long as the mathematical theory retains its present general form, this suggested interpretation leads to precisely the same results for all

physical processes as does the usual interpretation. Nevertheless, the suggested interpretation provides a broader conceptual framework than the usual interpretation, because it makes possible a precise and continuous description of all processes, even at the quantum level. . . . [5]

As a matter of fact, whenever we have previously had recourse to statistical theories, we have always ultimately found that the laws governing the individual members of a statistical ensemble could be expressed in terms of just such hidden variables. . . . [6]

The usual interpretation [in its finality and completeness] . . . presents us with a considerable danger of falling into a trap, consisting of a self-closing chain of circular hypotheses, which are in principle unverifiable if true.[7]

Let us now see just how these claims of increased understanding are cashed out in a class of actual experiments.

5.2 Neutron interferometry experiments

There are several variations of a basic neutron interference experiment in which an incident beam of neutrons is first Bragg-diffracted by a crystal plane (to the left in figure 5.1), each part of this split beam is then dif-

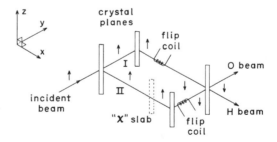

Figure 5.1 A neutron interferometry experiment

fracted by other planes (middle of figure 5.1) of the *same* single crystal, and these two beams are finally recombined by another plane (right of figure 5.1) of the single crystal.[8] The properties of neutrons that are relevant for my discussion are that a neutron is a massive particle having no electric charge but possessing (in the usual application of the formalism of quantum mechanics) an intrinsic "spin" (s) and a corresponding magnetic moment (μ).[9] The incident beam of neutrons is completely polarized with the spin aligned along the positive z-axis of figure 5.1. The basic idea is that the neutron's spin can be flipped, in one or the other of the diffracted beams, by means of magnetic fields established at right angles to the z-axis. The final, recombined beam will then in general consist of a superposition of different spin states.[10] Below I summarize the results of standard quantum-mechanical calculations for four different types of experiments that can be performed with the arrangement outlined in figure 5.1. As usual, the mathematical details are confined to an appendix (appendix 1). My presentation follows the now-classic work of Helmut Rauch and his collaborators in Vienna.[11]

5.2.1 Spin-flip coils

As indicated by the vertical arrow above the incident beam in figure 5.1, the incoming beam of neutrons is polarized along the positive z-axis (i.e., the spin of each incident neutron is aligned "up," in the positive z-direction). Things are so arranged that, were it not for the spin-flip coils, this incident beam would simply be split into two beams at the first crystal plane on the left, but the polarization of the beam would be unaffected. These upper and lower beams in the figure are denoted by I and II, respectively. The dotted, thin rectangle, labeled "χ" slab, in beam II is a phase shifter and, if inserted into this beam, will produce a change in the phase of the wave function associated with that beam. The flip coils shown in each beam are magnetic devices that can flip the spin of the beam passing through it. (Details of how they operate are contained in appendix 1). These coils are so tuned that they will exactly flip a spin up into a spin down and do so with (essentially) 100% efficiency. When the two subbeams (I and II) are recombined at the far right face of the interferometer, the emergent O beam can have an intensity different from that of the incident beam, if there has been a phase difference introduced between the two beams that have traveled along different paths through the interferometer. The polarization of the emergent beam can be different from that of the incident beam, if one or the other of the spin-flip coils has been in operation.

Here we need only appreciate that these coils allow us to flip the spin of the neutron at will. The coils can be operated in either of two modes. If we apply a steady (direct) current to a coil, then it just flips the spin, but exchanges no energy with the beam. In one type of experiment, a static flipper is placed in just *one* path (say, II). Then the recombined O beam is a superposition of a spin-up beam and a spin-down beam. This produces the quantum-mechanical result (which is completely unexpected classically) of an emergent beam polarized in the *x-direction*.[12]

The coils can also be operated with an alternating radio frequency (rf) current, in which case the neutron still has its spin inverted but now a quantum of energy is exchanged between the beam and the flipper.[13] This produces a *time-dependent* phase factor in the beam so that there can then be a time-dependent, or "beat," interference phenomenon when the two beams are recombined as they emerge from the interferometer. In this case, if an rf coil is placed in just one beam, the polarization of the emergent beam lies in the xy-plane and rotates in this plane with a frequency equal to that of the current driving the flip coil.

5.2.2 A modern "double-slit" experiment

Let me now apply these preliminary results to some actual neutron interferometry experiments. Once again, figure 5.1 will do to illustrate the class of experiments I am concerned with here. The four crystal planes (all actually part of *one* single crystal that has been machined to form this interferometer) perform the same function as the double-slit arrangement for the older optical-type experiments.[14]

I begin with a brief characterization of the incident neutron beam and of the dimensions of the single crystal and spin-flip coils.[15] The overall size of the single crystal is approximately 10 cm and the neutron wave packets are smaller than the size of the spin-flip coils or the separation between the diffracted beams. The de Broglie wavelength of the neutrons is $\lambda = 1.893$ Å $= 1.893 \times 10^{-10}$ m, the incident beam has a cross-sectional area of 2×4 mm², the size of the packet is about 0.03 mm $= 3 \times 10^{-5}$ m and the neutron's energy is 0.023 ev (electron volts) corresponding to a velocity of around $v = 2,000$ m/s.[16] The lateral separation of the subbeams is 20 mm $= 2 \times 10^{-2}$ m, the average separation between successive neutrons is 10 m and the length of the flip coils is about 20 mm. The transit time of a neutron through the crystal is 5×10^{-5} s and through the coil on the order of 1×10^{-5} s, while the time between neutrons is roughly 6×10^{-3} s. All of this makes it evident that just *one* neutron at a time is in the interferometer and that the wave packet does *not* overlap both (diffracted) beams and is smaller than the flip coil(s). "All of the experiments performed until now belong to the region of self-interference, where, at a certain time interval, only one neutron is inside the interferometer, if at all. The next one is usually not yet born and is still contained in the uranium nucleus of the reactor fuel."[17] That is, one is able to study the truly quantum-mechanical phenomenon of the *self-interference* of *single* neutrons. I consider here only cases in which the incident neutron beam is 100% polarized with its spin up, along the positive z-axis.

In the first type of experiment, a phase shifter is placed in the first leg of path II and a dc spin-flip coil in the second leg of path II, but path I has no such device in either leg (cf. figure 5.1 again). In this case, the emerging O beam has its polarization vector contained in the xy-plane and the direction of this vector is *fixed*, making an angle χ with the *x*-axis.[18] Next, consider the case in which just a phase shifter is placed in the first leg of path II and an rf flipper in the second leg of path II, but path I still has no such device in either leg (cf. figure 5.1). Again, the polarization of the O beams turns out to be perpendicular to that of the original beam and rotates in the *xy*-plane with the radio frequency of the flip coil. This phenomenon of two beams of different energy "beating" against each other can be observed stroboscopically in real time, as I discuss below for a different case.

If we have *two* rf flippers (one in each path, as actually shown in figure 5.1), driven at the *same* frequency and in phase with each other, then *both* beams are flipped. This gives an intensity that is *constant* in time and that can be varied by adjusting the phase shift χ. The polarization of the O beam is, as expected, directed along the negative *z*-axis. This arrangement was suggested in advance of the experiments and the predicted results have been observed.[19]

Finally, the most interesting case for the purposes of my general discussion here is that illustrated in figure 5.1 in which the two rf flippers are driven at *slightly* different (nearly resonant) frequencies. The incident beam of neutrons is completely polarized, which means here that the neutron spin points up along the *z*-axis. The magnetic spin-flip coils in beams I and II flip (with essentially 100% efficiency) the spin of the transmitted neutrons (to spin down along the *z*-axis). With just *one* neutron at a time in the interferometer, each neutron can interfere only with *itself*. The radio-frequency spin-flip coils operate by exchanging a *single* photon of energy $\hbar\omega_r$ with the neutron (where ω_r is the resonant frequency at which the coil is driven). Each coil is driven at slightly different resonant frequencies, ω_{r_1} and ω_{r_2}. The two beams, when recombined at the crystal plane on the far right of figure 5.1, have slightly different energies (or "frequencies") and so will exhibit a "beat" phenomenon in the intensity of this recombined beam (labeled the O beam). A straightforward application of the *formalism* of quantum mechanics (independent of any particular interpretation) has led to this prediction.[20] The actual experimental results show this interference or beat effect (cf. figure 5.2).[21] Even

Figure 5.2 The experimental interference curve

though the "neutron" (via beam I or beam II) has exchanged a definite quantity of energy with *one* of the flip coils, no "collapse" of the wave function has taken place, since interference effects are exhibited in the recombined O beam.

5.3 The "story" told by Bohm

On the Copenhagen interpretation, we are effectively stranded with the formalism and its predictions, leaving as a mystery just how the neutron (wave? particle?) interacts with *one* flipper and yet produces interference. The exchange of a photon with one flipper would seem to indicate the neutron behaves as a *particle,* yet subsequently it behaves as a *wave* (part traveling along path I and part along path II within the interferometer) to produce interference in the emerging O beam. Copenhagen intuition (perhaps only a folklore gloss, though) typically leads one to expect wave *or* particle, depending upon the environment or experimental arrangement. On the other hand, the causal interpretation, assigns *both* a wave *and* a particle to the neutron.[22] One can model the particles in such a way that calculations using the quantum potential show how energy and angular momentum are transferred by the quantum potential between the environment and a *localized* neutron (a *particle*) that passes through just one coil.[23] Whether or not such experiments and analyses of them take the discussion of rival interpretations beyond the level of mere preference and empty, heated debate may still depend upon one's prior predilections pro or con Copenhagen. The results of these experiments *invite* (but certainly do not *demand*) a conceptualization of an actual particle that exchanges energy with just one coil.[24] Without this, it is difficult to imagine how localized particles propagate through the apparatus.[25]

5.3.1 The "double-slit" experiment

Bohmian dynamics represents the neutron as a particle that has, in addition to its position, additional parameters affecting its motion. The wave function is now governed by the Pauli equation.[26] There are two basic options for modeling these parameters. One is to stay with positions *only* and simply modify the guidance conditions (eqs. [4.2] or [4.29]) to incorporate these additional degrees of freedom.[27] This is the most conservative approach and the one most in keeping with the spirit of all measurements ultimately being position measurements. Such a treatment of the additional degrees of freedom for spin is consistent and leads to the usual results for spin measurements (via, say, a Stern-Gerlach analyzer).[28]

It is possible to go one step further, though, and endow the particles with some internal structure.[29] The advantage is that this provides a graphic ontology for the interpretation. The additional parameters that are introduced can be thought of as a set of axes that define an orientation (of, say, a spin vector).[30] This causal interpretation of the Pauli equation yields a dynamics that governs the motion of a particle with an intrinsic spin. This model can be applied to the neutron-scattering experiments and it represents the center of mass of the particle following a trajectory $x(t)$ while its spin vector **s** precesses in space. The Euler angles (ϕ, θ, ξ) that specify the orientation of the spin become dynamical variables whose time evolution is governed by an extended version of the guidance condition.[31] The quantum potential becomes spin dependent so that "quantum torques" can act. Here I simply state the results of a causal analysis of the neutron interferometry experiments.[32]

In the setup with *no* spin-flip coil in either arm of the interferometer and a phase shifter in one, the initial polarization of the incident beam (say, spin up) is maintained for each split beam and the spin-dependent quantum-potential and the quantum-torque terms vanish, but an interference pattern persists.[33] That is, the polarization remains aligned in the positive z-direction. The intensity of the emerging beam does have the required harmonic variation with χ.[34] This is basically the same type of behavior we would expect by analogy with the case of a particle with no spin. That is reasonable, since here the spin of the neutron has played no role in the physical interactions.

If an rf spin flipper now operates in one arm and a phase shifter is placed in the other, then the Euler angles θ and ϕ have spatial variation and the spin-dependent quantum-potential and -torque terms come into play. The quantum-torque interaction aligns the spin so that the particles in the emerging beam have spin orientations $\theta = \pi/2$ and $\phi = \pi/2 - \omega_{rf} t - \chi$.[35] Angular momentum and energy are conserved on an event-by-event basis as each particle goes along one path or the other and undergoes energy/angular momentum exchanges via the quantum potential/quantum torque.

5.3.2 The EPRB experiment[36]

One can also give a causal discussion of the EPRB experiment in which the quantum potential (or background "field") properly accounts for the necessary exchanges of angular momentum to produce the observed results.[37] In the arrangement of figure 2.1, a source emits two particles in a state with total angular momentum zero.[38] The subsequent motion of these particles is governed by the two-body Pauli equation via a guidance condition.[39] When the initial spins of the two particles are calculated, one finds $s_1 = s_2 = 0$.[40] As the two wave packets (traveling in opposite directions) enter their respective Stern-Gerlach apparatuses, the spin variables become coupled to the position variables of *both* apparatuses.[41] This produces a measurement of the component of the spin of each particle along the direction of the analyzing field in the apparatus.[42] As each particle leaves its Stern-Gerlach apparatus, it enters one of the two emerging packets (much as in figure 4.1). The spin vectors do not at first lie along the directions of the analyzing fields, but approach them as $t \to \infty$.[43] In principle, the initial positions of the two particles (plus the configuration of the analyzing fields) uniquely determine where each particle will emerge and what spin projection it will have.

In general terms, what happens in an EPRB experiment is that *both* apparatus settings have an effect in producing the observed values of the spin at the two distant stations. If we change just *one* apparatus setting (i.e., the direction of its analyzing field), we (immediately) change the wave function (everywhere) and this (through the quantum potential) varies the behavior of *both* particles. It is this nonlocal interaction that *produces* the observed long-range correlations. Even though this action at a distance couples the *spin* of one particle to the *position* of the other (distant) one, this direct action at the level of individual particles cannot be used to signal.[44] The reason is that, since we cannot control precisely the initial positions, we have no way to predict what *would* have happened to one particle had the other not been disturbed by being subjected to a measurement.

The success of these and similar exercises demonstrates that one *can* consistently produce a quite visualizable picture of these microprocesses. At the beginning of this section I indicated that we do not have to commit ourselves to the *actual* existence of the spin vector of these particles, since, ultimately, we measure these spin projections by, say, splitting a beam (e.g., with a Stern-Gerlach apparatus) and detecting where the particle emerges (e.g., a position measurement in the "up" leg or the "down" leg of the detector). We have two options. One is a spinning-particle model that, when extended to many-body systems, implies that intrinsic properties (e.g., the spin of *a* particle) become a relational property of an entire system.[45] This could be seen as undercutting much of the motivation for a definite ontology.[46] There are also difficulties in extending this model to the relativistic domain.[47] Or we can take position as the only possessed value, and all other properties are then simply constructs for convenience of description. In any event, possessed values that are context independent cannot be extended beyond the position variable.[48] We may find it convenient to speak in terms of an intrinsic "spin" of the electron, but we do not observe that spin directly.

Appendix 1 The quantum mechanics of spin flipping

I begin with the case of a *static* magnetic field (B_1) *suddenly* applied to a neutron. In figure 5.3 the neutron beam is incident from the left (moving

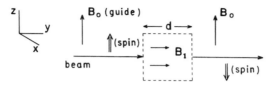

Figure 5.3 A static (dc) spin-flip coil

in the positive y-direction) and the spin is initially aligned "up" (along the positive z-axis). The guide field B_0 is a constant magnetic field (pointing along the positive z-axis) whose function is to keep the beam polarized. The beam enters a region (of length d along the y-axis) in which there is a uniform magnetic field B_1 aligned at right angles to the z-axis. In terms of the Pauli spin matrix σ, the spin s of the neutron is given as

$$s = \frac{\hbar}{2}\sigma. \tag{5.1}$$

We define its magnetic moment μ as

$$\begin{aligned}\mu &= -\mu\sigma \\ &= \frac{2\mu}{\hbar}s.\end{aligned} \tag{5.2}$$

Here μ is the *magnitude* of the magnetic moment of the neutron. The equation of motion for the spin s of the neutron is just

$$\begin{aligned}\frac{ds}{dt} &= \mu \times B_1 = -\frac{2\mu}{\hbar}s \times B_1 \\ &= -\frac{2\mu B_1}{\hbar}s \times \hat{n} \equiv -\omega_L s \times \hat{n},\end{aligned} \tag{5.3}$$

where ω_L is the Larmor precision frequency

$$\omega_L = \frac{2\mu B_1}{\hbar}, \tag{5.4}$$

and $\hat{n}$ is a *fixed* unit vector in the direction of B_1.[49] The corresponding Hamiltonian for this system is

$$H_B = -\mu \cdot B_1. \tag{5.5}$$

By adjusting the distance d, we can arrange to have the time of flight Δt of the neutron through the spin flipper

$$\Delta t = \frac{d}{v} \tag{5.6}$$

be such that the polarization is changed from the $+z$-direction to the $-z$-direction. That is, the state goes from the eigenstate $|+z\rangle$ to the eigenstate $|-z\rangle$, where

$$\sigma_z |\pm z\rangle = \pm |\pm z\rangle . \tag{5.7}$$

This means that, for spin flip, we arrange to have the Larmor precision β equal π

$$\beta \equiv \omega_L \Delta t = \frac{2\mu B_1}{\hbar} \frac{d}{v} = \pi . \tag{5.8}$$

Since there is no *explicit* time dependence in the Hamiltonian of eq. (5.5), the Schrödinger equation

$$i\hbar \frac{\partial \psi}{\partial t} = H_B \psi \tag{5.9}$$

has the solution

$$\psi(t) = e^{-iH_B t/\hbar} \psi_0 = \exp\left(-\frac{i\mu \boldsymbol{\sigma} \cdot \boldsymbol{B}_1 t}{\hbar}\right) |+z\rangle . \tag{5.10}$$

The actions of σ_x and σ_y on $|+z\rangle$ and $|-z\rangle$ can be summarized as

$$\sigma_x |\pm z\rangle = |\mp z\rangle , \tag{5.11a}$$

$$\sigma_y |\pm z\rangle = \pm i |\mp z\rangle . \tag{5.11b}$$

This, plus the general identity[50]

$$e^{i\frac{\phi}{2} \boldsymbol{\sigma} \cdot \hat{\boldsymbol{n}}} = \cos\left(\frac{\phi}{2}\right) + i \sin\left(\frac{\phi}{2}\right) \boldsymbol{\sigma} \cdot \hat{\boldsymbol{n}} , \tag{5.12}$$

shows that, when the Larmor precession angle β has the value π (as given in eq. [5.8]), then the state $|+z\rangle$ is transformed, or "rotated," into the state $|-z\rangle$. Since this is true for *any* direction of $\boldsymbol{B}_1$ orthogonal to the z-axis, for simplicity of illustration I take $\boldsymbol{B}_1$ to point along the positive y-axis.[51] In that case

$$\exp\left(-i\frac{\pi}{2} \sigma_y\right) |+z\rangle = |-z\rangle . \tag{5.13}$$

If in the experiment of figure 5.1, a static spin flipper is placed in just *one* path (say, II), the recombined O beam will have a state represented by the superposition[52]

$$\psi_O = |+z\rangle + |-z\rangle. \quad (5.14)$$

If we compute the *polarization P*, which is the expectation value of σ, we find

$$P = \frac{\langle\psi_O|\sigma|\psi_O\rangle}{\langle\psi_O|\psi_O\rangle} = \hat{\imath}. \quad (5.15)$$

This is a *very* nonclassical effect, since the composition of two beams of equal intensity and opposite states of polarization (along the $+z$-direction and $-z$-direction, respectively) does *not* lead to a mixed state of zero polarization, but to a superposition whose polarization is along the positive x-axis (at right angles to the polarization of the new beam). I return to this result below.

Next consider the effect of an alternating magnetic field $B_1(t)$

$$B_1(t) = B_1 \sin(\omega_L t)\,\hat{\jmath} \quad (5.16)$$

applied in the direction of the line of flight (cf. figure 5.4) of the neutron beam (i.e., along the y-axis in that diagram). Since the corresponding Schrödinger equation does not have a closed-form solution in this case,

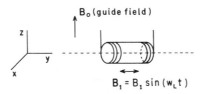

Figure 5.4 A radio-frequency (rf) spin-flip coil

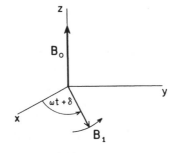

Figure 5.5 A rotating magnetic field

88 Chapter Five

we resort to a mathematical trick and to an approximation.[53] Suppose, first, that we had a *rotating* **B**-field as illustrated in figure 5.5.

$$\boldsymbol{B}(t) = B_1 \cos(\omega t)\,\hat{\boldsymbol{i}} + B_1 \sin(\omega t)\,\hat{\boldsymbol{j}} + B_0\,\hat{\boldsymbol{k}}. \tag{5.17}$$

We can go to the rotating frame of the B_1-field (where B_1 will then be a *constant* vector), solve the Schrödinger equation there, compute the Larmor precession, and then transform back to the laboratory reference frame of the actual experiment. The overall transformation that describes the time evolution of $\psi(0)$ is[54]

$$U(t) \equiv U_z(t + \Delta t)\, U_\beta(\Delta t)\, U_z^{-1}(t)$$
$$= \exp\left\{-i\,\sigma_z\,\frac{[\omega(t+\Delta t)]}{2}\right\} \exp\left\{-i\,\frac{\boldsymbol{\sigma}\cdot\boldsymbol{\beta}\,(\Delta t)}{2}\right\} \exp\left\{i\,\sigma_z\,\frac{(\omega t)}{2}\right\}. \tag{5.18}$$

The vector $\boldsymbol{\beta}$ defining the axis of the Larmor precession is given as

$$\boldsymbol{\beta}(\Delta t) = \omega_1\,\Delta t\,\hat{\boldsymbol{i}} - (\omega - \omega_L)\,\Delta t\,\hat{\boldsymbol{k}} \equiv \beta_x\,\Delta t\,\hat{\boldsymbol{i}} + \beta_z\,\Delta t\,\hat{\boldsymbol{k}} \tag{5.19}$$

and

$$\omega_L = \frac{2\mu B_0}{\hbar}, \tag{5.20a}$$

$$\omega_1 = \frac{2\mu B_1}{\hbar}, \tag{5.20b}$$

with Δt being the transit time of the neutron through the coil.

The efficiency for such a rotating magnetic field to flip the spin of the neutron is just the probability P (not to be confused with the polarization vector $\boldsymbol{P}$ of eq. [5.15])

$$P(-z, t; +z, t = 0) = |\langle -z|\psi_+(t)\rangle|^2. \tag{5.21}$$

It is the "chance" that, having begun in a state $|+z\rangle$ at $t = 0$, the neutron will be in a state $|-z\rangle$ at $t = t$ (i.e., after having passed through the coil). The result is[55]

$$P(-z, t; +z, t = 0) = \sin^2\left[\frac{1}{2}\,\omega_1\,\Delta t\left(1 + \frac{\beta_z^2}{\beta_x^2}\right)^{1/2}\right] \Big/ \left(1 + \frac{\beta_z^2}{\beta_x^2}\right) \tag{5.22}$$

$$= \sin^2\left\{\frac{\mu B_1 \Delta t}{\hbar}\sqrt{1 + \left[\frac{2(B_0 - \hbar\omega/2\mu)}{B_1}\right]^2}\right\} \Big/ \left\{1 + \left[\frac{2(B_0 - \hbar\omega/2\mu)}{B_1}\right]^2\right\}.$$

This probability reaches 100% (i.e., $P = 1$) when $\beta_z = 0$

$$\omega_r = \omega_L = \frac{2\mu B_0}{\hbar} \tag{5.23}$$

and when (simultaneously)

$$\omega_1 \Delta t = (2m + 1)\pi, \quad m = 0, 1, 2, \ldots. \tag{5.24}$$

Therefore, the frequency of the time-dependent field $B_1(t)$ is chosen to be the Larmor frequency ω_L and the length d of the coil (for a given velocity v of the neutrons in the beam) to satisfy eq. (5.24). Then, the neutrons will *definitely* be flipped from the state $|+z\rangle$ to the state $|-z\rangle$ as they pass through the coil.

In fact, the *oscillating* $B_1(t)$-field we are actually interested in, given by eq. (5.16), is (mathematically) equivalent to two counterrotating fields of *half* the amplitude because of the simple identity

$$B_1(t) = B_1 \sin(\omega t)\hat{j}$$
$$= \left[\frac{B_1}{2}\cos(\omega t)\hat{i} + \frac{B_1}{2}\sin(\omega t)\hat{j}\right]$$
$$+ \left[-\frac{B_1}{2}\cos(\omega t)\hat{i} + \frac{B_1}{2}\sin(\omega t)\hat{j}\right]. \tag{5.25}$$

For *either* of these rotating fields *separately*, we can solve the corresponding Schrödinger equation. Unfortunately, the equation for the field

$$B(t) = B_1(t) + B_0 = B_1 \sin(\omega t)\hat{j} + B_0\hat{k} \tag{5.26}$$

cannot be solved exactly for $B_0 \neq 0$.[56] However, an *approximate* solution, valid when $B_1/B_0 \ll 1$, can be developed. In this case, the approximate solution is essentially the superposition of these two counterrotating solutions.[57] I hereafter denote by ω_r the resonance frequency of the flipper (i.e., for 100% efficiency).

Since the Hamiltonian H_B entering into eq. (5.9) now becomes explicitly time dependent (cf. eq. [5.5]), the total energy of a neutron does not remain constant and an energy

$$\hbar\omega_r = 2\mu B_0 \tag{5.27}$$

is exchanged with the spin flipper.[58] From eq. (5.18) we see that if E is the energy of the neutron in the incident beam, then polarized states $|\pm z\rangle$ undergo the transition

$$|\pm z\rangle \to U(t) |\pm z\rangle \sim e^{-i(E/\hbar \mp \omega_r)t} |\mp z\rangle \qquad (5.28)$$

where I have, as usual here, omitted overall constant phase factors. Hence, if a spin flipper is placed in just *one* of the split beams, then

$$|+z\rangle \xrightarrow{\text{incident beam}} e^{-iEt/\hbar}|+z\rangle + e^{-i(E+\Delta E)t/\hbar}|-z\rangle$$

$$= e^{-iEt/\hbar} \begin{pmatrix} 1 \\ e^{i\omega_r t} \end{pmatrix}, \qquad (5.29)$$

where

$$\Delta E = \hbar \omega_r. \qquad (5.30)$$

If we now compute the polarization vector **P** defined in eq. (5.15), we find

$$\boldsymbol{P}(t) = \cos(\omega_r t)\, \hat{\boldsymbol{i}} + \sin(\omega_r t)\, \hat{\boldsymbol{j}}, \qquad (5.31)$$

which is again a superposition.

When the phase-shifting plate is inserted into a beam, it simply produces a phase shift $e^{i\chi}$ (with no loss of intensity) in the beam passing through it as[59]

$$|\psi\rangle \to e^{i\chi}|\psi\rangle. \qquad (5.32)$$

If a phase-shifting plate and a static spin flipper are both inserted into the beam along path II of figure 5.1 (but nothing is placed in path I), then from the results of eqs. (5.14) and (5.32) we see that the final state ψ_O for the recombined (O) beam is

$$|+z\rangle \to |\psi_O\rangle = |+z\rangle + e^{i\chi}|-z\rangle. \qquad (5.33)$$

From eqs. (5.7), (5.11), and (5.33), we find the polarization vector **P** to be

$$\boldsymbol{P} \equiv \frac{\langle \psi_O | \boldsymbol{\sigma} | \psi_O \rangle}{\langle \psi_O | \psi_O \rangle} = (\cos \chi)\, \hat{\boldsymbol{i}} + (\sin \chi)\, \hat{\boldsymbol{j}}. \qquad (5.34)$$

This is a superposition, not a mixture, since $\boldsymbol{P}_{\text{final}}$ is perpendicular to the spins of beams I and II before they are recombined in the *last* plate. This, like the result of eq. (5.15), is a classically unexpected effect.

Now consider the case with a phase shifter and an rf flipper in the path labelled II in figure 5.1. From eqs. (5.29) and (5.32) we see that the state ψ_O for the recombined (O) beam is

$$|+z\rangle \to |\psi_O\rangle = |+z\rangle + e^{i(\chi + \omega_r t)}|-z\rangle = \begin{pmatrix} 1 \\ e^{i(\chi + \omega_r t)} \end{pmatrix}. \qquad (5.35)$$

The polarization $\boldsymbol{P}(t)$ then follows from eqs. (5.7) and (5.11) as

$$\boldsymbol{P}(t) \equiv \frac{\langle \psi_O | \boldsymbol{\sigma} | \psi_O \rangle}{\langle \psi_O | \psi_O \rangle} = \cos(\chi + \omega_r t)\, \hat{\boldsymbol{i}} + \sin(\chi + \omega_r t)\, \hat{\boldsymbol{j}}. \qquad (5.36)$$

For a χ slab and two rf flippers, driven in phase and at the same frequency, eqs. (5.29) and (5.32) imply that

$$|+z\rangle_{initial} \to |\psi_O\rangle = e^{-i(E-\hbar\omega_r)t/\hbar}\left[|-z\rangle + e^{i\chi}|-z\rangle\right]. \quad (5.37)$$

The intensity I_O of the merging O beam is given as

$$I_O \propto |\langle\psi_O|\psi_O\rangle| \propto (1+e^{-i\chi})(1+e^{i\chi}) \propto 1+\cos\chi \quad (5.38)$$

and the polarization is

$$\boldsymbol{P}_O \equiv \frac{\langle\psi_O|\boldsymbol{\sigma}|\psi_O\rangle}{\langle\psi_O|\psi_O\rangle} = -\hat{\boldsymbol{k}}. \quad (5.39)$$

Finally, let the two rf coils be driven at frequencies ω_{r_1} and ω_{r_2}, each near resonance. We require that, while $\omega_{r_1} \neq \omega_{r_2}$, it should nevertheless be true that

$$\Delta\omega \equiv |\omega_{r_1} - \omega_{r_2}| \ll \delta\omega_{1/2} \quad (5.40)$$

where $\delta\omega_{1/2}$ is the (half) width of the resonance curve for the rf flippers. Each flip coil remains about 98% efficient under these conditions.[60] If we denote by δ_1 and δ_2 the phases of the magnetic fields in each coil, then we have (again, from eq. [5.28])

$$|+z\rangle_{initial} \to |\psi_O\rangle = \exp\{-i[(E/\hbar - \omega_{r_1})t - \omega_{r_1}\Delta t/2 - \delta_1]\}|-z\rangle \quad (5.41)$$
$$+ e^{i\chi}\exp\{-i[(E/\hbar - \omega_{r_2})t - \omega_{r_2}\Delta t/2 - \delta_2]\}|-z\rangle$$

It then follows that the intensity of the recombined O beam should have the time variation[61]

$$I_O(t) \sim 1 + \cos[\alpha - (\Delta\omega)t] \quad (5.42)$$

while the polarization is still

$$\boldsymbol{P}_O = -\hat{\boldsymbol{k}}. \quad (5.43)$$

Appendix 2   A causal interpretation of the Pauli equation

In section 4.1.1 we saw how Bohm was able to recast the equations of quantum mechanics for spinless particles into a form similar to Newton's law of motion. In a slightly more general or abstract vein, his program can be represented as follows.[62] In classical particle mechanics one begins with a Hamiltonian $H(q_j, p_j)$ (say, for a one-particle system here)

$$H = T + V = \frac{p^2}{2m} + V(q) \quad (5.44)$$

from which the equations of motion follow as

$$\dot{q}_j = \frac{\partial H}{\partial p_j}, \tag{5.45a}$$

$$\dot{p}_j = -\frac{\partial H}{\partial q_j}. \tag{5.45b}$$

In the present simple case, these reduce to

$$\dot{q}_j = \frac{p_j}{m} \tag{5.46a}$$

$$\dot{p}_j = m\ddot{q}_j = -\frac{\partial V}{\partial q_j}. \tag{5.46b}$$

or to Newton's second law

$$m\,a = -\nabla V = F. \tag{5.47}$$

For my purposes here, eqs. (5.44), (5.45), and (5.47) are equivalent statements of the same dynamical laws. One can also seek a canonical transformation to new canonical variables *all* of which are *constants* of the motion.[63] This can be effected by finding a generating function $S(q_j, t)$ (termed Hamilton's principal function) as the solution of the Hamilton-Jacobi equation

$$H\left(q_j; \frac{\partial S}{\partial q_j}\right) + \frac{\partial S}{\partial t} = 0. \tag{5.48}$$

Then one has

$$p_j = \frac{\partial S}{\partial q_j}, \tag{5.49a}$$

$$E = -\frac{\partial S}{\partial t}. \tag{5.49b}$$

In the case of the Hamiltonian given by eq. (5.44), we have

$$\frac{1}{2m}(\nabla S)^2 + V + \frac{\partial S}{\partial t} = 0, \tag{5.50a}$$

$$p = \nabla S. \tag{5.50b}$$

These two equations together imply

$$\frac{dp}{dt} = \frac{d}{dt}(\nabla S) = -\nabla V = F,$$

which is just Newton's law, eq. (5.47) again.

We have seen that, when the dynamics of quantum mechanics are recast into this form, the total energy must include the quantum potential as (cf. eq. [4.23])

$$E \equiv -\frac{\partial S}{\partial t} = \frac{p^2}{2m} + V + U. \tag{5.51}$$

The Hamilton-Jacobi equations can be extended to include "spin" as follows. A unit spinor can be used to define the orientation of this rigid body in space.[64] In terms of the Euler angles ϕ, θ, ξ for a general rotation, we can express the spinor eigenfunctions as[65]

$$\Psi = R\, e^{i\xi/2} \begin{pmatrix} \cos(\theta/2) e^{i\phi/2} \\ i \sin(\theta/2) e^{-i\phi/2} \end{pmatrix}, \tag{5.52}$$

where R is a real quantity. The spin vector s is given as[66]

$$s = \frac{1}{2} \frac{\Psi^\dagger \boldsymbol{\sigma} \Psi}{\Psi^\dagger \Psi} = \frac{1}{2}(\sin\theta \sin\phi, \sin\theta \cos\phi, \cos\theta). \tag{5.53}$$

The spinor Ψ satisfies the Pauli equation

$$i \frac{\partial \Psi}{\partial t} = H\Psi, \tag{5.54}$$

where the Hamiltonian is given as[67]

$$H = -\frac{1}{2m}(\nabla - ieA)^2 + \frac{e}{2m}(\boldsymbol{\sigma}\cdot B) + V. \tag{5.55}$$

There is then a continuity equation

$$\frac{\partial \rho}{\partial t} + \nabla \cdot (\rho v) = 0 \tag{5.56}$$

for $\rho = R^2$ with[68]

$$\begin{aligned} v &= \frac{1}{2mi}\left(\frac{\Psi^\dagger \nabla\Psi - (\nabla\Psi^\dagger)\Psi}{\rho}\right) - \frac{e}{m} A \\ &= \frac{1}{2m}(\nabla\xi + \cos\theta \nabla\phi) - \frac{e}{m} A. \end{aligned} \tag{5.57}$$

This is the modified "guidance" condition. The Hamilton-Jacobi equation (cf. eq. [5.48]) becomes[69]

$$\frac{1}{2}\frac{\partial \xi}{\partial t} + \frac{1}{2}\cos\theta\,(v\cdot\nabla)\,\phi + \frac{1}{2}mv^2 + U + V + H_s = 0, \quad (5.58)$$

with the usual quantum potential

$$U = -\frac{1}{2m}\frac{\nabla^2 R}{R} \quad (5.59)$$

and a spin-orbit coupling term

$$H_s = \frac{1}{8m}\left[(\nabla\theta)^2 + \sin^2\theta\,(\nabla\phi)^2\right] + \frac{e}{m}s\cdot B. \quad (5.60)$$

The equation of motion for the spin vector s is[70]

$$\frac{ds}{dt} = -\frac{s}{\rho}\times\sum_j\frac{\partial}{\partial x_j}\left(\rho\frac{\partial s}{\partial x_j}\right) - \frac{e}{m}s\times B. \quad (5.61)$$

The first term on the right of eq. (5.61) is a "quantum torque" that appears in addition to the classically expected $s\times B$ torque. (Compare this with eq. [5.3].) The equation of motion for the center of mass $x = x(t)$ turns out to be[71]

$$m\frac{dv_j}{dt} = -\frac{\partial}{\partial x_j}(V + U) -$$
$$\frac{1}{\rho}\frac{\partial}{\partial x_k}\left[\rho\left(\frac{\partial\theta}{\partial x_j}\frac{\partial\theta}{\partial x_k} + \sin^2\theta\,\frac{\partial\phi}{\partial x_j}\frac{\partial\phi}{\partial x_k}\right)\right]. \quad (5.62)$$

The energy of the particle is[72]

$$E(x, t) = -\frac{1}{2}\left(\frac{\partial\xi}{\partial t} + \cos\theta\,\frac{\partial\phi}{\partial t}\right). \quad (5.63)$$

These equations can now be applied to the neutron interference experiments described in the text and the usual quantum-mechanical results follow.[73] For example, in the setup with *no* spin-flip coil in either arm of the interferometer and a phase shifter in one, the initial polarization of the incident beam (say, spin-up) is maintained for each split beam and the spin-dependent quantum-potential and spin-dependent quantum-torque terms vanish, but a spatial interference pattern persists.[74] That is, the polarization remains

$$P = \hat{k} \tag{5.64}$$

and the intensity of the emerging beam varies with χ as

$$I_o \propto 1 + \cos \chi . \tag{5.65}$$

This, of course, is just the result obtained from the standard quantum-mechanical calculation (cf. eq. [5.32], where $I_o \propto \langle \psi_o | \psi_o \rangle$).

SIX
Opposing Commitments, Opposing Schools

If neither empirical (in)adequacy nor logical (in)consistency provides a sufficient explanation for a choice in favor of the Copenhagen over the causal interpretation of quantum mechanics, we might profitably consider other factors. Criteria such as fertility, beauty, and coherence, while often important, can have a Whiggish aspect to them if they are defined in terms of the successful, victorious, or accepted theory and then applied to a competing theory. I turn now to the influence of contingent historical events on the development of quantum theory. Could the philosophical outlooks and backgrounds of the creators of the Copenhagen hegemony have been important factors?

6.1 Cultural milieux circa 1900–1925

What are some of the relevant factors that may have affected the emergence of quantum theory and our view of the world based on it? Indeterminism (or the fall of determinism) in modern times had its roots in the nineteenth century and grew gradually.[1] Thermodynamics was one of the origins for the consideration and eventual acceptance of indeterminism. Also by the late nineteenth century there were significant philosophical precedents for the concept of indeterminism in nature, as opposed to the straightforward determinism often associated with classical physics.[2] For example, Charles-Bernard Renouvier (1815–1903) questioned the causality principle for physical processes, challenged the Kantian doctrine that acceptance of causality was a precondition for human understanding, and held that an object and its representation cannot be divorced even in principle.[3] Émile Boutroux (1845–1921) stressed contingency and inherent (i.e., not just representing *our* ignorance) chance in nature's actions.[4] Their thinking opposed belief in a completely rational universe and proposed instead an element of irrationality in a nature having contingent laws.[5] In a given situation, there would be equally possible alternatives to what does, *in fact*, occur. Henri Poincaré (1854–1912) was influenced by Renouvier's and Boutroux's works, while Louis de Broglie and other founders of the quantum theory studied the writings of Poincaré.[6] It has been argued that Søren Kierkegaard's (1813–1855) philosophy made an impression on Niels Bohr through the teachings of Harald Høffding (1843–1931).[7] Not only had Bohr attended Høffding's lectures as a student, but he also read his works and corresponded with him later in life. Bohr explicitly acknowledged the influence of Høffding's philosophy on his own formulation of complementarity.[8] One of Høffding's tenets was that in life decisive events proceed through sudden "jerks" or discontinuities. Philosophical trends alone did not determine the course of quantum theory in the early part of this century, but these concepts were available to the creators of quantum theory. Speculative thought, even in science, takes place amidst the ebb and flow of larger intellectual currents.

6.2 The Forman thesis

It has been over twenty years now since Paul Forman advanced his dramatic and controversial thesis[9] that acausality was embraced by German quantum physicists in the Weimar era (1919–1933) as a reaction to the hostlie intellectual and cultural environment that existed there prior to and during the formulation of modern quantum mechanics (around 1925).[10] In beginning his own article with the following passages from Max Jammer's often-cited book on the development of quantum theory, Forman himself acknowledged an intellectual debt.[11]

> Documentary material, if studied with scrutiny, will show that certain philosophical ideas of the late nineteenth century not only prepared the intellectual climate for, but contributed decisively to, the formation of the new conceptions of the modern quantum theory.... [12]
>
> The philosophical schools we have mentioned, contingentism, existentialism, pragmatism, and logical empiricism, rose in reaction to traditional rationalism and conventional metaphysics whose ultimate roots were in the philosophy of Descartes. Their affirmation of a concrete conception of life and their rejection of an abstract intellectualism culminated in their doctrine of free will, their denial of mechanical determinism or of metaphysical causality. United in rejecting causality, though on different grounds, these currents of thought prepared, so to speak, the philosophical background for modern quantum mechanics. They contributed with suggestions to the formative stage of the new conceptual scheme and subsequently promoted its acceptance.[13]

6.2.1 The basic theme

While Jammer had made a sweeping and, to some, an appealing claim, he had not, in Forman's opinion, provided the detailed historical argument necessary to establish this case. This is the task Forman set for himself. He hoped to provide a causal analysis of how, in the German cultural sphere after 1918, "scientific men are swept up by the intellectual currents."[14] Forman wanted to establish a *causal* connection between this social and intellectual milieu and the content of science, in this case quantum mechanics. In a later essay, contrasting the reactions in Germany and in Britain to the acausal aspect of the new quantum mechanics, Forman in summary listed an impressive array of physicists and mathematicians who, in the years between the end of the First World War and Heisenberg's 1925 paper on matrix mechanics, had announced their abandonment of the doctrine of causality.[15] These included Franz Exner, Hermann Weyl, Richard von Mises, Walter Schottky, Walther Nernst, Erwin Schrödinger, Arnold Sommerfeld, and Hans Reichenbach.

In broad outline, the structure of his argument is the following. Causality for physicists in the early twentieth century "meant complete lawfulness of Nature, determinism [(i.e., event-by-event causality)]."[16] Such lawfulness was seen by scientists as absolutely essential for science to be a coherent enterprise. A scientific approach was also taken to be necessarily a rational one. When, in the aftermath of the German defeat in World War I, science was held responsible (not only by its failure, but even more because of its spirit) for the sorry state of society, there was a reaction against rationalism and a return to a romantic, "irrational" ideal. The prevailing chain of thought might be represented as something like

$$\text{irrationalism} \to \text{no science} \to \text{no causality}.$$

"If the physicist were to improve his public image, he had first and foremost to dispense with causality, with rigorous determinism, that most universally abhorred feature of the physical world picture. And this [dispensation], of course, turned out to be precisely what was required for the solution of those problems in atomic physics which were then at the focus of the physicist's interest."[17]

In a subsequent publication on this same theme, Forman argued that the new quantum mechanics, which was overwhelmingly a German creation, was taken as supporting certain views (consonant with the prevailing German cultural milieu) that it did *not*, in fact, support.[18] These three characteristics are acausality, intuitiveness (or picturability), and individuality. These were valued characteristics of the milieu, Forman claimed, and quantum theory either exhibited, or in some cases was made to *appear* to exhibit, these same features by its creators. The new quantum mechanics was readily enough taken as rejecting causality, although one could (as Einstein did) see it merely as a statistical theory.[19] Only by a virtual inversion of the meaning of the term *anschaulich* as it had apparently applied to his own matrix mechanics was Heisenberg able, in his influential indeterminacy paper, to make the formalism (really to *define* it as) "intuitive."[20] Heisenberg discussed the intuitive or picturable *content* (however little or restricted) of quantum mechanics. It would still seem stretching matters to claim that quantum mechanics is overall, as a theory or representation of the world, either intuitive or picturable in any ordinary sense of those terms. Similarly, Bohr spoke of the individuality of atomic processes and even of "the 'indestructible individuality' of material particles."[21] Here too, in Forman's view, there was accommodation to a cultural milieu. These physicists made the most of the acausality in quantum mechanics—which was easy enough—and then contorted any natural readings of that theory's lessons in claiming it to be intuitive (or picturable) and to support a type of individuality.[22]

Forman did not take the "internal" demands of science (i.e., the phenomena) to be either irrelevant or wholly malleable (with regard to just *any* theory or interpretation).[23] It is unclear, though, whether the demands of society and of phenomena both required acausality by coincidence or whether there intrinsically existed a wide degree of latitude for such accommodations. Forman spoke of his claim "that the German physicists' predisposition toward acausal laws of nature likewise arose as a form of accommodation to their intellectual environment."[24] This can be taken as a predisposition for choosing or accepting a theory already on offer (whether "demanded" by the facts of nature or whatever), rather than as forcing the creation of such a theory. The crisis in the old quantum theory, which could not have been precipitated by internal factors alone, was possible only, in Forman's view, because of "the physicists' own craving for crises, arising from participation in, and adaptation to, the Weimar intellectual milieu."[25] He concluded "that substantive problems in atomic physics played only a secondary role in the genesis of this acausal persuasion, that the most important factor was the social-intellectual pressure exerted upon the physicists as members of the German academic community."[26]

6.2.2 *Some variations*

Central to Forman's methodology and his claims is a distinction between "psychological" and "sociological" analyses.[27] The former studies the "private" or individual makeup as background and motivation of a particular scientist (e.g., Bohr), while the latter relies on "public" or collective external social factors that are determinative of the action of members of that group. Forman's analysis, by his own declaration, is strictly sociological. However, personal, or psychological, factors may be even more important for the origin of central concepts in individuals who are key to generating and promulgating new theoretical structures.[28] This is illustrated in Jan Faye's study of the influence of Harald Høffding on Niels Bohr in the latter's formulation of complementarity.[29] More reasonable, to me, is the position that psychological factors play a larger (but not an exclusive) role in the specific formulation and interpretation of a theory, while sociological ones can be crucial for the acceptance and propagation of an already-formulated theory. Such "external" psychological or social factors are not *solely* responsible for the content of science. Science has its own "internal" demands and constraints that must also be accommodated. I argue that "internal" factors were most important for the emergence of the formalism of quantum mechanics, "external" ones for the nature of the interpretation that was accepted.

While Forman concentrated on the immediate social background as a key determining factor in the formulation of, or at least in the ready acceptance of, acausality in quantum theory, one can instead emphasize the longer historical process of the variation of the cultural background between the extremes of "Romantic" and "Realist" periods.[30] Probability and randomness came to play an important role in theoretical physics (e.g., statistical mechanics and radioactive decay) by the early part of this century. Consonant with this, there are grounds to claim that "indeterminism, and the denial of independent reality to atomic properties, are not unique to quantum mechanics, but emerged as the culmination of historical trends begun in the nineteenth century."[31] Einstein's resistance to the Copenhagen interpretation can be seen as based as much on his commitment to realism as on his dislike of indeterminism. An independent examination of the origins, in the political and social backgrounds of the culture, of the concept of statistical causality in Germany during the period between 1870 and 1920 has also led to conclusions that are consistent, in broad outline, with those of Forman.[32]

Forman's thesis has not been without its critics.[33] Most of the criticism is aimed at what I might term a *strong* Forman thesis, which would claim a major causal role of the cultural milieu in determining the very form and content of a scientific theory, as opposed to a *weak* Forman thesis, which would see the cultural milieu as sometimes playing an important role in the acceptance and propagation of an already-formulated scientific theory. This weak Forman thesis can remain agnostic on the internal or external nature of the factors that are responsible for creating and shaping the theory itself. This is the position I have advocated previously and the present case study, I believe, further supports such a view.[34] *If* the final decision between two theories were a wholly rational one, *independent of* these prior metaphysical commitments (and this is by no means obvious in the case of quantum mechanics), then the origins of and reasons for those commitments, while perhaps interesting in themselves, would actually be irrelevant for the ultimate theory choice and for its "logical" justification. This is just the usual distinction between the logic of discovery and the logic of justification.

6.3 Predilections—not uniquely/solely determining

I now examine the *prior* philosophical or metaphysical commitments that the key figures in the formulation of quantum theory brought with them to their study of problems in physics. As an illustration of cultural influences, Einstein's general position on foundational questions in physics—relativity, his stance on quantum theory, and his long-standing commitment to a unified field theory—can be seen as a search for the God of Spinoza. Einstein believed that God manifests himself in the rational structure of the external physical world, which the physicist tries to capture in a causal space-time theory.[35] Einstein's basic weltanschauung was that of a rational, causal world that could be comprehended in terms of an objective reality. At the same time, he was also subject to the currents of positivism—both Mach's view that the laws of science are essentially just summaries of experimental facts and Poincaré's conventionalism.[36] Einstein respected Poincaré's work in philosophy and he acknowledged explicitly the philosophical influence on him of Poincaré and of Mach.[37]

Similarly, a case has been made for the influence of Harald Høffding's philosophy on Niels Bohr's formulation and interpretation of quantum mechanics.[38] This can be better appreciated against the backdrop of the pro and con opinions on the issue. Many years ago, a book on the history of quantum physics sketched the philosophical currents present when quantum theory was formulated in the first quarter of this century and intimated a significant influence of Høffding's teaching on Bohr's own work in the foundations of quantum mechanics.[39] On the other hand, a coworker, perhaps better a disciple, of Bohr's, in a 1963 encomium published shortly after Bohr's death, in emotional tones categorically *denied* any external influence of earlier philosophy on Bohr's views of atomic phenomena.[40] Rather remarkably, the author claimed that as a young man Bohr had a strong interest in philosophy and was going to write a book on the subject, but essentially had no time to do so because he devoted his energies to constructing a theory of atomic phenomena. When, years later, Bohr had finally formulated his principle of complementarity and teased out many of its profounder philosophical and epistemological consequences, it so happened that these views *precisely* coincided with, and hence supported quite objectively, Bohr's own earlier philosophical views. This earlier philosophy was not seen as plausibly one causal factor in the doctrine of complementarity, but instead as a happy coincidence of harmony between philosophy and physics. In like vein, a recent biography of Bohr dismisses any significant influence of Høffding's, or anyone else's, philosophy on the end products of Bohr's ruminations about the philosophy of quantum physics.[41] Bohr is labeled "one of the major twentieth century philosophers" and we are told flatly that "there is no evidence of any kind that philosophers played a role in Bohr's discovery of complementarity."[42]

In contrast to such a position, a recent study of Bohr's philosophy holds "that there was a close connection between Niels Bohr's approach to the study of the atom and the philosophical influences which shaped his outlook from childhood and youth onwards."[43] The claim is not, of course, that these early philosophical influences *alone* completely determined Bohr's later position on the interpretation of quantum mechanics, but that they were a major factor. An "interpretation will normally be colored by the philosophical assumptions of the interpreter. This also applied in the case of Niels Bohr."[44] Bohr's interest in the problem of free will and his conception of biology are seen to have had their roots in Høffding's work, as did any (indirect) influence of Kierkegaard's or Kant's ideas on Bohr. "No persons other than Høffding and Einstein have meant so much to Bohr from an intellectual point of view."[45] Høffding himself "considered Bohr's interpretation of quantum mechanics as something growing out of his [Høffding's] philosophical and psychological ideas."[46] Bohr's own writing on causal versus space-time descriptions and on complementarity have counterparts in Høffding's epistemology, while Høffding's position on causal explanation and intelligibility are consonant with Bohr's equivalent use of causal and rational accounts.[47] Bohr's peculiar gloss on space, time, and causality as (Kantian) forms of perception has been attributed to Høffding, who passed it on to the young Bohr.[48] In formulating his approach to quantum mechanics, Bohr used dialectic relationships of whole and parts, continuity and discontinuity, rationality and irrationality, which Høffding had emphasized.[49]

The intersection of general predilections like those discussed here with the puzzles presented by physical phenomena results in a definite theory or research program. To illustrate this, I sketch how the opposing positions—wave versus matrix mechanics—were arrived at and what the arguments were for each view. The one traces its roots back to the nature of electromagnetic phenomena, the other to a study of discrete spectral lines. Not only were the general philosophical outlooks of these two groups of key players quite different, but it was also with very different classes of physical phenomena that each group began: for one the paradigm phenomena consisted of (continuous) radiation, while for the other they were taken to be (discrete) spectral lines. The discontinuity-versus-continuity dichotomy was contingently rooted in philosophical commitments and in physical phenomena.

6.4 The wave-mechanics route[50]

In 1909 Einstein used the Planck blackbody radiation law

$$\rho(v, T) = \frac{8\pi v^2}{c^3}(hv)\left[\exp\left(\frac{hv}{kT}\right) - 1\right]^{-1} \tag{6.1}$$

and Planck's energy quantization condition

$$\varepsilon = hv \tag{6.2}$$

to compute the mean square energy fluctuation for blackbody radiation in thermal equilibrium with another system as[51]

$$\langle \varepsilon^2 \rangle = (Vdv)\left[hv\rho + \left(\frac{c^3}{8\pi v^2}\right)\rho^2\right]. \tag{6.3}$$

He pointed out that the first term would be expected if radiation were composed of independent particles (his "photons") and the second term if radiation consisted of waves that could interfere with each other. At a conference in Salzburg in 1909, Einstein offered the opinion that a kind of fusion of the wave and emission theories of light was necessary for progress.[52] In retrospect, we tend to see this as an early flirtation with the concept of wave-particle duality.

In a paper on the quantum theory of radiation published in 1917, Einstein showed that a consistent and empirically adequate theory of the interaction of radiation with matter obtained only if, when molecules emit and absorb radiation under the influence of an external radiation field, momentum and energy are both conserved in each individual interaction.[53] Such directional radiation was termed "needle radiation" by Debye.[54] Compton's experimental work on the scattering of radiation by electrons gave support to the hypothesis of free electromagnetic quanta.[55] In 1922 Einstein and Ehrenfest analyzed the implications of the Stern-Gerlach experiments and asked by what physical mechanism the observed spins manage to take up their final (discrete, quantized) orientations.[56] They argued that, whether the atom is *always* in a quantized spin-orientation state or whether rapid variations occur during which states exist that violate the quantum rules, conceptual inconsistencies result (e.g., either energy conservation fails or only radiation-emitting systems can be quantized).[57] This was an early indication of the conceptual difficulties encountered when one attempts to construct (mechanical) pictures of atomic events. It would sharpen into a conflict between a representation in a continuous space-time background ("visualizability" of microevents) versus strict (event-by-event) energy conservation (and causality). In his 1917 paper, Einstein stated that the recoil direction of the molecule, which has emitted radiation, is "only determined by 'chance', *according to the present state of the theory*" and that "the weakness of the theory lies . . . in the fact . . . that it leaves the duration and direction of the elementary processes to 'chance'."[58] Here, as later, Einstein took this to be a shortcoming of the theory, a provisional fault to be overcome in the future.

The next key figure in this "continuity" school is, of course, Louis de Broglie. What would prove to be a recurrent theme in his work was not (classical) determinism, but rather the possibility of a precise space-time representation for a clear picture of microprocesses.[59] In this, his expectations were similar to Einstein's. Although de Broglie had been much impressed with Poincaré's writing on the philosophy of science, he parted company on the thesis of radical underdetermination. De Broglie did believe that *one* theory should best conform to nature.[60] He felt that classical Hamilton-Jacobi theory provided an embryonic theory of the union of waves and particles, all in a manner consistent with a realist (continuous) conception of matter.[61] This strong commitment by de Broglie to a realistic interpretation of the quantum formalism is consistent with his own high estimate of the work of Émile Meyerson, as stated by de Broglie in a preface to Myerson's *Essais*.[62] Meyerson, an influential French philosopher of science, attempted to dispel the positivist bias and held that the goal of science is an *ontological* one.[63] The Copenhagen interpretation would effectively sidestep the difficult issue of delineating a coherent ontology. Earlier, de Broglie had, under the influence of Bergson, turned from the positivistic trend in science, although he never embraced Bergson's radical disdain for positive science and his criticism of mathematical method in science.[64] For Bergson, continuity at the most fundamental level was an essential feature of nature.

Not only was there little interest in quantum physics in France after World War I (and France and Germany remained scientifically insulated from each other then), but even earlier in the century, theoretical physics there had fallen very much behind the times.[65] Léon Brillouin recalled there was no regular course of applied mathematics when he attended the Sorbonne, with very few people working in theoretical physics: "Theoretical physics was really at a low level when I was a student."⁶⁶ There was certainly no existing tradition of research in quantum theory in France when Louis de Broglie was a student. What theoretical research there was was dominated by classical physics. It was Louis's brother, Maurice, who was actively involved in experimental work on the photoelectric effect and who introduced the younger de Broglie to quantum physics. Louis de Broglie was poor in foreign languages and rarely went abroad.⁶⁷ In 1923 de Broglie initiated the theory of wave mechanics in attempting to understand the dual nature of Einstein's photon. In his early youth, Louis de Broglie had been impressed by the well-known formal (mathematical) analogy between wave optics and classical particle mechanics.⁶⁸ Building on this analogy and on some of his own previous work on Einstein's light quanta, de Broglie proposed a model of a particle that follows the trajectory of its associated phase wave.⁶⁹ There was a great affinity of views between Einstein and de Broglie. In 1925 Einstein stated his opinion that de Broglie's ideas "involve more than merely an analogy."⁷⁰ This is not surprising since Einstein had previously, in the context of general relativity, attempted to treat "particles" as the singularities in an underlying field.

This notice by Einstein drew attention to de Broglie's work. Walter Heitler recalled it was widely studied in Germany. Sommerfeld's group in Munich discussed de Broglie's paper, but everyone had objections and no one took the idea seriously.⁷¹ This applied to most theoretical physicists—except Schrödinger, who later said that "my theory was stimulated by de Broglie's thesis and by short but infinitely far-seeing remarks by Einstein."⁷² Uppermost for Schrödinger was a concern for conceptual matters and for physical models.⁷³ Important for him too were "allegorical" pictures of the physical situation and a clear physical interpretation.⁷⁴ An even more long-standing and central concern of his was physical pictures that undergird the formal mathematics of physical theories.⁷⁵ In the fall of 1926 Schrödinger wrote to Wilhelm Wien that "the point of view [using] visualizable pictures, which de Broglie and I assume, has not been carried through nearly far enough in order to render an account of the most important facts [of atomic theory]."⁷⁶ He took the comprehensibility of the external processes in nature to be an axiom. On the other hand, Bohr's Copenhagen school tenaciously held to the position that accounts of atomic phenomena must be formal rather than realistic.⁷⁷

In a paper on Einstein's gas theory, Schrödinger concluded that photons can be seen as the energy levels of the éther oscillators and that cavity radiation need not "correspond to the extreme light-quantum representation."⁷⁸ For him this meant "taking seriously the de Broglie–Einstein wave theory of moving particles, according to which the particles are nothing more than a kind of 'wave crest' on a background of waves."[79] A standard account has Schrödinger, in the same year (1926), exploiting Hamilton's analogy between mechanics and optics to obtain his wave equation.[80] An examination of Schrödinger's first two papers and of his early work suggests, however, that use of de Broglie's ideas applied to the study of ideal gases and analogies with other known wave equations in physics first led Schrödinger to his wave equation.[81] The derivation in the beginning of the second paper (via Hamilton's analogy) was a later construction on his part. In any event, once the wave equation was in hand, quantization was implemented by boundary conditions imposed on a continuous wave function. It is interesting to note in this connection that Schrödinger had early on acquired a mastery of eigenvalue problems in the physics of continuous media and that he was, by the time of his wave-mechanics papers, familiar with Courant and Hilbert's book on mathematical methods in physics (in which such eigenvalue problems are treated in detail).[82] Schrödinger worked out the *relativistic* case for hydrogen first (no spin, of course) and obtained an answer different from the Balmer formula and then put this work aside for some months before returning to the nonrelativistic case.[83] It would appear that his motivation and results owed nothing to Heisenberg's formulation of matrix mechanics or to Pauli's solution of the hydrogen atom with it. This is consistent with Schrödinger's own comment in a note in his paper on the equivalence between wave and matrix mechanics: "I did not at all suspect any relation to Heisenberg's theory at the beginning. I naturally knew about his theory, but was discouraged, if not repelled, by what appeared to me as a very difficult method of transcendental algebra, and by the want of perspicuity."[84]

The last thread in this part of my story is de Broglie's proposal in 1927 of a principle of the double solution in which the basic entity is a wave, but a wave that has a singularity whose trejectory would correspond to a particle path.[85] Because of mathematical difficulties with that proposal, at the 1927 Solvay congress he presented a simplified pilot-wave model. I return to this subject in section 7.2.

What I have sketched above is one (historical) route to *a* formulation of quantum mechanics—namely Schrödinger's wave mechanics. This small group (Einstein, de Broglie, and Schrödinger) of the creators of that theory stressed the continuous aspect of electromagnetic radiation and shared a commitment to a continuous wave as the basic entity subject to a realistic and, at least for Einstein and de Broglie, causal description. Visualizability and self-consistency had become accepted hallmarks of classical physical theories. To them it seemed preferable to "have an intelligible classical wave theory with flaws than a totally unintelligible 'dual' theory."[86]

A continuity/discontinuity divide separated this group from the one that shaped matrix mechanics. In his attempts to find an interpretation, Heisenberg took for granted the existence of discontinuities in the atomic domain, as did the other creators of matrix mechanics.[87] An examination of Born and Heisenberg's contribution to the 1927 congress shows that they believed the new mechanics was based on the idea that atomic physics is essentially different from classical physics on account of the existence of discontinuities.[88] Certainly the position taken by the wave-mechanics school was the more "natural" one relative to the then-accepted concepts of classical physics (i.e., it represented a less radical departure). The other program, based fundamentally upon discontinuity (and a lack, even in principle, of event-by-event causality), finally carried the day in spite of its much more radical nature. Since one might expect that the scientific community would have been inclined to take the more conservative of the alternatives on offer, it is all the more important to examine the arguments that were seen as convincing for the relatively drastic choice actually made.

6.5 The matrix-mechanics route

The small number of central players (Bohr, Heisenberg, Pauli, Jordan, and Born) involved in the program that led to matrix mechanics suggests that we ask whether this, too, was a closed group. That it was becomes all the more plausible when we realize that Pauli and Heisenberg were both Ph.D. students with Sommerfeld at Munich and that each in succession was then Born's assistant at Göttingen and later worked with Bohr at Copenhagen. Each of these young students was greatly impressed by Bohr's 1922 Göttingen lecture.[89] Jordan was also a student at Göttingen at this time.

I have already indicated philosophical factors in Bohr's own background that inclined him toward, or at the very least made him receptive to, discontinuous structure in nature at the most fundamental level and, eventually, to a doctrine of complementarity between opposites. An element of discontinuous transitions is a central feature in his 1913 "semi-classical" model for the hydrogen atom.[90] This was certainly the current language for discussing atomic phenomena in Sommerfeld's school. Prima facie, there is a case that Pauli and Heisenberg as young students were impressionable and naturally accepted this central tenet of atomic theory. Throughout his life, the ever-critical Pauli remained deferential toward his old teacher, Sommerfeld.[91] However, this *inclination* toward credence is scarcely sufficient to account for the prevailing strength of this conviction on the discontinuity-versus-continuity issue. Discontinuities, not causality as such, were initially the key issue in this formulation of quantum mechanics.[92]

Largely due to the *failure* of certain classical approaches, the main players took positions on what was and was not *possible* in principle. In his contribution to the 1949 Schilpp volume on Einstein's thought, Niels Bohr asked "whether the renunciation of a causal mode of description of atomic processes . . . should be regarded as a temporary departure from ideals to be ultimately revived or whether we are faced with an irrevocable step towards obtaining the proper harmony between analysis and synthesis of physical phenomena."[93] What Bohr wrote in this essay can reasonably be taken as his mature view on quantum mechanics. There he stated categorically that a causal analysis of microphenomena *must* be abandoned and emphasized the "inability of the classical frame of concepts to comprise the peculiar feature of indivisibility, or 'individuality', characterizing the elementary processes."[94] Bohr recounted the *failures* of classical attempts to account for the observed quantum phenomena and offered the opinion that a renunciation of continuity and causality "*appeared* to be the only way open to proceed with the immediate task of co-ordinating the multifarious evidence regarding atomic phenomena."[95] This is not a *proof* of the impossibility of a causal program. He later changed the modality of expression and stated that the "renunciation of the visualization of atomic phenomena is *imposed* upon us."[96] To support this claim Bohr discussed the Einstein-Podolsky-Rosen paper and concluded that a study of their argument "emphasize[s] how far, in quantum theory, we are beyond the reach of pictorial visualization."[97] What Bohr actually did, though, was to offer a description in terms of complementary pictures to resolve the difficulties raised. That is a *consistent* story, but it does not eliminate, in principle, a causal account. In his own summary of the argument in this essay, Bohr claimed that we "are not dealing with an arbitrary renunciation of a more detailed analysis of atomic phenomena, but with a recognition that such an analysis is *in principle* excluded."[98] We have, ultimately, a declaration of faith that consistency arguments constitute a refutation of even the *possibility* of an alternative point of view. These were not logical or in-principle refutations, but strong, practical *beliefs* that became dogma.[99]

Let me now turn the clock back a bit and see how these beliefs were fashioned. Bohr's own Ph.D. dissertation argued that the failure of the classical electron theory of metals was attributable to a fundamental insufficiency of the classical principles themselves.[100] Pauli, in his work on general relativity and related field-theory generalizations convinced himself (again, because of a *failure*) that a continuum field theory, with the particles as singularities, was not possible.[101] In his famous 1921 *Theory of Relativity*, Pauli already was of the opinion that "there is no point in discussing ... quantities [that] cannot, in principle, be observed experimentally."[102] In the same vein, he suggested that quantities that we cannot conceive of any method of observing should be considered "unobservable by definition, and thus be fictitious and without physical meaning."[103] This certainly has a strong operationalist air about it.[104] By 1923, in a letter to Arthur Eddington, Pauli required operational definitions of—really, descriptions for being able to measure, in principle—anything used in physics and the replacement of continuous concepts by discrete ones.[105] Observations are essentially localized and instantaneous, with measurements being discrete and, he felt, this structure should be carried over into the foundation of any theory that accounts for such observations and measurements. This would require, he believed, major conceptual revisions.

Both Pauli and Heisenberg had been involved in Born's program of attempting to apply the old quantum theory, with its orbitals, to molecular systems, and the utter failure of this approach convinced them that electron orbitals were meaningless.[106] A major logical difficulty of the old quantum theory with its visualizable orbitals was that, due to the quantization requirement for orbital angular momentum, the orbitals had to be oriented in certain special ways in any field, *however weak*.[107] If a field in one direction were reduced slowly to zero and then gradually increased to some small value in another, it was not at all clear how the orbit could properly align itself (rather than simply precessing about the initial axis). The Stern-Gerlach experiments (1921/22) supported quantum predictions (notice the name—*space quantization*—originally used for this effect), rather than a continuous (classical) magnetic moment.[108] Pauli's success with the Zeeman effect (1924) in terms of a classically nondescribable two-valuedness in the quantum-theoretical properties of the electron further strengthened his belief in nonvisualizability.[109] He was convinced as well that the exclusion principle could not follow from classical mechanics or from the old quantum rules.[110] Heisenberg began to be converted to Bohr's and Pauli's views on the inadequacies of mechanics.[111] The failure of the Bohr-Kramers-Slater theory (in which energy conservation had been given up) in 1925 indicated to Bohr that a complete renunciation of the usual space-time methods of visualization of the physical phenomena would be necessary for further progress.[112]

Not everyone who went to study at Copenhagen was taken with Bohr's style of argument. John Slater, who held a traveling fellowship from Harvard in 1923/24, spent part of this time at Bohr's institute. In his recollections we find:

> Bohr would always go in for this remark, "You cannot really explain it in the framework of space and time." By God, I was

determined I was going to explain it in the framework of space and time. In other words, that was Bohr's point of view on everything, and that was the fundamental difference of opinion between us.[113]

Before he went to Copenhagen, he had thought that "Bohr's papers looked like handwaving," but he hoped that it was simply Bohr's style that was "just covering up all the mathematics and careful thought that had gone on underneath."[114] Within a month, he had become convinced that "it was all just handwaving."[115] Slater found that Bohr was contemptuous of any attempt to work out a physical picture and that he would not even look at such work.[116]

Heisenberg's matrix mechanics provided just the type of renunciation that Bohr had demanded. Heisenberg is often taken at face value in his claim to have been motivated to construct this theory so as to banish unobservable quantities from it.[117] Born, in his influential lectures during the 1925/26 winter term at MIT, presented this positivistic approach as a basic axiom for any acceptable physical theory.[118] A more careful analysis of Heisenberg's actual work indicates "that the positivistic philosophy of elimination of unobservables was not used as a guiding principle in the emergence of the new quantum theory, but rather mostly as a post facto justification."[119] This is particularly interesting since Heisenberg, in his later recollections, claimed that, shortly after his formulation of matrix mechanics, he tried out an "observables only" line on Einstein in Berlin in April of 1926.[120] Because electron orbits inside the atom cannot be observed, Heisenberg told Einstein, and because a good theory should be based on observables only, such orbits should not enter into the theory. Einstein expressed surprise that anyone would believe that a theory could be based on observables only. When Heisenberg responded that he was simply following Einstein's own lead from the theory of relativity, Einstein said (so Heisenberg reports) that such an approach would be nonsense nevertheless and then made an oft-quoted statement: "But on principle, it is quite wrong to try founding a theory on observable magnitudes alone. In reality the very opposite happens. It is the theory which decides what we can observe. . . . Only theory, that is, knowledge of natural laws, enables us to deduce the underlying phenomena from our sense impressions."[121] This is really rather paradoxical, since an entire generation of young German physicists was influenced by what they took to be Einstein's "positivistic" approach to physics—an approach that Einstein himself here disavowed. Similarly, around 1932 Philipp Frank learned with discomfort that Einstein did not have positivist convictions and refused to accept "the new fashion" that was arising in physics. Frank protested:

"But the fashion you speak of was invented by you in 1905?" Einstein rejoined: "A good joke should not be repeated too often."[122]

Dirac was an exception in this group in that he had no particular interest in philosophical questions.[123] He recalled that, while he appreciated very much the beauty of de Broglie's work, he could not take those waves seriously. "I was so much imbedded in the [old] Bohr theory, I took the Bohr orbits very literally—we had the electrons as real particles, and the de Broglie waves seemed to me to be just a mathematical fiction of no importance to physicists."[124] Dirac cared only about the equations and what could be calculated with them. Later he took from Bohr the rejection of mental pictures in space-time. Like many, perhaps most, physicists, Dirac was willing to leave the philosophical considerations to someone else. He made the interesting observation that, in his opinion, part of the reason for the great impact that general relativity made in 1919 after the end of World War I was the need to forget the old and to focus on something wonderful and new.[125] This desire for a radical conceptual revolution was prevalent in the general cultural milieu of the time.[126] It was also dominant in Pauli's and Heisenberg's expectations about quantum theory.

It was the "collision" between matrix mechanics and wave mechanics that provided the impetus for the formulation of a consistent interpretation of quantum mechanics. Although it is not uncommon for scientists to believe that there is just *one* correct law or theory, Bohr even as a child believed in the uniqueness (i.e., the necessity) of natural laws.[127] Such a belief would justify one in looking for, or attempting to formulate, *the* correct version of quantum mechanics. Heisenberg's faith in the finality of quantum mechanics was essential for his struggle to fashion the Copenhagen interpretation via his uncertainty relations. Beller tells us:

> In his recollections, Heisenberg repeatedly stressed that his belief in the completeness of this mathematical scheme of quantum mechanics led him to assume that nature works only in such a way as not to violate the quantum mechanical formalism. Heisenberg's recollections regarding his belief are supported by correspondence at the time. This belief in the completeness of the mathematical scheme was essential—without it, Heisenberg, in the case of a discrepancy between "nature" and "formalism", would seek to improve the formalism rather than to reinterpret nature. It is in this context . . . that Einstein's dictum ("it is the theory which decides what we can observe") "suddenly" approached the status of a guiding principle.[128]

Born believed that microscopic coordinates were unmeasurable and therefore irrelevant.[129] Through his analysis of scattering processes with

Schrödinger's formalism, Born came to the opinion that even *perfect* initial information still led to uncertainty in the result and this implied, for him, a lack of causality.[130] Why was the complementarity principle taken as being complete and the final word in forbidding even the in-principle possibility of a description of microphenomena that is both causal and pictured in a continuous space-time?[131] One response is that (thus far) experience has shown the validity of complementary pairs of descriptions and that belief in the ultimate *necessity* of complementarity rests on the *subjective* epistemological criterion of the need for classical concepts and on the indivisibility of atomic phenomena (i.e., Bohr's act of faith).[132] Nathan Rosen summarized Bohr's position as "[physical] reality is whatever quantum mechanics is capable of describing."[133] For Bohr 'causality' meant the applicability of the exact laws of energy and momentum conservation.[134]

What is actually established by Bohr's arguments is that precise measurements of space-time properties and the verification of energy-momentum conservation require mutually exclusive arrangements. This really constitutes a *consistency argument* for, rather than a *proof* of, Bohr's principle of complementarity. We might say that the Copenhagen interpretation defined itself as true and strengthened its hold on physics, rewriting history so that Einstein, de Broglie, and Schrödinger largely fade from view, thus leaving Copenhagen as the only intelligible version of quantum mechanics.[135] John Heilbron, a historian of quantum physics, gives us some insight into the "external" pressures under which the Copenhagen group labored.

> The Pauli letters testify to the tremendous psychological *Zwang* ("constraint," another term of the Copenhagen art) under which the quantum physicists around Born and Bohr labored. Born himself oscillated between modest self-confidence and deep self-doubt; Bohr was frequently ill; Pauli hovered on the verge of breakdown; while Heisenberg, maintaining a healthy flippancy, succeeded in rescuing them all. Their correspondence is filled with strong words expressing despair, misery, and resignation, or joy, hope, and elation. . . . These romantic expressions . . . suggest the frame of mind that soon generated a religion of complementarity.[136]

I now turn to some of the personal and sociological factors that lent urgency to the Copenhagen school's establishing the hegemony of its interpretation.[137]

SEVEN
Competition and Forging Copenhagen

With this as background for the historical development that led to two opposing interpretations of the formalism of quantum mechanics, I can now ask how the Copenhagen interpretation was formulated under the challenge that Schrödinger's wave mechanics presented to the Göttingen-Copenhagen matrix-mechanics program and how this latter interpretation established its hegemony.[1] Since matters of logic and of empirical adequacy alone were insufficient for rejecting out of hand the possibility of a causal interpretation, it can be little surprise that other factors must be considered to provide an adequate explanation.

7.1 The challenge of wave mechanics

I return to the early days of the quantum formalism and the Solvay congress of 1927. In 1923 Louis de Broglie suggested a concept of wave-particle duality as extended to electrons and further developed these duality arguments in his doctoral dissertation.[2] Although, as we have seen, Einstein was impressed with de Broglie's thesis, and Schrödinger's own work on wave mechanics was influenced by de Broglie's insight, de Broglie nevertheless had the reputation of being an unorthodox theoretician. That did not condition the scientific community at large to consider his subsequent speculations seriously.[3] The French scientific community had isolated the Germans for reasons of nationalism after World War I and this was not conducive to a flow of ideas between the Germans and the French.[4] But this can scarcely have been an absolutely determining factor since, after all, Schrödinger was Austrian and *was* receptive to de Broglie's ideas. We have seen that a split in philosophical outlook, largely along generational lines, was relevant: the "older," essentially classical, worldview of people like Einstein, Schrödinger, and de Broglie versus a radically different, eventually indeterministic conception of physical processes engendered by the younger generation including Heisenberg, Pauli, Jordan, and Dirac. This tension between the Copenhagen group (e.g., Bohr, Born, Heisenberg, Jordan, and Pauli, with Dirac being an "honorary" member in this classification) and the likes of Einstein, Schrödinger, and de Broglie is a key element in appreciating how Copenhagen gained the ascendancy. Nor were the differences between these two groups merely philosophical. As will become clear, there was a very practical "social" or "professional" dimension to it as well.

7.1.1 Heisenberg's views on formalism and interpretations

Acceptance of an empiricist-operationalist philosophical tendency as an ideal among Heisenberg, Pauli, and Bohr can be traced in part back to Einstein's 1905 relativity paper. This operationalist approach seems to have exerted a profound influence upon young German physicists. This helps us in understanding the vehemence of the Copenhagen school's reaction against Schrödinger's wave mechanics. Matrix mechanics had been formulated by Heisenberg, and developed by other members of the Copenhagen school, as an essentially abstract mathematical formalism with no physical interpretation.[5] Heisenberg believed that a successful mathematical formalism of a physical theory, such as classical mechanics, was of a piece or whole and that it could not be modified in any essential way without destroying the entire structure.[6] This is what he meant by a 'closed theory'.[7] When such a formalism encounters difficulties (as classical mechanics did with quantum phenomena), it is not possible to modify that formalism successfully. A radically new formalism must be found to accommodate these new features of the physical world. (In his later years Heisenberg came to see "Kuhnian" revolutions in science as a natural outcome of this sharp break between formalisms.) Not only did Heisenberg see a successful formalism as unique and of a piece, but he also held a remarkable view on the relation between a formalism and its interpretation.[8]

In an interview in Munich on 22 February 1963, for the *Archive for the History of Quantum Physics* (AHQP), Thomas Kuhn discussed with Werner Heisenberg Heisenberg's own reaction (in 1926) to Erwin Schrödinger's development of wave mechanics.[9]

> *Heisenberg*—You just mentioned this attitude of (mine) toward the development of wave mechanics of Schrödinger and especially toward the paper of Born on collision processes. There the actual psychological situation for myself was that I felt, after we had written our *Drei Männer Arbeit,* at that time the mathematical scheme did make quite definite statements about how to calculate energy levels and to calculate amplitudes and intensities and so on. So I felt that from now on it is just a problem of

working things out, if one wanted to get the correct interpretation also for collision problems or for whatever else is to be found in atomic physics. Therefore, when Schrödinger's things came out, I found it very interesting. But I felt from the very beginning that either the thing is identical—is just a mathematical transformation of our own quantum mechanics—in which case it is extremely interesting, but it does not really contain new things except for new ways of talking about things; or it is different from our own attempt, in which case either or both must be wrong.

Therefore, I did not like so much that Born went over to the Schrödinger theory. Either that was just a new mathematical tool instead of an old one for the same thing, in which case all right, why not do it from the old mathematical tool. Or it really meant that Born doubted that our scheme was right, and thought rather Schrödinger's scheme was right. . . .

But I was so much afraid that by means of the Schrödinger mathematical scheme, a new interpretation of the thing would be brought in. Just because the interpretation was not perfectly clear at that time, I was very much afraid that now entirely wrong ideas could enter into the thing and actually have entered. Schrödinger, as you know, wanted to throw all the quantum jumps away and to say that there is no quantization, it's just all wave pictures and so on.[10]

Kuhn then pointed out that matrix mechanics was formulated and used without any interpretative commentary at all.

Heisenberg—Now I always felt that this [i.e., the meaning of the energy levels and amplitudes from Heisenberg's (1925) first quantum-mechanics paper] should already be sufficient to get the interpretation in all other cases, I knew that it had not been carried out, but I felt, "Well, that is now just a matter of doing things properly, then you will find out." . . .

In the Born paper on collisions, it looked as if he made now the new hypothesis, that we have to interpret the transformation matrices, or the Schrödinger waves, as a probability. I just felt that there was absolutely no room for any new hypothesis. Either the hypothesis follows from our own interpretation, then it's all right; or it does not follow, then either or both must be wrong. Perhaps our whole scheme is wrong.[11]

Kuhn then pressed Heisenberg on the possibility that, if it had taken longer to establish the equivalence between the two formulations of quantum mechanics, then physics might have developed quite differently.[12]

> *Heisenberg*—Yes, for a while, yet it might. Yes, that's quite true, yes. . . . I would say that however this story would have gone, at the end this would have been the result and then people would have realized, "Yes, we must use the words with this restriction."[13] . . .
> We didn't want to go back to the old line, and that was a disappointment with Schrödinger. . . . I felt, "Now Schrödinger puts us back into a state of mind which we have already overcome, and which has certainly to be forgotten."[14]

Heisenberg believed that one simply had to examine the formalism (here, of matrix mechanics) to find its proper interpretation. This view that a formalism gives uniquely its own proper interpretation made the appearance of Schrödinger's apparently very different formalism (theory) quite disturbing for Heisenberg. The situation was even worse after the two formalisms had been shown to be equivalent, since there then appeared to be two conflicting interpretations based on just *one* formalism.[15] This possibility, coupled with Born's "defection" to the enemy camp when he wrote his paper on the statistical interpretation of Schrödinger's wave function, produced in Heisenberg (so he told Kuhn in this interview) the fear that the physics community might take a wrong fork in the road and be led into error. Before any equivalence between the two formalisms had been established (and perhaps even more acutely afterwards), the possibility of a wrong choice was a major concern (according to Heisenberg here) so that it remained essential to find *the* correct interpretation of matrix mechanics as quickly as possible. This became a major undertaking in Copenhagen with Bohr, Heisenberg, and Pauli. When asked what might have happened if the equivalence (between matrix and wave mechanics) had taken longer to establish and if the theoretical physics community had gone the ("wrong") Schrödinger route, Heisenberg allowed that it might have *for a while*. But he was certain that things would have come back to where they in fact have ended (to the truth? to matrix mechanics?). Heisenberg affirmed a belief in a *unique*, correct theory—one truth.

7.1.2 Schrödinger's successful formalism: a threat
While Heisenberg's responses to Kuhn in this interview remained on the lofty plane of objective truth and concern with physics not being misled

into a wrong turn, there were other, much more practical and mundane, factors that lent urgency to the dispute/choice between matrix mechanics and wave mechanics. The formalism of matrix mechanics had *not* had many successful applications and it appeared to be bogged down in a mathematical morass before Schrödinger's wave mechanics allowed theorists to make a stunningly wide variety of well-supported calculations.[16] (In spite of Pauli's mathematical tour de force to obtain the Balmer spectrum for hydrogen, this overall practical failure of matrix mechanics remained a fact.)[17] Wave mechanics, not matrix mechanics, was the formalism employed by most theorists. This danger of losing the war on the calculational front threatened further consequences. Not only did Heisenberg have personal ambitions for advancement, but several chairs in theoretical physics were opening up in Germany. There was a conscious realization by members of the Copenhagen school that control of the future direction of theoretical physics was at stake. This conflict can be characterized as one over superiority and professional dominance (i.e., the interpretation and those physicists that would determine the future direction of research).[18] The victors, in Bohr's own characterization, would be able to realize their "'wishes' for the future of physics." [19]

The Copenhagen group had the talent, organization, and drive to carry the day in establishing the hegemony of its view.[20] Heisenberg's uncertainty-relation paper was a major step in accomplishing this. The Copenhagen group worked in concert, while its opponents (Einstein, Schrödinger, de Broglie) pulled each in his own direction.[21] Paradoxically, it was largely Schrödinger's techniques that were employed to establish the dominance of the interpretation associated with matrix mechanics.[22] Heisenberg and his coworkers unabashedly combined matrix methods (for electron spin) with Schrödinger's differential equation to solve several of the outstanding problems of atomic spectroscopy. The Bohr Institute in Copenhagen had an enormous influence on an entire generation of leading theoretical physicists who passed through it (e.g., most of those who later played dominant roles in establishing theoretical physics in the United States). As Ralph Kronig recalled, Bohr and his close colleagues were *authority* figures and a young person did not go against them (a sad example being Landé and Kronig himself when they abandoned the idea of electron spin because Pauli, who wanted no *classical* pictures of the electron, ridiculed it). Papers from Bohr's institute were accepted by *Zeitschrift für Physik*, Kronig claimed, without question.[23] It is also interesting to note that *no* French physicist ever worked at the Bohr Institute (at least not for any extended period of time).[24]

It was some of the more mathematically inclined contributors who maintained an openness on the question of the physical interpretation of the formalism. Dirac, for one, recognized that the formalism and an interpretation are separate so that a probabilistic interpretation was not necessarily demanded.[25] Similarly, Hilbert recognized a certain freedom of choice in the interpretation of a formalism.[26] The interpretation and the formalism were, in fact, intertwined in the *historical* development of matrix mechanics. This contingency made it easy to accept a particular interpretation as an essential part of the formalism.

7.2 Copenhagen succeeds

In an amazingly short period the Copenhagen interpretation gained ascendancy as *the* correct view of quantum phenomena. I now discuss some aspects of an event that was decisive for that outcome.

7.2.1 The 1927 Solvay Congress

A crucial encounter occurred at the 1927 Solvay congress.[27] In 1927 Louis de Broglie proposed a "principle of the double solution," which was a synthesis of the wave and particle aspects of matter.[28] The existence and nature of these singular solutions to his wave equation, in all but the simplest cases, proved quite complex mathematically. Under pressure to present a paper on the interpretation of quantum mechanics at the fifth Solvay congress, de Broglie decided, as an interim measure, to accept, or simply postulate, the existence of a particle accompanied by its phase wave ψ and he "contented [himself] with a presentation of the pilot-wave view" at the congress.[29] A physical particle was pictured as being guided by the pilot wave.

In discussion at that congress, Wolfgang Pauli criticized de Broglie's theory on the basis of the example of the inelastic scattering of a plane wave by a rigid rotor.[30] Although de Broglie felt he understood the general outlines of a suitable response to Pauli's objection, his rebuttal in fact appeared ad hoc and was not convincing.[31]

In addition to Pauli's negative reaction, neither Einstein nor Schrödinger gave positive support to de Broglie's ideas. De Broglie presented a conceptual mixture of waves and particles, which did not incline Schrödinger kindly toward it since, at this time, he wanted an interpretation based *wholly* upon the wave concept. One might have expected Einstein to be receptive to de Broglie's theory, since it allowed a picture of actually existing physical particles following well-defined trajectories—an observer-independent, objective physical reality. Einstein was at that time, and remained thereafter, quite partial toward a belief in such a reality. In the general discussion at the congress, Einstein did address the questions of causality, determinism, and probability. He claimed that any truly fundamental theory, including quantum mechanics, should be a *complete* theory of individual processes (as opposed to yielding information about the statistics of ensembles *only*). He said de Broglie was right to search for a complete theory.[32] Still, Einstein remained distrustful of this particular model of de Broglie's. This was likely related to his own abortive attempt at a hidden-variables theory for quantum mechanics, to which I return in section 8.3.

The people who were producing results with the hybrid wave- and matrix-mechanics formalism for problems involving spin (e.g., Heisenberg and Born, who also spoke at the 1927 Solvay congress) strongly favored the indeterministic or noncausal picture.[33] Bohr was for a long time against the concept of the photon, so that de Broglie's ideas had never spread rapidly in the Copenhagen school.[34] The institute at Copenhagen was a very closed community and those invited there were identified as the "respectable" theorists. De Broglie was never a member of this group. By 1930, when he wrote a *very* standard quantum-mechanics book, de Broglie had himself changed his mind about the pilot-wave theory: "It is not possible to regard the theory of the pilot-wave as satisfactory."[35] In that same book he rehearsed other arguments against the pilot-wave theory, both general conceptual ones and a specific thought experiment involving the reflection of light from an imperfect mirror.[36] John von Neumann's 1932 impossibility "proof" for hidden variables theories further confirmed de Broglie's position against his own previous theory.[37]

7.2.2 Pauli's objection

Let me now examine just what Pauli's objection was. In de Broglie's pilot-wave model, the wave function ψ played a dual role: the square of its norm gave the *probability* of finding a particle at a given location and its phase (ϕ) determined the trajectory followed by the particle according to the guidance condition

$$v = \frac{1}{m} \nabla \phi . \qquad (7.1)$$

This assigns particle trajectories, or "flow lines," to *individual* microsystems. In this conception of microprocesses, a particle always follows some *specific* trajectory during its motion. For a scattering process, say, a particle enters the scattering region with a fixed and definite energy moving along its entrance trajectory and leaves the scattering region with a final (perhaps different) energy as it travels on its exit trajectory.

Pauli considered an *inelastic* scattering process and argued that the guidance formula could *not* assign a definite energy (or a single trajectory) to the scattered particle. Pauli used as his example the scattering of an incident particle (or beam) by a quantum-mechanical rotator. Not long before the Solvay congress, Enrico Fermi had produced an explicit solution to this problem.[38]

Pauli's point was the following. In the standard (later, Copenhagen) interpretation, the final superposition state ψ allows one to calculate the *probabilities* of various outcomes.[39] However, the actual physical system always ends up in just *one* specific state upon observation (by what would become known as "reduction of the wave packet"). But, Pauli objected, use of the guidance condition of eq. (7.1) required that this ψ be written as

$$\psi = Ae^{i\phi/\hbar}, \qquad (7.2)$$

where ϕ would be given by a complicated expression (since ψ is a superposition) involving *all* of the possible final energy states of the system, not just a *particular* one corresponding to a definite outcome of an observation.[40] The resulting v would not correspond to *one* rotator energy and *one* (definite) energy for the scattered particle. Such a result, Pauli claimed, would contradict the experimentally well-confirmed fact that the scattered particle *always* emerges in *some* well-defined energy state, *not* in a superposition.

De Broglie began his response with an observation to the effect that Fermi's problem and his own were essentially different since Fermi worked in configuration space and he in ordinary space.[41] He claimed this to be an important difference between his approach and that of Schrödinger (whose Ψ-function did exist in configuration space).[42] However, the relevance of such a distinction for Pauli's example was perhaps not clear and it may have appeared that de Broglie had missed the point of Pauli's criticism. De Broglie argued that this difficulty was an artifice of the plane-wave approximation in which the wave extends over *all* space so that the incoming and diffracted waves will *always* interfere.[43] Pauli had attempted to block this move by pointing out that, since ψ is periodic in an angular variable in *configuration* space, even a ψ_0 initially representing a confined wave packet would correspond to a state ψ spread over an extended region in configuration space.[44] We shall see that David Bohm (but only *much* later, in 1952) properly dispatched Pauli's objection. At the time of the 1927 Solvay congress, though, de Broglie did *not* respond convincingly to what appeared to be a mortal blow struck by Pauli against the pilot-wave model.

In summarizing the reaction of the congress participants to his theory, de Broglie recalled that a few of the "old guard" (Lorentz, Einstein, Langevin, Schrödinger) insisted on the necessity of finding a causal interpretation of wave mechanics, but did not support his theory. Meanwhile, Bohr, Born, Heisenberg, and Dirac categorically favored the purely probabilistic interpretation that they had developed and would not even discuss his work.[45]

7.2.3 Victory and hegemony

De Broglie himself appears not to have been wholly confident in his own rejoinder to Pauli. Heisenberg's 1927 uncertainty-relations paper had a considerable influence on de Broglie prior to the Solvay congress, but de Broglie still had not abandoned his "classical" worldview.[46] The indifference and negative reaction to his pilot-wave theory at the congress, Heisenberg's uncertainty paper, and the generally favorable attitude toward and acceptance of the Copenhagen view all took their toll on de Broglie. After he returned to the Sorbonne, he still taught the theory of the pilot wave for a while, but he no longer genuinely believed it. By early 1928 he had decided to adopt the views of Bohr and Heisenberg. Invited to deliver some lectures at the University of Hamburg in the spring of 1928, he "there for the first time declared in public [his] formal adherence to the new ideas."[47] By 1927–1928, the issue of the Copenhagen interpretation versus a causal one had been *effectively* settled. Ralph Kronig recalled: "I mean, the idea of Bohr and the interpretation of Heisenberg, and so on, in 1928, already, was not a subject one discussed very much any more."[48]

So, there were two types of factors operative: "external" pressures that lent urgency to the need to find *the* correct interpretation of the formalism of quantum mechanics and rational but, as I have argued, not wholly compelling reasons (even though many *took* them to be so) for settling on Copenhagen. What were really consistency arguments (for complementarity, for acausality, etc.) in favor of Copenhagen often became transmuted into statements about the in-principle impossibility of alternative viewpoints.[49] The general acceptance of the Copenhagen interpretation (even if still somewhat ill defined) precluded, for most, any consideration of causal interpretations. Any possibility of a causal interpretation was generally *believed* to be and *accepted* as a dead issue by 1928. In the reconstruction of the history and in the propagation of the new orthodoxy (in, for example, Heisenberg's influential 1929 lectures on quantum mechanics), "de Broglie, Einstein, and Schrödinger disappear from Heisenberg's story, which includes as creators of quantum mechanics only those who understood it in the Copenhagen sense."[50]

Once the Copenhagen interpretation had established its hegemony in Europe, this dominance spread almost of necessity to the United States. Prior to 1935, the major connection of young American quantum physicists (e.g., David Dennison, Robert Lindsay, John Slater, Harold Urey) was to the research centers in Copenhagen, Göttingen, and Cambridge (where the orthodoxy had already spread).[51] Many American physicists were first alerted to the "new quantum theory" by Max Born when he visited MIT and other U.S. institutions in the winter of 1925/26.[52] Of thirty-two American visitors to Europe, all went to Copenhagen-doctrine centers (and none, for example, to France).[53] America was still in a learning, catch-up phase prior to 1930 and the pragmatic American approach to quantum mechanics led to its acceptance there, by and large without philosophical qualms. Most American physicists simply did not care about the question of an interpretation. Later, notable achievements in quantum theory in the United States were in applications of a largely established formalism.[54]

We can perceive a parallel (and not just in America) between the rapid spread of orthodox quantum mechanics (once the formalism had been fixed and the Copenhagen interpretation forged) and Newtonianism. There was an emphasis, by most of the rapidly growing band of practitioners, on computation as opposed to thinking about foundational questions. It is simplest and most efficient for disciples to follow the path of the master and use the scheme to calculate, carrying along without much reflection a certain amount of philosophical "baggage."

Appendix Pauli's 1927 criticism of de Broglie's theory

While the physical insight and mathematical artifices that Fermi employed to solve the relevant quantum-mechanics problem are admirable enough to make a fascinating story in their own right, here I simply state those of his *results* needed to follow Pauli's argument.[55]

A rotator (or rigid "dumbbell") is located at the origin and rotates about a fixed axis that is perpendicular to the scattering plane. Its moment of inertia is $I = MR^2$ and its dimensions are taken to be negligible. A particle is incident along the z-axis and leaves ("scatters") along that same direction. If z is the coordinate of the particle and θ is the angular coordinate of the rotator, then the interaction potential $V(z, \theta)$ between the rotator and the particle was taken by Fermi to vanish except in a small region around $z = 0$. The initial energy of the particle is

$$E_1 = \frac{1}{2} m v_0^2 \qquad (7.3)$$

and that of the rotator

$$E_2 = \frac{1}{2} I \omega_0^2 = \frac{n^2 \hbar^2}{2I} \qquad (7.4)$$

so that the total energy is

$$E = E_1 + E_2. \tag{7.5}$$

The initial wave function, before any interaction (or "collision"), is just

$$\psi_0 = A \exp\left[\frac{i}{\hbar}(Et - \sqrt{2mE_1}\, z - \sqrt{2IE_2}\, \theta)\right]. \tag{7.6}$$

Using an analogy with the equation describing optical diffraction by a grating, Fermi was able to show that the final wave function after the interaction (or "scattering") consisted of the superposition

$$\psi = \sum_j A_j \exp\left[\frac{i}{\hbar}(Et - \sqrt{2mE'_{1j}}\, z - \sqrt{2IE'_{2j}}\, \theta)\right]. \tag{7.7}$$

Here

$$E = E_1 + E_2 = E'_{1j} + E'_{2j} \tag{7.8}$$

and the final rotator energy E'_{2j} is given by

$$E'_{2j} = \frac{\hbar^2}{2I}(n + j)^2. \tag{7.9}$$

That is, if $j \neq 0$, then an *inelastic* collision has taken place.

Pauli observed that the guidance condition, eq. (7.1), applied to the wave function of eq. (7.7) would not yield a velocity corresponding to just *one* of the terms in that sum (since the various terms overlap). This, he claimed, contradicted the requirement that an observation must produce a *definite* state of motion (i.e., corresponding to just *one* value of j in the sum of eq. [7.7]).

EIGHT

Early Attempts at Causal Theories:
A Stillborn Program

In this chapter I begin to trace the origins and eventual fate of the causal quantum-theory program. This begins with de Broglie's theory of phase waves, includes various hydrodynamical and velocity-field (essentially "hidden variables") interpretations, such as those due to Madelung and Einstein, and progresses through de Broglie's theory of the double solution and his (provisional) pilot-wave model, the encounters with the Copenhagen group at the 1927 Solvay congress, the impact of von Neumann's "impossibility" proof, an interpretation by Rosen, Bohm's 1952 paper and, finally, Nelson's 1966 work on a stochastic basis for the Schrödinger equation.

After his fundamental paper on wave mechanics in 1926, Erwin Schrödinger at first attempted to give a *realistic* interpretation to the ψ-function.[1] One objection to this was that, while a real wave might exist in the physical *three*-dimensional space (or possibly in a four-dimensional space-time) in which we exist, it would make little sense to speak of a physical wave existing and propagating in a $3n$-dimensional *configuration* space (where n is the number of particles in the system).[2] Such a ψ must be merely a mathematical construct, useful perhaps for calculating probabilities, but surely not to be assigned actual physical reality.

Max Born in his successful probability interpretation of $|\psi|^2$ did not initially *categorically* rule out the possibility that quantum mechanics, as formulated at that time, might be a statistical theory as a matter of *practical* necessity, rather than as a matter of absolute principle.[3] In an article in *Nature* in early 1927, Born even allowed the possible existence of microscopic atomic coordinates that are averaged over in practice. "Of course, it is not forbidden to believe in the existence of [microscopic] coordinates."[4] Born was at least agnostic on the question of whether, in later terminology, a hidden-variables version of quantum mechanics might indeed exist.

However, Pascual Jordan, in a later issue of the same volume of *Nature,* was on the whole less open to the possibility of a basically classical, continuous, and picturable view of microphenomena.[5] Jordan did not offer proofs or even compelling arguments for foreclosing this *possibility* of continuity, but rather referred to the *opinions* of most scientists and to the *difficulty* of conceiving of such an alternative. Born later came to believe in the impossibility of microscopic coordinates, mainly as a result of Heisenberg's "uncertainty" paper.[6]

The previously dominant worldview of classical physics had been based on continuity and picturability for physical processes, and even Born was not at first hostile toward the possibility of a *largely* classical ontology being compatible with the new quantum theory. How positions hardened against this a priori intuitively appealing understanding of the basic physical phenomena is the question I now address.

8.1 Madelung

In the fall of 1926, Erwin Madelung suggested a hydrodynamical interpretation of quantum mechanics.[7] He began with the Schrödinger equation for the wave function ψ and made a set of mathematical transformations (similar to the type Bohm would make decades later). All of the mathematical details (appendix 1 to this chapter) aside, Madelung suggested interpreting the Schrödinger equation as representing a physical "fluid" (of identical particles of mass m) of density ρ and with a velocity field v. This ideal fluid had no viscosity. One difficulty of this interpretation, since Madelung was considering a fluid consisting of a continuous distribution of *charge,* was that the equations contained a "quantum force" term (to use a later terminology here) that depended only upon the *local* density ρ ($= |ψ|^2$), but not upon the total charge distribution. The physical meaning or significance of the additional ("quantum potential") term in his Newtonian equation of motion will be a recurrent theme of, or actually a problem for, various causal interpretations I discuss. Although Madelung claimed that this model gave an intuitively clear picture of quantum phenomena, it is not wholly evident, at least to a modern reader, just *what* this ideal fluid was, how it represented an atom in some state, or how emission and absorption phenomena were to be envisioned.[8] In spite of conceptual difficulties that remain, it is clear that Madelung was attempting to provide a classical picture or explanation of quantum phenomena.

George Temple, in his 1934 book on quantum mechanics, derived the equations of motion for a charged fluid and then discussed Madelung's theory. He showed how, for certain cases, Madelung's transformation reduced the nonlinear hydrodynamical equations to the Schrödinger equation.[9] He observed that, to prevent radiation from these moving charges, the fluid flow must remain steady and this cannot occur under the influence of electromagnetic forces *alone*. Hence, there was need for the *quantum force* (or *quantum potential U*).[10] That is, Temple attempted to give some motivation for introducing the quantum potential in order to account for the lack of electromagnetic radiation in stationary states.

8.2   De Broglie

The similarity of the Schrödinger equation to classical hydrodynamical equations, as well as the analogy of the Hamilton-Jacobi formulation of classical mechanics to classical wave optics, led de Broglie to his own attempts at a largely classical formulation of quantum mechanics. He suggested two different approaches: the hypothesis of the double solution and the theory of the pilot wave. Neither succeeded and these failures made him "see better the necessity for adopting entirely new ideas which were developed during the course of the same year by Bohr and Heisenberg."[11]

Before I sketch some of the details of de Broglie's attempts, let me indicate the significance of this quotation. The original (French) version of *Physics and Microphysics* (the source of this citation) was published in 1947. At that time, de Broglie was very much in the Copenhagen camp, having been converted (by 1930) after his bitter experience at the 1927 Solvay congress. David Bohm's 1952 paper was to have a profound ("reconversion") effect on de Broglie. However, in 1947, de Broglie still repudiated his former ideas. In a note to the 1955 translation of *Physics and Microphysics*, he acknowledged Bohm's paper (which he had seen only in preprint at the time of these comments) and essentially rejected it as just a version of his pilot-wave theory.[12] Shortly thereafter, de Broglie admitted that Bohm had successfully overcome the original objections to a pilot-wave model. I return to this part of the story later.

De Broglie's main goal was to unify the wave and particle dualism into a *single* coherent picture or model. His hope was to treat the "particle" as a (mathematical) singularity in the center of an extended wave. As outlined in 1927, this *extended*, continuous part of the wave would "sense" the environment (obstacles, slits, etc.) and thus vary the motion of the singularity accordingly. He wanted two related wave solutions, one for the singularity and the other for the continuous wave. This latter would account for the statistical behavior of a collection of particles. A classical conception of actually existing entities in a continuous space-time background was to underlie his theory or worldview.[13] De Broglie's solution to the wave-particle duality was a synthesis of wave *and* particle, versus the wave *or* particle of the (eventual) Copenhagen interpretation. For him, the (continuous, extended) wave aspect was to be represented by the function ψ and the singularity by a function u.

His basic motivation had been Planck's fundamental relation for light ($\varepsilon = h\nu$) which united a particle aspect (localized energy ε) and a wave aspect (the frequency ν).[14] In his Nobel Prize acceptance speech in 1929, de Broglie recalled that he had been dissatisfied with Planck's relation because it defined the energy of a light corpuscle by a relation that contains a frequency ν. De Broglie felt that a purely corpuscular theory should not contain a frequency. On the other hand, the stable motions of the electrons in the atom are characterized by whole numbers that are typically associated in physics with interference and standing waves. This suggested to him "that electrons themselves could not be represented as simple corpuscles either, but that a periodicity had also to be assigned to them, too."[15]

In early 1927 de Broglie attempted to generalize these intuitive arguments (cf. appendix 2 to this chapter) into his "principle of the double solution." To every continuous solution $\psi = Re^{i\phi/\hbar}$ of the wave equation there was to correspond a singularity solution $u = fe^{i\phi/\hbar}$ having the same phase ϕ as ψ, but whose amplitude f represented a moving singularity.[16] On the basis of this principle, he was able to show that the velocity v of the singularity in u (the "particle") was to be determined by the "guidance formula" ($v = \nabla\phi/m$).[17] It was this phase ϕ, rather than the amplitude of ψ, that determined the motion of the singularity representing the particle.[18] Only for the case of a *free* particle was de Broglie able to carry through these pilot-wave ideas explicitly.[19] He also suggested that $\rho = |\psi|^2$ represented the probability of finding a particle ("singularity") at a point in space.[20] The ψ-wave gave statistical information about the behavior of an ensemble of particles (or, equivalently, probabilistic information about the behavior of a single particle whose initial location—the "hidden variable"—is unknown or uncertain). At the end of his 1927 paper on this double-solution theory, de Broglie observed that one might simply postulate the existence of two distinct realities, particle and wave, with the motion of the particle determined by the phase of the wave. He considered such a move to be not really satisfactory and only provisional.[21] His pilot-wave theory is mathematically the same as Madelung's formalism.[22]

It should be evident that de Broglie's style of doing physics was an intuitive one in which general insights played the major role. He was not a formalist possessing great mathematical power. The existence and nature of the singular solution u to the wave equation for the general case of motion in the presence of a force field proved quite complex mathematically. In later years de Broglie himself suggested that a *nonlinear* equation may be required for u.[23] In the face of these severe mathematical difficulties, de Broglie decided, as an interim measure, to accept, or simply *postulate,* the existence of a particle accompanied by its phase wave ψ.[24] He presented this hybrid model, his pilot-wave theory, at the fifth Solvay congress in October of 1927. I have already indicated why Schrödinger was not enthusiastic about this theory and what Pauli's reaction to it was. Although I have also alluded to what I believe was a factor in Einstein's dismissing de Broglie's theory, I now consider that in more detail.

8.3 Einstein

A fascinating (unpublished) manuscript in the Einstein Archives is titled "Does Schrödinger's Wave Mechanics Determine the Motion of a System Completely or Only in the Sense of Statistics?"[25] In it, he tells us that to each solution of the wave equation there corresponds the motion of an *individual* system that is determined unambiguously and uniquely. In 1927 Einstein wrote to Born about this work and indicated that it would soon be published.[26]

Conceptually, but not in its mathematical details, this theory was very much in the spirit of Madelung's hydrodynamical model. In an addendum to this manuscript, Einstein mentioned that Walther Bothe had pointed out a difficultly with this scheme for compound systems whose overall state may be represented by a *single* product of the wave functions of each of the subsystems. Such a system is made up of *independent* (i.e., noninteracting) subsystems. It turns out that the motions of the compound system will not be simply combinations of motions for the subsystems, as Einstein required on physical grounds. This showed that Einstein's particular recipe for determining the "flow lines" was not tenable (not, of course, that there could not exist another one that would be tenable). Einstein suggested that it might be possible to overcome this difficulty, but nothing specific followed. The fact that this paper, originally presented orally at a meeting of the Prussian Academy in early 1927, was never published indicates that this "entanglement" problem remained grounds, for him, to reject this particular "classical" attempt at interpreting quantum mechanics.[27] Years later, Born commented that he himself could no longer "remember it now; like so many similar attempts by other authors, it has disappeared without trace."[28]

The general, at least conceptual, similarities among this attempt by Einstein, the Madelung hydrodynamical model, and the de Broglie pilot-wave theory probably account for Einstein's lack of interest in de Broglie's 1927 Solvay congress presentation. All three of these attempts suffered from apparently bizarre, physically unacceptable properties: Madelung's (of which Einstein surely already knew) had a peculiar "inner" force of the continuum (cf. eqs. [8.14] and [8.15] in appendix 1 to this chapter), Einstein's own had the entanglement feature I have just discussed, and de Broglie's pilot-wave theory would be made to seem incoherent by Pauli's objection at the 1927 Solvay congress itself. Whether it was these strange features or the general nonlocality of quantum theory that left Einstein cool toward de Broglie's pilot-wave presentation is unclear.[29] We do know that by the spring of 1927 (*prior* to this Solvay conference) Einstein was already critical of quantum mechanics (in either the matrix or the wave formulation) because, among spatially separated systems, there existed correlations that seemed to violate a principle of action by contact.[30]

8.4 Kennard, Rosen, Fürth

In 1928, Earle Kennard published in *The Physical Review* a discussion of the formalism and application of Schrödinger's wave mechanics. His opening sentence had a sense of finality on the interpretation question.[31] Early in the paper Kennard cited Madelung's work and arrived at a "Newtonian" equation of motion and stated:[32]

> *Thus each element of the probability moves in the Cartesian space of each particle as that particle would move according to Newton's laws under the classical force plus a "quantum force"* given by the h-term in (Newton's second law of motion. eq. [8.14]).
>
> The motion here considered occurs in a space of n dimensions. We can also, however, replace the n-dimensional packet by n separate packets, one for each particle, all moving in the same ordinary space.[33]

He went on to point out that, in spite of this similarity with Newton's second law of motion, there is a profound difference since the motion of any one particle depends, in general, on the instantaneous location and velocities of all the other particles as well.[34] Here, again, is nonlocality.

I now step somewhat out of the time sequence of events and mention another example of a hidden-variables interpretation because it nicely focuses on those central features of such theories that most found objectionable on physical grounds. In a somewhat obscure journal, Nathan Rosen in 1945 published a paper whose purpose was to explore "the extent to which classical concepts can be carried over into the quantum theory and the conditions under which they conflict with the formalism of this theory."[35] He cited Madelung's and Kennard's papers and obtained the same equations of motion as they did. With regard to the dynamical equation of motion (Newton's second law, as modified by the quantum potential), Rosen observed "that the motion of each particle of the ensemble depends on the density with which all the members of the ensemble [of *possible* representatives of the *actual* particle under consideration] are distributed, so that effectively we have an interaction among the [virtual] particles."[36] The basic point that Rosen is making here is that the quantum potential U depends upon $\rho = |\psi|^2$, the density of particles in an ensemble in which only *one* particle (the *actual* one) is real and the rest are merely *possible*. This is worse than mere nonlocality among distant, actually existing particles and, interpreted as Rosen does, calls for a bizarre influence of possible systems on actual ones. This would scarcely help a realistic interpretation of the phenomena. On the other hand, Bohm's 1952 paper made a virtue of the quantum potential by interpreting it in terms of a nonlocal influence of the (actual) environment upon the (actual) particle under consideration—all very realistic.

There was also an early precursor to the Brownian-motion type of stochastic mechanics that I return to in the next chapter. Schrödinger noted the *formal* (mathematical) similarity between the diffusion equation and his own wave equation, but he saw the *physical* differences between these two cases as more significant than the formal resemblances.[37] In 1933, Reinhold Fürth, citing an earlier review by Schrödinger on the interpretation of quantum mechanics, discussed this formal analogy between the equations for the position probability of a classical statistical mechanical system and quantum mechanics.[38] Unsharp observables resulted and there were inherent (practical) limitations on the accuracy of measurements, from which "Heisenberg"-like inequalities followed. A relation was established between diffusion equations for real density functions and the Schrödinger equation for complex functions and this was related to reversible and irreversible natural processes. Fürth examined some specific cases and then argued/conjectured that in general an "uncertainty" relation $\Delta x \Delta v \approx D$ held, where D is the diffusion coefficient in the diffusion equation.[39] (In section 9.3.2 we shall see that in 1966 Edward Nelson obtained a similar result and made the identification $D = \hbar/2m$.)[40] Fürth also referred to previous literature on the inherent limitations that Brownian motion placed on the accuracy of measurements.[41] There was considerable discussion of such limitations in the years 1925 to 1935.[42]

Similar stochastic-mechanics approaches to quantum phenomena would recur several times after the Second World War.[43] In section 9.3.2 I discuss this program at some length. I mention these early attempts here to indicate not only that such an approach was *possible* just after the 1927 Solvay congress, but also that *in fact* it was considered. However, there soon appeared to be a reason to believe that all such attempts *must*, ultimately, fail.

8.5 Von Neumann's "proof"

We have seen how, by 1927–1928, the issue of the Copenhagen versus a causal interpretation had been essentially decided. The story has included both "external" factors and rational arguments that were taken at the time as convincing for choosing Copenhagen. Even by September of 1927, *prior* to the October Solvay congress, Bohr's Como lecture had to a large extent solidified the matrix of what would become the Copenhagen interpretation. This general acceptance of the Copenhagen interpretation effectively precluded any consideration of causal interpretations. With the appearance in 1932 of von Neumann's *Mathematical Foundations of Quantum Mechanics*, however, there appeared to be a logically irrefutable proof that any type of hidden-variables theory that gave *all* of the same predictions as standard quantum mechanics was *impossible*.

8.5.1 The impact of the theorem

In spite of the considerable mystique that has surrounded von Neumann's theorem in recent decades, that proof was probably *not* the decisive reason that hidden-variables theories were not actively pursued in the 1930s. Most people already *believed* (for reasons that, in retrospect, appear considerably less convincing now than they did at the time) that such extensions were ruled out on physical or experimental grounds. Although the proof was cited by proponents of the Copenhagen interpretation, that was usually done as a nod to mathematical purity or as a put-down "clincher," rather than as *the* central element in a refuting argument.[44] Of course, it is not possible to tell what effect that theorem *may* have had in diverting people from seriously pursuing hidden-variables theories.

Von Neumann's proof further confirmed de Broglie's position against his own causal theory.[45] While that may be, de Broglie had *already* been converted to Copenhagen. Similarly, in a letter of 2 July 1935 to Pauli on the recent EPR paper and Bohr's response to it, Heisenberg included a long addendum titled "Is a Deterministic Completion of Quantum Mechanics Possible?"[46] Although Heisenberg does refer approvingly to von Neumann's book there, he does not single out the "impossibility" proof specifically as a key element in his lengthy discussion of why such a completion is not possible.

In 1936 Jordan spoke out against causality in atomic phenomena, but as was characteristic of him, much more categorically than many of his colleagues. In his popular lectures on physics in this century, Jordan discussed the example of individual photons passing through a polarizer and then told his reader that a "denial of the classical concept of causality is not to be understood as a temporary imperfection of our knowledge, but is inherent in the nature of the thing—again showing how incorrect our previous, classical concepts were."[47] In his more technical treatise on quantum mechanics, Jordan handled this same example in virtually the same way.[48]

In a 1939 conference report in *New Theories in Physics,* Bohr referred to von Neumann's proof, but only as the most clear and elegant demonstration of the already well known fact "that the fundamental superposition principle of quantum mechanics logically excludes the possibility of avoiding the non-causal feature of the formalism by any conceivable introduction of additional variables."[49] While he reported on von Neumann's own discussion of the "impossibility" proof at the conference, Bohr also claimed that this result is already evident from more elementary considerations.[50]

As time passed, the theorem gained in importance. Born's Waynflete lectures at Oxford in 1948 drew up a case against a causal interpretation of quantum mechanics and von Neumann's result was cited as being one of the more important elements in the argument. Born first used historical precedent to argue against the likelihood of a reversion to a primitive conception, such as determinism, and offered his opinion that the attendant mathematical difficulties could not be overcome. He then cited von Neumann's *Mathematische Grundlagen der Quantenmechanik* and assured his audience "that the formalism of quantum mechanics is uniquely determined by [a few plausible] axioms; in particular, no concealed parameters can be introduced with the help of which the indeterministic description could be transformed into a deterministic one [so that] . . . if a future theory should be deterministic, it cannot be a modification of the present one but must be essentially different."[51]

The *limitations* of the theorem were rarely stressed. For instance, Pauli in 1948 referred to "von Neumann's well known proof that the consequences of quantum mechanics cannot be amended by additional statements on the distribution of values of observables, based on the fixing of values of some hidden parameters, without changing some consequences of the present quantum mechanics."[52] Even many years later, in a special 1958 issue of the *Bulletin of the American Mathematical Society* dedicated to the memory of John von Neumann, as reflective a theoretical physicist as Léon van Hove would state that "von Neumann could show that hidden parameters with this property [of reinstating causality] cannot exist if the basic structure of quantum theory is retained."[53] Van Hove does imply that von Neumann may have allowed the possibility of certain types of hidden-variables theories as long as the formalism of quantum mechanics was *modified.*[54]

As I now show, von Neumann was able to obtain his result only with the assumption that one of ordinary quantum theory's rules for statistical ensembles could be *extended* to the dispersion-free ensembles of any hidden-variables theory. This assumption turns out to be unwarranted (although it was *many* years before John Bell pointed this out with great clarity in the mid-1960s).

8.5.2 The actual theorem

The hope central to the class of causal interpretations I consider is that there exists (at least *in principle*) a set of hidden variables (say, the actual microscopic coordinates of the particles) that, if known or controllable by the experimenter, would *completely* and *uniquely* determine the motion of individual microsystems. The results of experiments would be determined on an event-by-event basis. In practice, this hope would hold, these hidden variables are unknown and must be averaged over to obtain predictions for an ensemble of particles. I denote these assumed hidden variables collectively by the symbol λ (i.e., λ may stand for an entire set $\lambda_1, \lambda_2, \lambda_3, \ldots$).[55] There should then be subensembles such that the value of the observable in question should have a *definite* value for all members of the subensembles (i.e., the measured values of this observable should be *dispersion free*). Von Neumann's proof shows that there will necessarily be situations in which such dispersion-free states cannot exist, provided one is to maintain *all* of the predictions (for experimental outcomes) of the standard formalism of quantum mechanics.[56] This, it turns out, is really equivalent to the question of the *completeness* of the Copenhagen interpretation of quantum mechanics.

The problem as actually posed by von Neumann involves the possible decomposition of an ensemble E into subensembles.[57] He put this as follows:

> We could attempt to maintain the fiction that each dispersing ensemble can be divided into two (or more) parts, different from each other and from it, without a change of its elements. . . . This is the question: is it really possible to represent each ensemble $[E_1, E_2, \ldots, E_N]$, in which there is a quantity $\mathcal{R}$ with dispersion, by the superposition of two (or more) ensembles different from one another and from it?[58]

After the details of his proof, he concluded that all ensembles have dispersions.[59] He told his reader that "the present system of quantum mechanics would have to be objectively false, in order that another description of the elementary processes than the statistical one be possible."[60] Von Neumann denied the possibility of fine-grained versus coarse-grained knowledge being the source of the quantum-mechanical dispersions that would be produced in an averaging process.

Von Neumann was able to produce a mathematical contradiction only by assuming that, even for dispersion-free subensembles, the expectation value of the sum of two (noncommuting) operators is simply the sum of the expectation values of each separately (as *is* the case in quantum mechanics). This was widely accepted as establishing that general dispersion-free states cannot exist so that a hidden-variables extension of quantum mechanics is impossible *in principle*. With the aid of hindsight provided by later work, we can see today that what has actually been established is that the sum rule is inconsistent with a hidden-variables extension.[61] Another way to parse this is that such a hidden-variables extension requires a *more* complete specification of the state of a microsystem than that possible with the quantum-mechanical state vector ψ.[62]

While some philosophers seem to have been explicitly aware of this logical point (even in the 1930s), this caveat was not emphasized by physicists in most references to von Neumann's proof.[63] The sociological and psychological reasons for the largely uncritical acceptance by the physics community of the implications of von Neumann's proof have been studied in detail.[64] Von Neumann had tremendous intellectual prestige among scientists, including such leaders in quantum theory as Wigner.[65] He had been David Hilbert's favorite when, as a young man, von Neumann attempted an axiomatization of mathematics in the 1920s. That project, like his no-hidden-variables proof a decade or so later, nearly succeeded. It was shut down by Gödel's incompleteness theorem. These were both brilliant, but ultimately failed, undertakings. It was only with Bell's work in the mid-1960s that this proof's irrelevance for many types of hidden-variables theories became widely understood.[66]

In Bohm's theory, how a microsystem behaves depends upon its environment (i.e., an observed value is contextual). That makes it evident why a set of hidden variables of the microsystem *alone* could not fix definite values for outcomes of incompatible measurements. The quantum potentials (or, equivalently, the wave functions) for these different experimental arrangements would be different. This contextuality allows Bohm's theory to escape von Neumann's theorem.[67]

We see that von Neumann's theorem is fine as a mathematical theorem (i.e., as a correct exercise in deductive logic, *given* his axioms), but that it is simply irrelevant to a large class of hidden-variables theories (one of which is Bohm's). The theorem just does not imply everything it was taken to.

Appendix 1 Madelung's derivation

Madelung accepted Schrödinger's wave equation

$$-\frac{\hbar^2}{2m}\nabla^2\psi + V\psi = i\hbar \frac{\partial \psi}{\partial t} \qquad (8.1)$$

as his starting point and made the substitution[68]

$$\psi = R \exp(i S/\hbar) . \qquad (8.2)$$

If this is substituted into the Schrödinger equation and real and imaginary parts are separated, two equations result (just as, of course, happened in Bohm's theory in appendix 1 to chapter 4):

$$\frac{\partial R}{\partial t} = -\frac{1}{2m}[R\, \nabla^2 S + 2\nabla R \cdot \nabla S] \qquad (8.3)$$

$$\frac{\partial S}{\partial t} = -\left[\frac{(\nabla S)^2}{2m} + V - \frac{\hbar^2}{2m}\frac{\nabla^2 R}{R}\right]. \qquad (8.4)$$

If eq. (8.3) is multiplied by R, the result can readily be rewritten as

$$\frac{\partial}{\partial t}(R^2) = -\frac{1}{m}\nabla\cdot(R^2\, \nabla S) . \qquad (8.5)$$

If we identify

$$\rho = R^2 \qquad (8.6)$$

as a fluid density and the phase

$$\phi = S \qquad (8.7)$$

as a velocity potential so that the velocity field v of this fluid is given as[69]

$$v = \frac{1}{m}\nabla\phi = \frac{1}{m}\nabla S , \qquad (8.8)$$

then eq. (8.5) becomes the continuity equation

$$\nabla\cdot(\rho v) + \frac{\partial \rho}{\partial t} = 0 \qquad (8.9)$$

expressing the conservation of mass of the fluid. So, eq. (8.3) is basically the continuity equation. The other separation equation, eq. (8.4), becomes, with the use of eq. (8.8),

$$\frac{\partial \phi}{\partial t} + \frac{1}{2m}(\nabla\phi)^2 = -V + \frac{\hbar^2}{2m}\frac{\nabla^2 R}{R}. \qquad (8.10)$$

If we take the gradient of eq. (8.10), divide by m and use eq. (8.8), we obtain from the left-hand side of eq. (8.10)

$$\frac{\partial \boldsymbol{v}}{\partial t} + \frac{1}{2}\nabla(v^2) \equiv \frac{d\boldsymbol{v}}{dt}. \qquad (8.11)$$

This is just the acceleration of a volume element of fluid as it moves along a flow line defined by $\boldsymbol{v}(x, t)$ since, for *any* $f(x, t)$ along such a flow line [where $x = x(t)$],

$$\frac{df}{dt} = \sum_{j=1}^{3}\frac{\partial f}{\partial x_j}\frac{dx_j}{dt} + \frac{\partial f}{\partial t} = \boldsymbol{v}\cdot\nabla f + \frac{\partial f}{\partial t}. \qquad (8.12)$$

If we use eq. (8.8), as $v_j = \partial\phi/\partial x_j$, and set $f = v_k$ in eq. (8.12), we find[70]

$$\frac{dv_k}{dt} = \sum_{j=1}^{3}v_j\frac{\partial v_k}{\partial x_j} + \frac{\partial v_k}{\partial t} = \sum_{j=1}^{3}v_j\frac{\partial v_j}{\partial x_k} + \frac{\partial v_k}{\partial t} \qquad (8.13)$$

$$= \frac{1}{2}\frac{\partial v^2}{\partial x_k} + \frac{\partial v_k}{\partial t}.$$

Therefore, the gradient of eq. (8.10) (multiplied by $1/m$) is just

$$\frac{d\boldsymbol{v}}{dt} = -\frac{1}{m}\nabla V + \nabla\left(\frac{\hbar^2}{2m^2}\frac{\nabla^2 R}{R}\right). \qquad (8.14)$$

This last "force" term can be written as $-\nabla U$, where

$$U \equiv -\frac{\hbar^2}{2m}\frac{\nabla^2 R}{R} \qquad (8.15)$$

(later called a "quantum potential" by de Broglie).[71]

Appendix 2 De Broglie's guidance argument

De Broglie's basic 1923 argument was the following.[72] Consider a particle of rest mass m_0 and denote its rest frame by S'.[73] In this frame let there be a periodic phenomenon associated with the particle and characterized by the frequency v_0

$$E_0 \equiv m_0 c^2 = hv_0 \qquad (8.16)$$

and represented by the "wave"

$$\psi_0 = A_0 \exp(-2\pi i v_0 t_0). \qquad (8.17)$$

Notice that ψ_0 does *not* depend upon the spatial coordinates in frame S'. There is a frame S in which an observer (at rest) sees this particle moving with a velocity $v = \beta c$ along the positive x-axis. The corresponding phase wave ψ as seen by S is

$$\psi(x, t) = A_0 \exp\left[-2\pi i \nu_0 \gamma\left(t - \frac{\beta x}{c}\right)\right], \qquad (8.18)$$

since the time coordinate t_0 in S' is related to the space-time coordinates (x, t) in S via the Lorentz transformation

$$t_0 = \gamma\left(t - \frac{\beta}{c} x\right), \qquad (8.19)$$

$$\gamma = \frac{1}{\sqrt{1 - \beta^2}}. \qquad (8.20)$$

Equation (8.18) represents (in frame S) a phase wave

$$\psi(x, t) = A_0 \exp\left[-2\pi i \nu\left(t - \frac{x}{V}\right)\right] \equiv A_0 e^{i\phi/\hbar}, \qquad (8.21)$$

$$\nu = \gamma \nu_0, \qquad (8.22)$$

$$V = \frac{c^2}{v}. \qquad (8.23)$$

Here ν is the frequency of this phase wave and V is its (phase) velocity.[74] The quantity $\phi/\hbar$ is the phase of ψ. The frequency ν is just what one would have expected from

$$h\nu = E = mc^2, \qquad (8.24)$$

$$m = \gamma m_0, \qquad (8.25)$$

which is the analogue of eq. (8.16) once the relativistic value for m in frame S is used (in place of m_0 for S'). For reference later note that since

$$\phi = 2\pi \hbar \nu \left(\frac{x}{V} - t\right), \qquad (8.26)$$

it follows that

$$\frac{1}{m}\frac{\partial \phi}{\partial x} = \frac{2\pi \hbar \nu}{mV} = \frac{h}{m}\frac{\gamma \nu_0}{V} = \frac{h \nu_0}{m_0}\frac{v}{c^2} = v, \qquad (8.27)$$

which is just the velocity of the particle. That is, the gradient of the *phase* of ψ gives the *velocity* of m. Equation (8.27) is the same as eq. (8.8).

Furthermore, de Broglie observed that the phase of the periodic phenomenon associated with the particle in frame S' remained everywhere the same as the phase of ψ in S. To see this consider a collection of "clocks" in S' each having a period τ_0 in that frame. One of these is to be associated with the particle of rest mass m_0. The frequency

$$\nu_0 = \frac{1}{\tau_0}, \qquad (8.28)$$

originally defined in S', becomes for S

$$\nu_S \equiv \frac{1}{\tau} = \frac{\nu_0}{\gamma} \qquad (8.29)$$

because of the time dilation effect

$$\tau = \gamma \, \tau_0 \,. \qquad (8.30)$$

Now, de Broglie reasoned, if S follows the "particle" along for a time dt (and, as it travels, by definition, $dx = v \, dt$), the observed *change* of phase (for S) of this "intrinsic" phenomenon will be just

$$- 2\pi \hbar \nu_s dt = - 2\pi \hbar \frac{\nu_0}{\gamma} dt \,. \qquad (8.31)$$

However, if $\phi_1 = \phi(x_1, 0)$ was the phase of ψ at the initial position x_1 of m at $t = 0$, then the phase of ψ at the new location of m a time dt later will be (cf. eq. [8.21])

$$\phi_2 = \phi(x_1 + dx, dt) = \phi_1 + 2\pi \hbar \nu \left(\frac{\beta}{c} dx - dt \right)$$

$$= \phi_1 - 2\pi \hbar \nu (1 - \beta^2) \, dt = \phi_1 - 2\pi \hbar \frac{\nu_0}{\gamma} dt \,. \qquad (8.32)$$

The "intrinsic" phase of the "particle" (of mass m) moving with velocity v in S remains *fixed* relative to the phase wave ψ moving with (phase) velocity V in S. In other words, the phases of the "particle" and of the wave ψ are *always* equal to each other at the location of the particle as it moves along its trajectory in frame S. Starting from such a congruence, one can also argue back to the (by now familiar) guidance condition[75]

$$v = \frac{1}{m} \nabla \phi \,. \qquad (8.33)$$

Appendix 3   Einstein's 1927 hidden-variables theory

Einstein's basic idea was that the time-independent Schrödinger equation

$$\frac{\hbar^2}{2m}\nabla^2\psi + (E - V)\psi = 0 \tag{8.34}$$

can be used to find the kinetic energy $K = E - V$ for any given wave function solution ψ defined on an n-dimensional configuration space.[76] He used the quantum-mechanical expression for the kinetic energy

$$K = -\frac{\hbar^2}{2m}\frac{\nabla^2\psi}{\psi} \tag{8.35}$$

to define an equivalent kinetic energy in point-particle mechanics as

$$K = \frac{1}{2}mv^2 = \frac{1}{2}m\,g_{\mu\nu}\dot{q}_\mu\dot{q}_\nu, \tag{8.36}$$

where $g_{\mu\nu}$ is the metric tensor for the configuration space and $\dot{q}_\mu$ is the velocity component of the particle.[77] These $\dot{q}_\mu$ are functions of the configuration-space coordinates (that is, they define a velocity field, the tangents to which are the "flow lines" or possible particle trajectories). Specifically, having set

$$\nabla^2\psi = g^{\mu\nu}\psi_{\mu\nu}, \tag{8.37}$$

where $\psi_{\mu\nu}$ (which Einstein termed "the tensor of ψ-curvature") is the covariant derivative, he then sought a "unit" vector A^μ

$$g_{\mu\nu}A^\mu A^\nu = 1 \tag{8.38}$$

that would render

$$\psi_{\mu\nu}A^\mu A^\nu \equiv \psi_A \tag{8.39}$$

an extremum. This is the normal curvature of the differential geometry of surfaces.[78] A hermitian quadratic form like eq. (8.39) is rendered an extremum by those vectors A^μ that are the solution to the eigenvalue problem[79]

$$(\psi_{\mu\nu} - \lambda g_{\mu\nu})A^\nu = 0. \tag{8.40}$$

In terms of these A^μ and their eigenvalues $\lambda_{(\alpha)}$, Einstein was able to give an expression for *uniquely* assigning the $\dot{q}_\mu$ in terms of a given ψ. (The details of the recipe need not concern us here.)[80]

The essence of Bothe's objection is that the (covariant) derivative $\psi_{\mu\nu}$ for such a product wave function $\psi = \psi_1\psi_2$ is not zero when μ is an index for referring to the first subsystem and v one for the second subsystem. That is why the motions of the compound system will not be simply combinations of motions for the subsystems, as Einstein demanded that they be on physical grounds.[81]

Appendix 4 Von Neumann's unwarranted assumption

To illustrate what is at issue in von Neumann's theorem, let me consider some physical observable A for which there are possible measurement outcomes a_k, $k = 1, 2, 3, \ldots$. The idea is that for some particular value λ_0, every single observation (of a specified type) would with certainty yield a value $a(\lambda_0)$ (i.e., a definite and fixed one of the a_k). As usual, the average value $\langle A \rangle$ for a set of N observations is defined as

$$\langle A \rangle = \frac{1}{N} \sum_{j=1}^{N} a_j. \tag{8.41}$$

If the hidden variables λ are distributed according to some density function $\rho(\lambda)$, then the ensemble average is obtained as

$$\langle A \rangle = \int \rho(\lambda) \langle A \rangle_\lambda \, d\lambda = \int \rho(\lambda) a(\lambda) \, d\lambda. \tag{8.42}$$

It is these $\langle A \rangle$ that would have to agree with the predictions of standard quantum mechanics, while the $\langle A \rangle_\lambda$ might remain inaccessible in practice. For λ restricted to the *fixed* λ_0, we would have $a_j = a(\lambda_0)$ for *all j* so that

$$\langle A \rangle_{\lambda_0} = \frac{1}{N} \sum_{j=1}^{N} a_j(\lambda_0) = a(\lambda_0). \tag{8.43}$$

Similarly, $\langle A^2 \rangle_{\lambda_0}$ would have the value

$$\langle A^2 \rangle_{\lambda_0} \equiv \frac{1}{N} \sum_{j=1}^{N} a_j^2 = a^2(\lambda_0). \tag{8.44}$$

Now the *dispersion* of A (here restricted to the subensemble λ_0) is defined as

$$\Delta A_{\lambda_0} \equiv \sqrt{\langle A^2 \rangle_{\lambda_0} - \langle A \rangle^2_{\lambda_0}} \tag{8.45}$$

and is generally taken as a measure of the speed or "scatter" of the individual values of a_j observed. However, if the system is in a *dispersion-free* state specified by the λ_0, then

$$\Delta A_{\lambda_0} = 0. \tag{8.46}$$

Suppose we consider two noncommuting hermitian operators representing observables A and B and define another operator C as

$$C = A + B. \tag{8.47}$$

If λ_0 represents an in-principle possible dispersion-free state of an ensemble, then

$$\langle A \rangle = a(\lambda_0), \langle B \rangle = b(\lambda_0), \langle C \rangle = c(\lambda_0). \tag{8.48}$$

If—and this is von Neumann's critical assumption—the same linearity rule for expectation values that holds in standard quantum mechanics

$$\langle C \rangle = \langle A \rangle + \langle B \rangle \tag{8.49}$$

can be extended to dispersion-free ensembles then, for this λ_0,[82]

$$c(\lambda_0) = a(\lambda_0) + b(\lambda_0). \tag{8.50}$$

However, there are many simple counterexamples to eq. (8.50).[83] For instance, begin with the spin operator $\boldsymbol{\sigma}$

$$\boldsymbol{\sigma} = \sigma_1 \hat{\boldsymbol{i}} + \sigma_2 \hat{\boldsymbol{j}} + \sigma_3 \hat{\boldsymbol{k}}, \tag{8.51}$$

where $\hat{\boldsymbol{i}}, \hat{\boldsymbol{j}}$, and $\hat{\boldsymbol{k}}$ are unit vectors along the x-, y-, and z-axes respectively, and the σ_j are the 2×2 Pauli spin matrices. With the choices

$$A = \sigma_1, B = \sigma_2, C = \boldsymbol{\sigma} \cdot \boldsymbol{n}, \tag{8.52}$$

where $\boldsymbol{n} = (1, 1, 0)$, we have $C = A + B$. Since all of the Pauli matrices have eigenvalues $+1$ and -1 *only*, it follows, in this case, that $a(\lambda_0) = +1$ or -1, $b(\lambda_0) = +1$ or -1, and $c(\lambda_0) = +\sqrt{2}$ or $-\sqrt{2}$. Equation (8.50) would then require that

$$\pm\sqrt{2} = \pm 1 \pm 1, \tag{8.53}$$

which is impossible for *any* choice of the signs in eq. (8.53).

Perhaps a detailed physical model that actually accomplishes what the theorem is often taken as forbidding will be helpful for the reader.[84] For the "spin" operator[85]

$$\hat{\sigma} \equiv \frac{1}{\sqrt{2}} (\sigma_x + \sigma_y) = \frac{1}{\sqrt{2}} \begin{pmatrix} 0 & 1-i \\ 1+i & 0 \end{pmatrix} \tag{8.54}$$

one easily verifies that the eigenvalues of $\hat{\sigma}$ are ± 1 and that the corresponding eigenvectors are

142   Chapter Eight

$$\chi_+ = \begin{pmatrix} \frac{1}{\sqrt{2}} \\ \frac{1+i}{2} \end{pmatrix}, \chi_- = \begin{pmatrix} \frac{1}{\sqrt{2}} \\ \frac{-1-i}{2} \end{pmatrix}. \tag{8.55}$$

Since direct calculation shows that

$$\langle \chi_+ | \hat{\sigma} | \chi_+ \rangle = 1, \tag{8.56a}$$

$$\langle \chi_+ | \sigma_x | \chi_+ \rangle = \frac{1}{\sqrt{2}} = \langle \chi_+ | \sigma_y | \chi_+ \rangle, \tag{8.56b}$$

we see that

$$\langle \hat{\sigma} \rangle_{\chi_+} \equiv 1 = \frac{1}{\sqrt{2}} \left(\frac{1}{\sqrt{2}} + \frac{1}{\sqrt{2}} \right) = \frac{1}{\sqrt{2}} (\langle \sigma_x \rangle_{\chi_+} + \langle \sigma_y \rangle_{\chi_+}) \tag{8.57}$$

with a similar relation (except for an overall minus sign) holding for the state χ_-. This is an example in which three operators A, B, and C are such that

$$A = B + C, [B, C] \neq 0, \tag{8.58}$$

but still

$$\overline{\langle A \rangle} = \overline{\langle B \rangle} + \overline{\langle C \rangle}. \tag{8.59}$$

Here the averages are to be taken over an *entire* ensemble (not just over one "pure" subensemble).

But, *need* this be so for dispersion-free (hidden-variables) states? It is, of course, true in quantum mechanics that there cannot be a simultaneous complete set of eigenstates $\{\psi_j\}$ of noncommuting hermitian operators, since the conditions

$$A = A^\dagger, B = B^\dagger, A\psi_j = \alpha\psi_j, B\psi_j = \beta\psi_j \tag{8.60a}$$

on such a set are necessary and sufficient for

$$[A, B] = 0. \tag{8.60b}$$

However, one *can* give a more complete, dispersion-free state description for the example of eq. (8.54). Consider an "electron" moving in a straight line and let its spin variable be $\boldsymbol{\lambda}$ and the direction along which this spin will be measured be denoted by the vector $\boldsymbol{a}$ (where both $\boldsymbol{\lambda}$ and $\boldsymbol{a}$ lie in a plane perpendicular to the direction of motion). The *observed* value of the spin is to be assigned according to the rule

$$\text{sgn }(\boldsymbol{\lambda}\cdot\boldsymbol{a}) > 0 \Rightarrow +1 \quad (8.61a)$$

$$\text{sgn }(\boldsymbol{\lambda}\cdot\boldsymbol{a}) < 0 \Rightarrow -1 \quad (8.61b)$$

Then, *by construction*, the "observed" values in any *fixed* state $\boldsymbol{\lambda}$ of the subensemble of those $\boldsymbol{\lambda}$ lying between the x- and y-axes of figure 8.1 are[86]

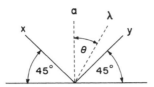

Figure 8.1 The "hidden variable" $\boldsymbol{\lambda}$

$$\langle\hat{\sigma}\rangle_\lambda = 1, \langle\sigma_x\rangle_\lambda = 1, \langle\sigma_y\rangle_\lambda = 1 ,$$

so that

$$\langle\hat{\sigma}\rangle_\lambda \neq \frac{1}{\sqrt{2}} (\langle\sigma_x\rangle_\lambda + \langle\sigma_y\rangle_\lambda) . \quad (8.62)$$

Therefore, these dispersion-free states do *not* satisfy the sum rule. It is also evident that

$$\overline{\langle\hat{\sigma}\rangle} \text{ (averaged over } all \text{ } \boldsymbol{\lambda} \text{ in the upper half plane)} = +1. \quad (8.63)$$

Now assume that the "hidden variable" $\boldsymbol{\lambda}$ is distributed over the *half* plane above the horizontal axis of figure 8.1 according to the "probability" $P(\boldsymbol{\lambda})$[87]

$$P(\boldsymbol{\lambda}) = \frac{1}{2} \cos(\theta) . \quad (8.64)$$

Then direct calculation yields

$$\overline{\langle\sigma_x\rangle} = \frac{1}{2}\int_{-\pi/2}^{\pi/4} \cos(\theta)\, d\theta - \frac{1}{2}\int_{\pi/4}^{\pi/2} \cos(\theta)\, d\theta = \frac{1}{\sqrt{2}} = \overline{\langle\sigma_y\rangle}, \quad (8.65)$$

so that

$$\overline{\langle\hat{\sigma}\rangle} = \frac{1}{\sqrt{2}} (\overline{\langle\sigma_x\rangle} + \overline{\langle\sigma_y\rangle}) . \quad (8.66)$$

This *local* hidden-variable model reproduces the *statistical* predictions of quantum mechanics.

NINE
The Fate of Bohm's Program

By the mid-1930s, the option of a causal quantum-theory program had generally been rejected as logically impossible and necessarily empirically inadequate when, in fact, it need not have been taken to be either. There matters essentially stood until 1952 when David Bohm gave a brilliant and detailed exposition of a hidden-variables theory based on the familiar quantum formalism.[1] This was a specific counterexample to the dogma that there could exist no causal version of quantum mechanics that was observationally equivalent to standard quantum mechanics. Given the presumed objectivity and impartiality of the scientific enterprise, one might expect that such an interpretation would be given serious consideration by the community of theoretical physicists. However, it was basically ignored, rather than either studied or rebutted. Just as external factors had played a key role in establishing the Copenhagen hegemony, so they once again contributed to keeping this competitor from the field. That a generation of physicists had been educated in the Copenhagen dogma made it all the more difficult for Bohm's theory.

9.1 The initial reception of Bohm's 1952 paper

One finds no extensive examination of this new theory in the technical literature.[2] For example, in 1952 a note containing a technical criticism of Bohm's paper was published, but Bohm successfully responded, essentially pointing out that the objection was based on a misunderstanding.[3] As for the one positive reaction in the physics literature, Bohm actually took exception to some of its suggestions.[4] There were, however, a few sharp and quite negative responses. In the main these came from men who were by then seen as the central, if dated, figures in the debates over the foundations of the Copenhagen interpretation.

9.1.1 *The intellectual environment*
A summary of the major attitudes prevalent towards quantum mechanics around 1950 will provide a picture of the arena into which Bohm's theory

entered. A convenient source for this sampling is a 1948 issue of *Dialectica* devoted to the concept of complementarity, a topic chosen by Pauli, who edited that issue of the journal. The contributors included Bohr, Einstein, de Broglie, Heisenberg, and Reichenbach. In his editorial preface, Pauli wrote that any gain in knowledge about one aspect of an atomic system must come at the cost of knowledge about some other aspect of that system. For him this lesson of the loss of our ability to predict the results of individual observations was the most fundamental result of such complementarity. He denied the possibility of an objective reality that is independent of the act of observation, contrary to Einstein's apparently reasonable expectations, and cited von Neumann's proof that quantum mechanics cannot be completed by the addition of hidden variables.[5]

Bohr gave a brief exposition of causality and complementarity, telling his reader that "the individuality of the quantum processes excludes a separation between a behaviour of the atomic objects and their interaction with the measuring instruments defining the conditions under which the phenomena appear."[6] He even extended complementarity to the realm of psychology. Einstein returned to the theme of the incompleteness of quantum mechanics and the specter of action at a distance. The completeness of a state description in terms of the wave function, he claimed, would "imply the hypothesis of action-at-distance, an hypothesis which is hardly acceptable."[7] Heisenberg discussed his ideas about "closed" physical theories, a topic covered elsewhere in this book.[8]

Finally, the philosopher Hans Reichenbach asked whether or not, in quantum mechanics, complementary parameters might possess precise values that are simply unknown to us and whether, if these values were known, future measured values could be predicted. Both questions were given negative answers. He formulated the lack of causality in terms reminiscent of Rosen's:

> If there is radiation going through a diaphragm with two slits, and a particle goes through one slit, its further travel will depend on whether the other slit is open or closed; the possibility that the particle might have gone through the other slit determines its actual behavior. Actuality is made a function of possibility.
> The peculiar relationship can also be interpreted as a violation of the principle of action by contact. The causality introduced by artificial interpolation of unobserved values, however these values be chosen, is of such a kind that what happens at one space point depends on conditions at distant space points, while there is no physical transmission of this effect through space. If there is a causality "behind" the observables of quantum mechanics,

it is not of the normal type, but contradicts the principle of action by contact.[9]

While Bohm would not disagree with much of what is stated here, he was willing to accept a "nonlocal" causality, unlike Reichenbach. It is perhaps worth recalling that Bohm had begun as an orthodox Copenhagen supporter and even wrote a widely praised and much-used 1951 textbook on standard quantum theory. He then decided that he did not find that theory really coherent and in 1952 formulated his own theory. It would not seem that Bohm held any a priori ideological brief against the Copenhagen view. At that time, though, the prevailing climate of opinion about the broader implications of the formalism of quantum mechanics was not conducive to an impartial consideration of Bohm's interpretation.

9.1.2 Einstein remains silent

Einstein was never enthusiastic about Bohm's theory.[10] In a letter of 12 May 1952, Einstein commented to Born on Bohm's attempt: "Have you noticed that Bohm believes (as de Broglie did, by the way, 25 years ago) that he is able to interpret the quantum theory in deterministic form? That way seems too cheap to me."[11] What did Einstein mean by this? One might have expected Einstein to be receptive to such a theory since it obviously saved an objective reality. Einstein's interest in an ensemble interpretation of quantum mechanics was largely as a means of arguing for the incompleteness of (Copenhagen) quantum mechanics since, in his opinion, a truly fundamental theory should, in principle, apply to actually existing *individual* particles and systems, not just to ensembles.[12] It was not the final picture.

In a 1948 letter to Born, Einstein expressed this view: "The concepts of physics relate to a real outside world, that is, ideas are established relating to things such as bodies, fields, etc., which claim a 'real existence' that is independent of the perceiving subject. . . . It is further characteristic of these physical objects that they are thought of as arranged in a space-time continuum."[13] On this basis, Bohm's program should surely have scored a plus.

However, Einstein had concerns about locality. That same letter continued:

> An essential aspect of this arrangement of things [in a space-time continuum] in physics is that they lay claim, at a certain time, to an existence independent of one another, provided these objects 'are situated in different parts of space.' Unless one makes this kind of assumption about the independence of the existence (the 'being-thus') of objects which are far apart from one another

in space—which stems in the first place from everyday thinking—physical thinking in the familiar sense would not be possible. It is also hard to see any way of formulating and testing the laws of physics unless one makes a clear distinction of this kind.[14]

In giving his reasons against Born's interpretation that took quantum theory to be complete, Einstein again stressed the need for locality to be "free from spooky actions at a distance."[15] The nonlocality of Bohm's theory was objectionable to Einstein. In section 10.2, I return to this question in light of Bell's theorem and of the no-signaling result for quantum-mechanical correlations. But in 1952, when Bohm's quantum-potential papers appeared, Einstein would still have felt that nonlocality told decisively against Bohm's theory.

Einstein had even deeper reasons for rejecting such hidden-variables approaches to a completion of quantum mechanics. Quite simply, they were not radical *enough*. For him the available formulations of quantum mechanics dealt with the wrong concepts and the general approach was fundamentally flawed.[16] In 1954 Einstein remarked that "it is not possible to get rid of the statistical character of the present quantum theory by merely adding something to the latter, without changing the fundamental concepts about the whole structure."[17] The existence of an observer-independent, objective reality (which he conceived of in a separable space-time with local interactions) was more of a concern for him than was causality itself.[18] In a 1954 letter to Born, Pauli gave a similar representation of Einstein's position.[19]

In this light, we could say that Bohm saved causality and (some type of) space-time, but at the price of nonlocality. However, in 1952 Einstein told his old friend Michelangelo Besso that "I reject . . . the assumption that a rigid coupling exists between parts of a system spatially arbitrarily far apart from each other (instantaneous action at a distance, which does not decrease with increasing distance)."[20] And again in a volume dedicated to de Broglie, Einstein focused on the inadequacy of basing an acceptable theory on the concepts of classical mechanics.

> I do not at all doubt that the contemporary quantum theory (more exactly "quantum mechanics") is the most complete theory compatible with experience, as long as one bases the description on the concepts of material point and potential energy as fundamental concepts. [The difficulties of quantum mechanics] are connected with the fact that one retains the classical concepts of force or potential energy and only replaces the laws of motion by something entirely new. The completeness of the mathemati-

cal mechanism of the theory and its significant success turn attention away from the difficulty of the sacrifice that has been made.

To me it seems, however, that one will finally recognize that something must take the place of forces acting or potential energy (or in the Compton effect, of wave fields), something which has an atomistic structure in the same sense as the electron itself. "Weak fields" or forces as active causes will then not occur at all, just as little as mixed states.[21]

In Einstein's view, only a completely different approach, such as a unified field theory, held any hope of producing a truly fundamental and complete theory of quantum phenomena. For him, Bohm's attempt truly did not go far enough.

9.1.3 De Broglie is reconverted

In 1949, de Broglie still accepted the Copenhagen view and spoke of it as the *only* interpretation compatible with the facts.[22] He referred to the "Bohr-Heisenberg interpretation", making no essential distinction between the positions of Bohr and of Heisenberg.[23] Initially, de Broglie was against Bohm's ideas (which were similar to his own pilot-wave theory of 1927) and he raised the same objections against Bohm's theory as had been raised against his own.[24] Bohm's work of the early 1950s eventually reconverted de Broglie to his former ideas.[25]

In a book written not long after the appearance of Bohm's 1952 paper, de Broglie recalled receiving in the summer of 1951 the draft of a paper from Bohm. At first it appeared to de Broglie that Bohm had simply repeated his own ideas of the 1927 pilot-wave theory—something already dispatched by Pauli. Around that time, Jean-Pierre Vigier, who worked at the Henri Poincaré Institute with de Broglie, pointed out a similarity between Einstein and Grommer's treatment in 1927 of the motions of particles in general relativity and de Broglie's own discussion of the motion of a singularity in the theory of the double solution.[26] For de Broglie, Bohm's paper presented an "analysis of the processes of measurement as seen from the point of view of the pilot-wave which would seem to permit removing the objections which had been directed against my ideas by Pauli in 1927."[27] About any possible charge of inconsistency in returning to those views he had abandoned twenty-five years previously, de Broglie responded: "To this, if I may joke, I can answer with Voltaire: 'A stupid man is one who doesn't change.'"[28]

It appears that, essentially as a response to Bohm's 1952 paper, de Broglie wrote a book in the nature of a priority claim (or "reminder") for the pilot-wave theory.[29] In it he collected several of his wave-particle duality, double-solution, and pilot-wave papers from the 1920s, as well as a series of more recent publications, from 1951 and 1952, that had been stimulated by Bohm's paper. There Vigier also discussed a possible extension of de Broglie's double-solution concept to general relativity.[30] In this program we see what was a recurrent theme for de Broglie: the demand for a clear picture of physical processes in space-time. With the quantum potential one could provide *some* model for fundamental processes.[31]

There is, however, a basic distinction that is at times relevant between two positions that are often conflated as the "de Broglie–Bohm" theory. In spite of some general conceptual similarities, de Broglie's pilot-wave theory and Bohm's (1952) causal theory are *not* identical. De Broglie himself has characterized the difference as his having a real wave in three-dimensional space while Bohm's wave (like Schrödinger's in 1927) exists in a ($3n$-dimensional) configuration space.[32] If one wants to take Bohm's ψ-wave as an independently, actually existing wave (rather than simply as a mathematical way to represent the influence of the environment), then this can certainly be done in a multidimensional configuration space, although the physical meaning of waves in configuration space is problematic.[33] Another really central difference between these two theories is evident in the way electromagnetic waves are treated. For de Broglie's theory, there are *both* a photon *and* an electromagnetic (pilot) wave, whereas for Bohm there is no separate photon "particle" (but the super wave field can localize electromagnetic energy—effectively a "photon").[34] The de Broglie theory appears to conflict with experiment, whereas Bohm's does not.[35]

9.1.4 Pauli is caustic

Pauli had been one of de Broglie's harshest critics at the 1927 Solvay congress and he maintained the same attitude toward Bohm's theory. Interestingly enough, when Bohm sent him a copy of the paper in which Bohm showed Pauli's objections to the causal interpretation to be specious, Pauli never responded.[36] In a letter to Born in 1953, Pauli referred to the retrograde efforts by Schrödinger, Bohm, and Einstein and confidently predicted that the statistical character of the wave function and of the laws of nature would determine the style of the laws for centuries to come. Any dream of a way back to the classical style of Newton and Maxwell seemed to him both hopeless and in bad taste.[37]

In a volume dedicated to de Broglie on his sixtieth birthday in 1952, Pauli gave his views on hidden variables and the pilot-wave theory, in which he included Bohm's theory.[38] He claimed that either such hidden parameters produce no observable effects, which "characterize this form of the theory as artificial metaphysics," or the hidden variables do lead to empirical consequences, in which case "the consequences of the theory will be in disagreement with the general character of our experiences."[39] He accused Bohm of not having really examined the old objections to de Broglie's pilot-wave theory before reviving it. Pauli then reviewed his criticisms from the 1927 congress, pointing out that de Broglie himself subsequently accepted the Copenhagen theory and that de Broglie had reiterated these objections to Bohm. He identified as a central flaw in Bohm's *deterministic* theory the assumption that $P = |\psi|^2$ (recall eq. [4.6]) for the probability distribution since P ought in principle, Pauli claimed, to be able to have *any* arbitrary form [e.g., $P = \delta(x - x_0)$].[40] In a letter to Einstein dated 26 November 1953, Born, referring to this article by Pauli, stated that "Pauli has come up with an idea . . . which slays Bohm not only philosophically but physically as well."[41] Bohm *did* respond to this challenge by showing that an arbitrary initial distribution for P would be driven, through interactions, to $|\psi|^2$.[42] Pauli also cited approvingly von Neumann's "proof" against hidden variables and emphasized the nonlocality of Bohm's theory as making a mockery of any claim to intrinsic properties for individual particles.[43] He categorically rejected the possibility of a meaningful hidden-variables extension of quantum mechanics. Interestingly enough, his arguments against Bohm's theory have nothing to do with relativity and covariance, but are general conceptual ones. In sum, Pauli's *logical* criticisms are not compelling. He did, though, offer *ideological* ones as well.

An extensive study of the Pauli-Fierz correspondence reveals that Pauli's interest in the philosophical possibilities opened up by uncertainty was one reason for his opposition to Bohm's attempt (and to hidden-variables theories generally).[44] Pauli worried that these ill-conceived efforts might, for ideological reasons, divert younger scientists from more fruitful (orthodox) avenues of research. His letter to Markus Fierz in 1952 was quite sarcastic and contemptuous of Bohm's work.[45] Fierz has offered the view that de Broglie was a Catholic and Bohm a Communist and that, to Pauli, this was the black and red alliance.[46] Pauli, so Fierz claims, felt that the de Broglie–Bohm theory was either prattle or an unnatural way of stating the same thing as standard quantum theory. Fierz also cited the objection that the wave function exists in a multidimensional configuration space. Both Pauli and Fierz expressed strong negative opinions in emotionally charged terms. For Pauli more was at stake than just physics. Whereas materialism considers the spiritual as a mere by-product of the material, "Pauli especially understood the 'epistemological lesson' of atomic physics as a signpost pointing in another direction. He, like Heisenberg, thought that quantum theory has clearly shown materialism to be unsatisfactory."[47]

Pauli was quite explicit in holding that it is wrong to imagine the existence of the material external world without including the observer and his psyche in the worldview.[48] The ideas of a detached observer and of realism were to be abandoned.[49] He saw the Copenhagen interpretation as not being compatible with a materialistic worldview.[50] On the downside for the causal program from Pauli's perspective was the fact that "psychological problems associated with observations are quite neglected by Bohm, not to speak of the manifestations of the unconscious [and] in his causal-deterministic universe Bohm does not see any 'quantum mechanical paradoxes'—but there is no place for free will either."[51] Pauli wanted no part of a deterministic, clockwork universe and felt that the trend of Western culture since the seventeenth century had been dangerous.[52] The "irrational influences" that statistical causality implied were important for him.[53] Pauli's concerns spilled over into the theological realm as well:

> Causality, *i.e.*, the possibility of describing natural phenomena in a rational way, is an expression of God's intelligence. The simple idea of deterministic causality must, however, be abandoned and replaced by the idea of statistical causality. For some physicists . . . this has been a very strong argument for the existence of God and an indication of His presence in nature.[54]

There were deep philosophical, moral, and religious issues involved here. I return to this theme when I consider the role of Soviet science in the criticism of the Copenhagen interpretation. An attack on the canonical version of quantum theory resonated quite negatively and strongly with Pauli's beliefs about God and his views on epistemology. It was inconceivable to him that anything like a return to a "classical" world with causality and picturable, continuous processes in space-time was either possible or anything less than a disgusting loss of nerve and a return to darkness.[55]

It is not wholly clear whether, in his philosophically reflective later years, it was fear of a return to an oppressive form of Laplacian determinism or belief in a necessary relationship between the observer and the observed (i.e., complementarity for him) that was the more fundamental for Pauli in resisting realistic interpretations of quantum theory.[56] Complementarity was underpinned by the essential role the observer played, both in interpreting data and through a link between deep-seated psychological structures (Jung's archetypes) and structures in the external world.[57] On either reading, though, there were commitments and speculations that lacked the clarity and rigor that we usually associated with the "conscience of physics," to use an epithet often applied to Pauli.

9.2 Continued hostility and general disinterest

Having sketched early reactions to Bohm's theory by some of the major figures, let me now turn to the dominant attitude that has prevailed since then.[58] A type of guilt by association soon came into play.

9.2.1 An aura of Marxism

By his own recollection, Bohm was stimulated to rethink his conventional position on quantum theory when, in 1951, he read a paper (by Blokhintsev or by Terletskii) critical of Bohr's approach.[59] A connection between interpretative questions in quantum mechanics and the work of physicists from Communist countries is interesting, given the views of Pauli noted earlier, and would appear to be more than mere coincidence.[60] Nearly two decades ago, Max Jammer suggested that the possible influence of social and political factors on the increased interest in Marxist ideology in the West during the 1950s should be pursued in as much depth as Paul Forman had done for the influence of Weimar culture on the formulation of quantum theory in the 1920s.[61] There is a tension between scientific materialism and the Copenhagen interpretation over the existence or nonexistence, respectively, of objective, mind-independent matter. In the Soviet Union, party philosophers effected a split among physicists there over the Copenhagen dogma.[62] The pro-Copenhagen school included Vladimir Fock, Lev Landau, and Igor Tamm, while those who attempted to fashion a materialistic quantum theory included Dimitri Blokhintsev and Jacob Terletskii.[63] The *motivation* for this critical approach to the Copenhagen orthodoxy was ideological, not scientific (in any narrow sense of that term). There is no reason to suppose that Bohm himself was motived to pursue hidden-variables models because of an interest in Marxism. It does appear, however, that the writings of some Soviet physicists may have started him thinking along those lines.[64] Bohm became a victim of the McCarthy crusade against un-American activities not because of this but because he had worked for Oppenheimer as a graduate student at Berkeley.[65] On the other hand, Vigier, who brought Bohm's theory to de Broglie's attention, was a Communist, active during World War II in the French underground. Some of the earliest enthusiastic reactions to Bohm's 1952 paper came from Marxist (though not always Soviet) scientists.[66]

Another physicist, Hans Freistadt, had, like Bohm, run afoul of the anti-Communist hysteria in the United States in the late 1940s and he later published several papers on Bohm's program and the philosophy associated with it.[67] His technical papers during the 1950s included a general Hamilton-Jacobi formulation of quantum field theory (as a development of Bohm's 1952 work) and a lengthy, positive review of Bohm's program.[68] Two philosophical pieces contrasted positivism (including operationalism) and agnosticism (akin, it appears, to the common use of the term 'instrumentalism') with dialectical materialism.[69] He found such materialism to be the proper setting for science and Bohm's interpretation quite conducive to it. In a general discussion of the crisis in physics, Freistadt claimed that "the vehemence of the criticism currently leveled at Bohm's approach illustrates the firm grip that positivist philosophies have on the thinking of modern scientists."[70] He argued that there was no necessary conflict between free will and determinism and that positivism was an overreaction attempting to restrict science only to correlating sensory data and forbidding science from speaking about an underlying causal reality. For him, determinism and objective reality were inextricably linked.[71] Since Rosenfeld had published a detailed explanation of complementarity and of the Copenhagen interpretation, along with a bitter diatribe against Bohm's program, Freistadt responded with a philosophical defense of materialism and determinism as necessary for the description of an objective (observer-independent) reality.[72]

9.2.2 *Materialism, mechanism, and mere metaphysics*
In the opening lines of the chapter of *Physics and Philosophy* devoted to criticisms of alternative interpretations of quantum mechanics, Heisenberg asserted that the "Copenhagen interpretation of quantum theory has led the physicists far away from the simple materialistic views that prevailed in the natural science of the nineteenth century."[73] He categorized opponents of the Copenhagen view as wanting to return "to the ontology of materialism," to a completely objective description of nature, rather than accepting Copenhagen's subjective element in the description of atomic events.[74]

> At this point we realize the simple fact that natural science is not Nature itself but a part of the relation between Man and Nature, and therefore dependent on Man. The idealistic argument that certain ideas are *a priori* ideas, i.e., in particular come before all natural science, is here correct. The ontology of materialism rested upon the illusion that the kind of existence, the direct "actuality" of the world around us, can be extrapolated into the atomic range. This extrapolation, however, is impossible.[75]

Here we sense that more is at stake than just physics. Heisenberg attempted to dismiss any program (such as Bohm's) that is based on the same *formalism* as is Copenhagen since "they only repeat the Copenhagen interpretation in a different language [and] from a strictly positivistic standpoint one may even say that we are here concerned not with counter-proposals to the Copenhagen interpretation but with its exact repetition in a different language."[76] This is consonant with Heisenberg's view that a correct formalism determines (or contains) its interpretation. The thrust of his discussion was that Copenhagen and Bohm differ only over matters of language, not of physics. That was a strange move since he also allowed that the issue between the two was a matter of ontology.

In 1954 Henry Margenau wrote a more philosophically oriented, more balanced review of the different interpretations of quantum mechanics. There he placed Bohm's theory in the "mechanistic" category, essentially dismissed it as anachronistic and espoused a more radical interpretation in which position would no longer be a primary quality, but would arise only in the act of perception.[77] (Margenau favored the terms 'possessed' and 'latent'.)[78] He then used philosophical and methodological criteria, such as simplicity, economy, and coherence, to argue for this latter view. He admitted that such an evaluation was certainly subjective and incomplete as regards scientific evidence.[79]

Mario Bunge shortly thereafter published a broad review on the interpretation question.[80] He was critical of the Copenhagen interpretation and gave Bohm a fair representation, but did not finally come out in favor of any particular interpretation. He took the multiplicity of interpretations to be a sign of crisis in physics and urged a freer climate of discussion of the various possibilities.[81] What is distinctive about Bunge's article is that it is open-minded on Bohm, rather than categorically negative.

Old-guard, loyal supporters of Bohr, such as Leon Rosenfeld, heaped scorn on Bohm's and other unorthodox approaches (much as Fierz would do later). An example of this was Rosenfeld's defensive and accusatory conference lecture delivered at the University of Bristol in the spring of 1957. There he declared that "quantum theory eminently possesses this character of uniqueness; every feature of it has been forced upon us as the only way to avoid the ambiguities which would essentially affect any attempt at an analysis in classical terms of typical quantum phenomena."[82] While it is not surprising that some "nonstandard" conference speakers, such as Fritz Bopp, David Bohm, and Jean-Pierre Vigier, disagreed with Rosenfeld, even so orthodox a physicist as Markus Fierz found some of Rosenfeld's claims too dogmatic.[83] At this conference, where roughly equal numbers of physicists and philosophers spoke, Bohm's ideas were not simply rejected out of hand. However, this was the exception. The general reaction to Bohm's theory in the West, and in the United States in particular, ranged from indifferent to strongly negative.[84] A decade after his first "interpretation" paper came out, Bohm published a lengthy article on the interpretation of quantum mechanics and argued in detail that his own hidden-variables theory was not only free from inconsistencies but also in several ways preferable to the standard interpretation.[85]

In his 1963 book, *The Concept of the Positron*, the philosopher Norwood Hanson defended the Copenhagen theory at length and argued against Bohm's program.[86] "There is now one, and only one, physical theory [orthodox quantum theory] which deals at all successfully with the perplexing phenomena which ... upset classical physics."[87] This was meant to be more than just a statement about what theories happened to be known *then*. On the basis of the noncommutativity of the operators that are at the heart of quantum mechanics, he concluded that "there is no intelligible way in this system of *speaking* of the exact position of an electron of precisely known energy."[88] According to Hanson, the wave-particle symmetry of quantum theory is an essential feature that cannot be broken while still maintaining the successful features of the theory, and this is guaranteed by von Neumann's theorem.[89]

It would at least seem arguable that Bohm, if given a fair reading, could be seen as doing just what is here forbidden. Hanson claimed that Bohm's theory "only added extra philosophical notions of heuristic value" and noted approvingly that Bohr and Heisenberg did not need such a hypothesis.[90] That is surely a normative judgment on his part and need scarcely be taken as a killing criticism. It is by Copenhagen's *own* standards that Hanson condemns Bohm. He stated, more than argued, that the Copenhagen *interpretation* itself was essential for the progress that was made in Dirac's theory of the electron in 1928, for example. Much of his case is built on the fact that there was no fully articulated alternative to Copenhagen in 1927. The thesis here really is that, given the way historical events *actually* occurred, only the Copenhagen interpretation was viable at the time. Even if this were granted, it begs the question as far as any logical or intrinsic superiority of the Copenhagen interpretation is concerned. Hanson further warned that scientists do not abandon a working theory unless it faces anomaly *and* another theory is available. They replace a successful theory with an entirely new one only when forced to do so by failure.[91] Again, even if correct, this is more a statement of psychology or sociology, or perhaps a sensible strategy for getting on with things, rather than a telling criticism of Bohm's program as such. When Hanson tells us that "Bohm ... treat[s] ... non-commutativity [of position and momentum operators] as constituting an experimental limitation only," he is being unfair by representing the influence of the environment (via the quantum potential) as though it were something that might in principle *in our actual world* be circumvented experimentally.[92] It is unclear that the 'Bohm' that Hanson refutes is the actual Bohm of the

1952 paper. His book is a retrospective reconstruction of a sweeping victory for *a* methodology and *an* interpretation. It does not, though, sufficiently separate historical fact from logical necessity.

Such a dismissive attitude is not confined to past decades. A recent review article on consistent interpretations of quantum mechanics contends that "there is no serious alternative to it [the Copenhagen interpretation], since the approach through hidden variables, whatever its interest, has not been developed to the point of giving a theory but only the preliminaries of a theory."[93] There is no mention of Bohm's 1952 paper or any discussion of specific causal quantum theory programs, in spite of the statement that an interpretation is "consistent if free from internal contradiction and complete if it provides precise predictions for all experiments."[94] It is difficult to see how Bohm would fail on either count. Another recent discussion of the interpretation question coauthored by a former student and later confidant of Heisenberg essentially rehearses Heisenberg's own position and relegates Bohm's theory to the "excess-baggage" category since it is observationally indistinguishable from Copenhagen and since it adds "ontological assumptions which go, as far as quantum theory can say today, beyond the realm of human knowledge, thus being neither provable nor refutable with our means."[95]

9.2.3 *A few typical encounters*

The lack of attention to Bohm's early work can be understood in terms of most scientists' being uninterested in what they perceive to be (mere) metaphysical questions, coupled with their belief that there can be only *one* correct theory and that they have found it (or are in the process of doing so).[96] A few firsthand, anecdotal illustrations of this, all involving Bohm's quantum-potential approach, may be enlightening here.

At a weekly theorists' journal club meeting in a major theoretical physics department of a university in the eastern United States in the early 1950s, someone suggested that Bohm's 1952 paper be reported on the following week. The theorists were so disdainful of any proposed attempt at a quantum theory different from the orthodox one that they assigned the paper to an experimentalist. (Any practicing physicist will appreciate this plausible but condescending move.) When the experimentalist reported the following week that he had found no logical flaw or contradiction in the paper, the theorists were upset. They were certain there *had* to be an inconsistency, but were unwilling to take the time themselves to locate it. The theory was simply rejected out of hand because it conflicted with Copenhagen.

A well-known theorist was teaching a graduate course in quantum mechanics at an American university in the Midwest in the early 1950s and one of his students asked him about Bohm's paper. Initially inclined to dismiss this attempt by Bohm, the theorist nevertheless read the work and concluded that, indeed, there was really nothing wrong with this alternative. Sometime later he was a visitor at Princeton's Institute for Advanced Study and he asked Robert Oppenheimer, the institute's director, what he and the other theorists there thought of Bohm's theory. In essence, Oppenheimer's response was that they were all certain Bohm was incorrect but he admitted, when pressed, that no one had seriously studied the work.

A few years ago in the United States at a national meeting of the American Physical Society, there was a symposium on chaos theory and the topic of chaos in quantum mechanics was discussed. The invited speaker, himself a distinguished theorist, pointed out the well-known problem that quantum chaos can be difficult to define concisely since the usual characterization in classical mechanics is in terms of trajectories that diverge rapidly in time. However, in quantum systems we cannot, he reminded his audience, speak of trajectories. A member of the audience asked whether Bohm's interpretation, in which there *are* particle trajectories that exhibit chaotic features, might be a useful point of departure. The question fell flat when the speaker replied that he had never *heard* of Bohm's theory.

Here, then, is what is likely to happen when one attempts to discuss Bohm's theory with typical physicists. First, most will never have heard of Bohm's theory. Those who have are certain it has to be wrong, although they're not sure exactly why (aside from the fact that it conflicts with Copenhagen). Those few who still remain to listen to a description of Bohm's theory are usually surprised when they learn that they can, if they wish, consistently believe in actual particle trajectories in a space-time background. After all, Copenhagen has told them this is impossible (even though they often speak of and conceptualize physical phenomena in terms of just such trajectories in their research). This they usually welcome. But when they realize that Bohm's theory is nonlocal, they reject it since it is *merely* as empirically adequate as Copenhagen and it came later. Pointing out to them that Copenhagen has forbidden *both* trajectories *and* locality and that, at least, Bohm has given them back half a loaf of bread is to no avail.[97] Copenhagen got there first, great minds, such as Bohr and Heisenberg, have thought about these issues and provided answers (*some* answers, whatever they may be), and that's good enough to foreclose the need to agonize over such metaphysical baggage. Science has to do with observations—period.

All this is not meant to imply that *no one* cares about such issues. But it does indicate that there are fundamental misconceptions about how conclusive some of the Copenhagen arguments are as well as about

Bohm's program. There is no interest in wasting time on matters of interpretation that seem to have no prospect for an immediate impact upon observations and experiments and thus upon the course of physics.

9.3 Subsequent work

As one might expect, the Bohm causal quantum theory program did not stop with the 1952 "existence theorem" (of a logically possible alternative to Copenhagen). Since that time there have been several attempts to underpin this and similar causal interpretations with models of physical processes to account for, say, the quantum potential. While Bohm's quantum-potential theory is observationally equivalent to the Copenhagen one, detailed dynamical models to go beyond it typically (but not always) have, in principle, some predictions that differ from those of the standard theory.[98] The following is not a survey of all such attempts, but rather a description of a few programs to provide a background for the last section of this chapter.

9.3.1 The Bohm-Vigier model (1954)

One early attempt at such a model was the work of David Bohm and Jean-Pierre Vigier in which a physically real field ψ (i.e., one based on actual physical processes in three-dimensional space) in many ways equivalent to a "Madelung fluid," having irregular and random fluctuations, provided a background medium that acted on particle-like inhomogeneities in this fluid.[99] In this model they were able to show that P, the density of this fluid, would necessarily be driven to the (stable) equilibrium distribution $P \to |\psi|^2$. This "fluid" would undergo random fluctuations about the mean "guidance condition" velocity v and about the mean density $|\psi|^2$ determined by the Schrödinger equation. Their proposal was tentative and *suggestive* of a research program. The interests of Bohm and of Vigier later diverged into two quite different research programs. Bohm's interests took a metaphysical bent in terms of the wholeness or interrelatedness of all physical phenomena, while Vigier's subsequent work (on a stochastic interpretation of quantum mechanics) concentrated on specific physical models based on the concept of a covariant ether.[100]

The idea of such an ether had been suggested previously by Dirac, who, in a series of papers, proposed a new classical theory of electrons in which an ether seemed to appear naturally.[101] In Dirac's theory, there is "no contradiction with relativity, because all the equations are Lorentz invariant."[102] He also pointed out that arguments using Lorentz invariance to rule out an ether for the vacuum of classical physics would no longer be valid in a vacuum subject to quantum mechanics. The basic reason for this, Dirac explained, is that the velocity of an ether would be subject to the uncertainty relations and a suitable wave function would make all values of this velocity equally probable.[103] This suggestion is, of course, very different from Einstein's previous identification (within the framework of general relativity) of the structure of space itself as a type of (immaterial) ether: "Therefore, instead of speaking of an aether, one could equally well speak of physical qualities of space."[104]

The "vacuum" of modern quantum field theory is hardly empty or structureless in any common sense of those terms. Nevertheless there is general resistance to introducing a "subquantum medium" to provide a possible explanation for quantum phenomena.[105] David Bohm and Basil Hiley have recently returned to such a stochastic model (in actual three-dimensional space) of quantum phenomena, perhaps to avoid the common objection to Bohm's earlier quantum-potential approach in which the ψ-wave exists in a configuration space that raises serious ontological questions.[106]

Neither of these programs—Bohm's "implicate-order" ontology or the "stochastic-fluid" model—has proven wholly satisfactory. Rather than pursue them here, I turn instead to a program that attempts to interpret quantum mechanics in terms of a (largely classical) stochastic mechanics. My reason for doing this is that this project is interesting in its own right and will be useful for my "counterfactual historical" thesis in the next chapter.

9.3.2 Nelson's stochastic model (1966)

In section 8.4 we have seen that the similarity between the Schrödinger equation and the diffusion equation was noted by Schrödinger himself as early as 1931. Several subsequent attempts were made to show that quantum phenomena could be interpreted in terms of (classical) stochastic processes and I begin here with the work of Edward Nelson.[107] It is my contention that, conceptually, Nelson's work is based on ideas current in the late 1920s. His early work has developed into a carefully articulated and mathematically precise program.[108]

The spirit of this project was concisely presented in the abstract to Nelson's now-classic 1966 paper:

> We examine the hypothesis that every particle of mass m is subject to a Brownian motion with diffusion coefficient $\hbar/2m$ and no friction. The influence of an external field is expressed by means of Newton's law $F = ma$. . . . The hypothesis leads in a natural way to the Schrödinger equation, but the interpretation is entirely classical. Particles have continuous trajectories and the

wave function is not a complete description of the state. Despite this opposition to quantum mechanics, an examination of the measurement process suggest that, within a limited framework [of observations that may be reduced to position measurements] the two theories are equivalent.[109]

While Nelson's theory is not a deterministic one, probability enters just as it does in a classical theory of, say, Brownian motion. It has long been appreciated that Brownian motion creates limitations on the accuracy obtained by measuring instruments.[110]

In Nelson's model, a particle of mass m is subject to a (random) Brownian-motion force and to the classical force. The equations of motion for m are formulated in terms of what Einstein called the *osmotic velocity* u [arising from the variations of the density ρ of the fluid (ether)] and the *current velocity* v.[111] If $x(t)$ is the position of the particle, then $dx(t)/dt$ is *not* a continuous function of t. The highly nonlinear dynamical equations that govern the motion of this particle can, by means of a mathematical transformation, be converted into a linear differential equation. This process is effectively the inverse of what Madelung and Bohm did and results in the Schrödinger equation. (The details are given in appendix 1). We have a *formal* mathematical equivalence between the dynamical equations for the Brownian motion of m and the Schrödinger equation.

Work on the equivalence between stochastic and Schrödinger mechanics has been extensive, beginning with the association of diffusion processes and solutions to the Schrödinger equation.[112] Depending upon the exact nature of the random process influencing the particle motion, one may or may not obtain *exact* (versus only approximate) equivalence between these two formalisms.[113] One can question whether Nelson's stochastic approach is completely equivalent to the Schrödinger equation because of problems of possible multiple-valuedness in the function S that appears in the "guidance" condition in stochastic mechanics (whereas $\psi = e^{iS/\hbar}$ is usually required to be single-valued in quantum mechanics).[114] Complete equivalence does obtain on multiply connected spaces (which are needed in certain physical situations).[115] For single-time observables (of the type encountered in standard quantum mechanics), there is no observational difference between the two formulations. However, there are naturally defined expectation values for combinations of noncommuting operators that need not (and in some cases cannot) agree in the two theories. It is unclear what the observational consequences of this are. Nelson's program has been extended (not without additional assumptions) to the relativistic (spinless-particle) case and to the Dirac equation.[116] This program of stochastic mechanics has survived a fair amount of scrutiny.

Nelson established the equivalence between his stochastic mechanics and the Schrödinger equation when all measurements are reducible to position measurements.[117] He has made an observation that is quite relevant to measurement theory in Bohmian mechanics.

> The first thing to say about measurements is that the outcome of any conceivable experiment may be expressed in terms of the positions of macroscopic objects. This is a triviality. It does not concern the physical observable being measured, but how we as human beings receive information through our senses. Furthermore, the outcome of any conceivable experiment may be expressed in terms of the position of macroscopic objects, including recording devices, at a single time. Consequently, if the laws of quantum mechanics apply both to the system being measured and to the measuring apparatus, then the predictions of quantum mechanics will be identical with those of Markovian statistical mechanics, since the positions of all constituents of the system plus apparatus are determined by the probability density $\rho = |\psi|^2$, where ψ is the wave function of the system plus apparatus.[118]

He ended his 1966 paper with the insightful comment, "It appears that the phenomena which first led to the abandonment of classical physics admit a simple classical interpretation which is only in a limited sense equivalent to quantum mechanics."[119] That is, the class of phenomena upon which quantum mechanics was based historically is susceptible to a (largely) classical interpretation. I return to this theme in chapter 10.

Nelson has pointed out that "stochastic mechanics is more vulnerable than quantum mechanics, because it is more ambitious: it attempts to provide a realistic, objective description of physical events in classical terms."[120] An obvious and important question is whether stochastic mechanics with its background field can account for the types of long-range correlations observed in the EPRB experiments. We know that Bell's theorem and the outcomes of experiments rule out any theory that is both deterministic and local. Nelson opted for a description that would have intrinsic randomness, but that would be local.[121] Such a move is still unable to account for these long-range correlations: "I have loved and nurtured Markovian stochastic mechanics for 17 years, and it is painful to abandon it. But its whole point was to construct a physically realistic picture of microprocesses, and a theory that violates locality is untenable."[122] His theory is *both* indeterministic *and* nonlocal, and Nelson found nonlocality totally unacceptable. These stochastic models also run into difficulty in producing *physical* accounts of spin and of the statistics of identical particles.

We have seen, on the other hand, that a (simple) Bohm quantum-potential theory pays *one* price—nonlocality—and then provides an account of all phenomena with no measurement problem and with no classical-limit difficulty. This allows one to have an at-base deterministic worldview with continuous space-time trajectories. In that case, we do have to face a certain type of nonlocality (yet with no *observational* violations of special relativity). This required adjustment of our worldview may be analogous to the one physicists faced with the lack of simultaneity that relativity itself brought.[123]

9.4 Some recent developments (since 1980)

Causal quantum theories fall into two main categories. One class follows de Broglie's pilot-wave idea, in which there are particles and a new physical field of force associated with the wave function $\psi(x, t)$.[124] Even though there are variations in this set, for my purposes here I take Bohm's (1952) *completely deterministic* quantum-potential version as representative of that approach.[125] The other broad class of theories consists of the *inherently indeterministic* hydrodynamical models, which are by and large the lineal descendents of Madelung's 1926 work in which $|\psi|^2$ represents the density of a fluid, $v = \nabla S/m$ is the stream velocity of the fluid, and inhomogeneities in the fluid exhibit particle-like behavior. As we have seen, Bohm and Vigier proposed such a model in 1954. Related to this hydrodynamical formalism are the stochastic mechanics of Nelson and, more recently, the stochastic interpretation of Bohm and Hiley.[126] In this section I address a basic problem for which both approaches attempt to provide a solution and I examine some of the recent work on Bohm's mechanics.

9.4.1 Quantum equilibrium via mixing

As we saw in section 4.1.2, Bohm's quantum-potential formulation and standard quantum mechanics are empirically equivalent only as long as

$$P = |\psi|^2. \tag{9.1}$$

In 1953 Bohm examined a few specific cases to show that, even if initially $P \neq |\psi|^2$, then random interactions with other systems would drive P to its equilibrium value, $P \rightarrow |\psi|^2$.[127] Once eq. (9.1) is satisfied, then the continuity equation for P and Schrödinger's equation for ψ guarantee that eq. (9.1) will continue to be satisfied.[128] So, it is reasonable to ask under what conditions a system will be driven to this equilibrium distribution. Bohm and Vigier offered a hydrodynamical model to underpin Bohm's quantum-potential approach, and they argued that random fluctuations would drive the fluid to an equilibrium distribution *in the mean*, but there would still exist quantum fluctuations about this mean.[129] Their model is not *completely* empirically equivalent to the standard formulation of quantum mechanics—there could be small, in-principle detectable differences. This is unlike Bohm's original quantum-potential approach and his argument for an equilibrium distribution for P.[130]

Recently, Antony Valentini has given an insightful discussion of the relation among the equilibrium distribution for P, the uncertainty relations, and the possibility of superluminal signaling, all within the framework of Bohm's 1952 theory.[131] There are no inherently indeterministic subquantum fluid fluctuations, but rather random subquantum interactions that drive the system to equilibrium. This is conceptually very similar to the way we typically envision the molecules in a large sample of gas being driven toward the Maxwell-Boltzmann equilibrium distribution through random interactions among the gas molecules themselves (rather than via interaction with some "background fluid" in which they find themselves). Even if all of the gas molecules in a room were initially squeezed into a small volume in one corner of the room, they would quickly (once released) diffuse throughout the room and reach an equilibrium distribution.

In the present case, the equilibrium distribution is characterized by eq. (9.1) for an ensemble. Valentini shows that the impossibility of instantaneous signaling ("signal locality"), the Heisenberg uncertainty principle and a statistical law of subquantum entropy increase are all related and turn on the equilibrium-distribution condition of eq. (9.1).[132] On this view, the world would be fundamentally nonlocal in its structure, yet possess signal locality as a *contingent* fact once equilibrium has been reached. Valentini's arguments build on Bohm's earlier ones and on Bohm's original insight that the uncertainty relations obtain *only* when eq. (9.1) is satisfied.[133] He also proves that once the wave function Ψ for a "universe" (which contains an ensemble of identical subsystems) satisfies the condition of quantum equilibrium (eq. [9.1]), then the wave function ψ for any individual subsystem drawn from the larger system will also satisfy this equilibrium condition *for measured values*.[134] Once this universe is itself in quantum equilibrium, then, for the ensemble of subsystems that eventually have a wave function ψ, the probability P of outcomes of this ensemble of observations must be given by eq. (9.1).

One can understand intuitively how "mixing" in configuration space produces a result like eq. (9.1) *on a coarse-grained level*. If $\wp$ is the initial distribution (for an ensemble of identical systems) defined over a multidimensional configuration space (whose coordinates are denoted collectively by X) and Ψ is the wave function defined on this space, then a function f is defined as the ratio of $\wp$ to $|\Psi|^2$.[135] Along any given system trajectory in configuration space, this f can be shown to have some specific *constant* value. (One may think of each trajectory as having its own particular constant associated with it.) These various (*different*) constants are fixed and need not (in fact, cannot) approach some *common* constant. However, for a sufficiently complicated system, the values of f (for various trajectories) will, as these trajectories time evolve and "mix," become randomly distributed over all accessible regions of configuration space. A collection of the threadlike fibers (or system trajectories) in any one cell will then yield about the same average value $\bar{f}$ (which can always be taken as unity by normalizing $\wp$ and $|\Psi|^2$) as for any other cell. Both $\wp$ and $|\Psi|^2$ are governed by a continuity equation and the dynamics, via the $\dot{X}$ of the guidance condition, determines the time evolution of these two quantities.[136] In general terms, one can say that the wave function Ψ generates the velocity field that, in turn through the continuity equation for $\wp$, drives $\wp$ to quantum equilibrium. In time, the two distributions become thoroughly "mixed" by interactions. Of course, there remain about such approach-to-equilibrium arguments deep questions that lie at the very heart of statistical mechanics. Valentini's work does not propose new resolutions of these issues. However, his mixing and H-theorem arguments (cf. appendix 2 of this chapter) do reduce an understanding of the fundamental quantum-equilibrium condition to the same status—no better, no worse—that equilibrium distributions have in classical statistical mechanics.

Considering in detail how one extracts a specific measured result q for a given particle (say "1") from the entire system represented by Ψ, Valentini demonstrates explicitly that $\dot{X}_1$ is determined *solely* by the term $\psi_q(X_1, t)$ corresponding to the measured value of q and that the density distribution $P(X_1, t)$ is given by $|\psi_q(X_1, t)|^2$. [Just as in Bohm's description of the measurement process given in section 4.2.1, once a specific result has been obtained in a measurement, only that term of the total superposition wave function containing *that* $\psi_q(X_1, t)$ continues to have any influence on the system. This is the *effective* "collapse" of the wave function.] The particle, or subsystem, we have singled out (by measurement or state preparation) effectively has the wave function $\psi_q(X_1, t)$ and the probability density $P = |\psi_q(X_1, t)|^2$, so that eq. (9.1) is satisfied. Therefore, once the probability distribution $\wp(X, t)$ for the "universe" has been driven to its equilibrium state, *any* subsystem will *necessarily* satisfy eq. (9.1). This "nesting" property is an important feature of Bohm's theory if that theory is to be consistent.

Finally, Valentini shows with specific calculations that, if we have two interacting systems, whose overall state is an *entangled* one, then a change on one subsystem (accomplished, say, by modifying the Hamiltonian for that subsystem) can produce an instantaneous change in the probability distribution P for the other system *if and only if* eq. (9.1) is *not* satisfied for the interacting system.[137] He establishes (again) a no-signaling theorem when $P = |\psi|^2$ (as well as, of course, when the state for the system factors—i.e., no entanglement), but illustrates how to signal *instantaneously* for an entangled state when $P \neq |\psi|^2$. He is also able to relate eq. (9.1) to the Heisenberg uncertainty principle. Bohm previously showed that eq. (9.1) leads to that principle, while Valentini exhibits a counterexample when eq. (9.1) fails.[138]

Valentini sees a scrambling of holistic information that may initially be present so that, eventually, it is no longer detectable at the coarse-grained level. As a complicated system approaches quantum equilibrium, "the possibility of instantaneous signalling fades away, and statistical uncertainty takes over [so that] the nonlocal connections of quantum theory may no longer be used for practical signalling [and must] be deduced indirectly, via Bell-type theorems."[139] The quantum-equilibrium distribution "has special features which are not fundamental to the underlying theory [such as] the symmetry of Lorentz covariance."[140] In this picture, the analogue of Boltzmann's heat death of the universe has actually occurred.[141] Notice that a single-variable system not in quantum equilibrium can *never* (left to itself) relax to quantum equilibrium. But any single-variable system is extracted from a larger system. If that larger system is in quantum equilibrium, then the extracted subsystem will also *necessarily* be in quantum equilibrium and no violent relaxation mechanism is required.[142]

Here we have a realistic ontology of actually existing particles and events in which several remarkable features of our world—the uncertainty relations, no instantaneous signalling, Lorentz covariance—all are given a unified and essentially understandable explanation. It would seem unreasonable to hold any of this *against* a causal interpretation.

9.4.2 *A new program of Bohmian mechanics*
A different approach to understanding the equilibrium condition of eq. (9.1) has been provided by Detlef Dürr, Sheldon Goldstein, and Nino Zanghi in a series of papers on Bohmian mechanics.[143] In a review article, Goldstein initially based his discussion on Nelson's stochastic-mechanics formalism.[144] While Nelson sought to underpin his mechanics with an underlying classical theory, Goldstein suggested taking stochastic mechanics itself as *the* fundamental theory of phenomena. This stochastic mechanics was based on inherent randomness in nature and was conceptually different from Bohm's completely deterministic mechanics. Goldstein recognized the justification of eq. (9.1) as a central problem for stochastic mechanics.[145] Aside from indicating how this natural extension

166 *Chapter Nine*

of classical mechanics to a world with intrinsic randomness accounts for (i.e., dissolves) the measurement problem, Goldstein also argued that a modified topology of the configuration space can produce the Aharonov-Bohm effect and can induce quantum statistics as well.[146] He pointed out that the Copenhagen reasoning, according to which it is *in principle* meaningless to posit a detailed description of what occurs on a microscopic level between observations "must be fallacious; it claims to establish the impossibility of just the sort of detailed description provided, in a very natural way, by stochastic mechanics."[147]

In subsequent work, Dürr, Goldstein, and Zanghi began directly with Bohm's deterministic mechanics as their basic formalism in which the *state* of a system is specified by its coordinates $(q_1, \ldots, q_N)$ and its wave function $\psi(q_1, \ldots, q_N)$.[148] They see "Bohmian mechanics" as radically non-Newtonian, but a priori just as "natural" as a theory of particle motion, as is ordinary classical Newtonian mechanics.[149] For them, the quantum potential appears as an artificial means to recast this essentially non-classical mechanics into a superficially Newtonian form. Dürr, Goldstein, and Zanghi are able to use the assumption that, in Bohmian mechanics, a complete state description consists of (q, ψ) (rather than of ψ *alone*), full Galilean invariance, and simplicity arguments for the functional $v_k^\psi \equiv v_k^\psi(q_1, \ldots, q_N)$ to arrive at the form[150]

$$\frac{dq_k}{dt} = v_k^\psi(q_1, \ldots, q_N) = \frac{\hbar}{m} Im\left(\frac{\nabla_k \psi}{\psi}\right). \qquad (9.2)$$

This is just the guidance condition of eq. (4.2).[151]

In an extensive analysis they addressed the question of why eq. (9.1) should hold for a system.[152] Their perspective is very different from Valentini's mixing approach to an equilibrium distribution.[153] Like Valentini, they begin with the wave function Ψ of the universe. In their view of Bohmian mechanics, for conceptual coherence one *must* start at that level and then *deduce* the dynamical laws governing subsystems. They ask what *meaning* $\wp$ would have as a probability for the entire *universe* at some initial time.[154] We have only *one* actual universe, not an *ensemble* of universes. In a sense (which I elaborate on below) they assume the equivalent of eq. (9.1) initially for the entire universe and then find eq. (9.1) to be also virtually necessary for any extracted subsystem, while Valentini uses a "mixing" proof to justify eq. (9.1) for an ensemble of identical systems, from which he then gets to eq. (9.1) for a subsystem.[155] Valentini uses a quantum equilibrium ensemble, while Dürr, Goldstein, and Zanghi make direct contact with an individual universe that has the properties typical for the quantum equilibrium ensemble.[156] They show that, for a large number (*M*) of identical subsystems, each having a wave function ψ, the *empirical* distributions of the configurations will be well approximated by eq. (9.1), which they refer to as Born's law.[157] In other words, eq. (9.1) will obtain for *every* ψ for the *overwhelming* majority of initial choices for q.[158] They then extend their analysis to show that Born's statistical law holds for experiments performed on one or more subsystems at different times. A result is that (for "global" quantum equilibrium) in Bohmian mechanics it is *in principle* impossible to know the configuration of a system with more precision than specified by eq. (9.1).[159]

A sketch of their argument is this. Much as in Valentini's discussion, an *effective* wave function ψ(*x*) for a subsystem is "extracted" from Ψ as[160]

$$\Psi(x, y) = \psi(x) \Phi(y) + \Psi^{\perp}(x, y) . \tag{9.3}$$

Here *x* stands for the coordinates of the subsystem of interest, *y* for the remaining coordinates of the rest of the universe. This $\Psi^{\perp}$, the remainder of Ψ, must have (essentially) no overlap *in the y-coordinates* with Φ(*y*). The actual values of the configuration variables *Y* then lie in the support of Φ. This is typically what happens in a "measurement" (cf. appendix 1.2 to chapter 4) so that the total wave function (here, Ψ) effectively factors into one part, ψ(*x*), for the subsystem and a remainder, Φ(*y*), for the rest. Just as for Valentini, this extraction, measurement, or preparation of the subsystem is a special type of "intervention" and ψ(*x*) represents the subsystem immediately thereafter.

Let the state (*Q*, Ψ) be governed by Bohmian dynamics as[161]

$$\frac{dQ}{dt} = \sum_{k=1}^{N} \frac{\hbar}{m_k} Im\left(\frac{\nabla_k \Psi}{\Psi}\right)\bigg|_{q=Q} , \tag{9.4a}$$

$$i\hbar \frac{\partial \Psi}{\partial t} = -\sum_{k=1}^{N} \frac{\hbar^2}{2m_k} \nabla_k^2 \Psi + V(q)\Psi , \tag{9.4b}$$

with the quantum equilibrium condition

$$\wp = |\Psi(Q)|^2 . \tag{9.4c}$$

To make sense of eq. (9.4c), suppose *provisionally* that there *were* an ensemble of universes all having the same initial wave function Ψ_0 but different initial configurations *Q*. These initial configurations are assumed to be distributed according to $\wp_0 = |\Psi_0|^2$ (for this *ensemble* of universes). That is, *Q* is assumed to have a random distribution given by $\wp_0(Q)$. Since this quantum equilibrium distribution is maintained at any future time t (a property termed *equivariance*), it follows that $\wp(Q_t) = |\Psi_t(Q_t)|^2$ so that Q_t is also distributed randomly.[162] We now make the division $Q_t = (X_t, Y_t)$ into subsystem (X_t) and the environment consisting of the rest of the universe (Y_t). For a *fixed* Y_t (i.e., for given "environment"), the (conditional) probability of X_t having a set of values x is[163]

$$P(x|Y_t) = |\Psi_t(Q_t)|^2 \equiv |\psi_t(x)|^2 . \quad (9.5)$$

If at some time t the x-system consists of M *identical* subsystems (each having the identical wave function ψ) with configurations $x_1, x_2, \ldots, x_M$ and wave function

$$\psi_t(x) = \psi(x_1) \psi(x_2) \cdots \psi(x_M) , \quad (9.6)$$

then[164]

$$P(x|Y_t) = |\psi(x_1)|^2 |\psi(x_2)|^2 \cdots |\psi(x_M)|^2 . \quad (9.7)$$

This means that each of the x_j ($j = 1, 2, \ldots, M$) are also independent random variables distributed according to $P(x) = |\psi(x)|^2$.

Of course, all of this is a fiction so far because we do not actually have an *ensemble* of universes (which would be necessary to give $\wp(Q)$ a coherent meaning as a *probability*), but only *one* actual universe from which we may extract samples. Dürr, Goldstein, and Zanghi argue that, if M is sufficiently large, then by the law of large numbers the *observed* or *empirical* distribution of the xs must be close to the distribution $P(x) = |\psi(x)|^2$.[165] "In other words, *for typical initial configurations of the universe* (consistent with the environment of the x-system at time t) *the empirical distribution is given by Born's law*."[166] The basic idea is that for a typical initial condition of a universe, Born's law for probabilities holds. If one is willing to accept that our universe "must," with overwhelming likelihood, have begun in a "typical" initial configuration, then the ensemble of universes can be dispensed with. The crucial question here is how convincing or applicable one finds probability arguments applied to a singular event (i.e., the actual and only initial configuration of our universe).

The result of this line of argument is that, if eqs. (9.4) hold for the universe, then any subsystem whose state is (X, ψ), where ψ is the effective wave function given by eq. (9.3), will be governed by the equations

$$\frac{dX}{dt} = \sum_{j=1}^{n} \frac{\hbar}{m_j} Im\left(\frac{\nabla_j \psi}{\psi}\right)\bigg|_{x=X} , \quad (9.8a)$$

$$i\hbar \frac{\partial \psi}{\partial t} = -\sum_{j=1}^{n} \frac{\hbar^2}{2m_j} \nabla_j^2 \psi + V(x)\psi \quad (9.8b)$$

and will have[167]

$$P = |\psi(X)|^2 . \tag{9.8c}$$

With overwhelming probability, eq. (9.8c) will well approximate the empirical distributions of actual configurations (X).[168]

In broad terms, both "derivations" of eq. (9.1) are based on probability arguments and involve the wave function of the universe. Central to each is the important distinction (also made long ago by Bohm) of the differences, in general, between the probabilities (or distributions) predicted by quantum mechanics for results *after* measurement and the distributions for quantities *before* measurement.

Dürr, Goldstein, and Zanghi feel that Bohmian mechanics has several desirable features lacking in standard quantum mechanics:

> Quantum randomness can best be understood as arising from ordinary "classical" uncertainty—about what is *there* but *unknown*. The denial of the existence of this unknowable—or only partially knowable—reality leads to ambiguity, incoherence, confusion, and endless controversy. What does it gain us? . . . [169]
>
> What does the incorporation of actual configurations buy us? A great deal! It accounts for:
>
> (1) randomness
> (2) absolute uncertainty
> (3) the meaning of the wave function of a (sub)system
> (4) collapse of the wave packet
> (5) coherent—indeed, familiar—(macroscopic) reality[170]

Bohmian mechanics leads to quantum chaos and to the usual operator formalism of quantum mechanics as a phenomenological description of measurement.[171]

Appendix 1 Nelson's equations

The variation in the position $x(t)$ of a particle subject to Brownian motion is given as

$$dx(t) = (v + u)dt + dw , \tag{9.9}$$

where $w(t)$ represents the random effects of the fluid on the particle (the Brownian motion). The *kinematical* equations relating density ρ, the osmotic velocity u, the current velocity v, and the mean acceleration a of the particle are the usual continuity equation

$$\frac{\partial \rho}{\partial t} + \nabla \cdot (\rho v) = 0 \qquad (9.10)$$

and the pair

$$\frac{\partial u}{\partial t} = -\nu \nabla(\nabla \cdot v) - \nabla(v \cdot u), \qquad (9.11)$$

$$\frac{\partial v}{\partial t} = a - (v \cdot \nabla)v + (u \cdot \nabla)u + \nu \nabla^2 u, \qquad (9.12)$$

where ν is the diffusion coefficient.[172] Incidentally, this is essentially the kinematics used by Einstein to describe Brownian motion.[173]

Since macroscopic bodies (i.e., those with large mass) do not exhibit Brownian motion, Nelson took the diffusion coefficient ν to be inversely proportional to m. Later identification showed that

$$\nu = \frac{\hbar}{2m}. \qquad (9.13)$$

The *dynamical* content of Nelson's theory is obtained by setting the mean acceleration a of the particle equal to that in Newton's second law of motion as

$$ma = F = -\nabla V. \qquad (9.14)$$

If we make the identifications

$$u = \frac{\hbar}{m}\frac{\nabla R}{R} = \frac{\hbar}{2m}\nabla \ln \rho, \qquad (9.15a)$$

$$v = \frac{1}{m}\nabla S, \qquad (9.15b)$$

$$\rho = R^2 = |\psi|^2, \qquad (9.15c)$$

where, as usual

$$\psi = Re^{iS/\hbar},$$

then eqs. (9.10) through (9.12) become equivalent to[174]

$$i\hbar\frac{\partial \psi}{\partial t} = -\frac{\hbar^2}{2m}\nabla^2 \psi + V\psi. \qquad (9.16)$$

Appendix 2 Valentini's H-theorem

Let Ψ be the wave function for a large system (which contains N particles) and $\wp$ to be the initial probability distribution, assumed not to satisfy the quantum equilibrium condition. Valentini asks how and in what sense the relation

$$\wp = |\Psi|^2 \qquad (9.17)$$

can come about over time. Below I sketch his arguments.

The wave function $\Psi = \Psi(X_1, \ldots, X_N, t)$ is denoted by $\Psi(X, t)$ and the corresponding probability as $\wp = \wp(X, t)$. This Ψ satisfies a Schrödinger equation and the velocities of the particles are, as usual, given as

$$m\dot{X}_j = \frac{\partial S(X, t)}{\partial X_j}, j = 1, 2, 3, \ldots, N, \qquad (9.18)$$

where, as before, S is the phase of Ψ. Valentini considers an ensemble of ("identical") systems, each with the same Ψ, but not necessarily the same X. Such an ensemble of systems is required in order that $\wp$ have meaning as a probability distribution (in configuration space). Initially, $\wp$ can be an arbitrary probability distribution on this configuration space so that Valentini, following Bohm's earlier work, writes (as a *definition* for f)

$$\wp(X, t) = |\Psi(X, t)|^2 f(X, t) . \qquad (9.19)$$

Since $\wp$ is a probability function, it satisfies a continuity equation

$$\frac{\partial \wp}{\partial t} + \nabla \cdot (\wp \dot{X}) = 0 , \qquad (9.20)$$

as does Ψ (as follows directly from the Schrödinger equation)

$$\frac{\partial |\Psi|^2}{\partial t} + \nabla \cdot (|\Psi|^2 \dot{X}) = 0 . \qquad (9.21)$$

Since $\dot{X}$ is given by eq. (9.18), Ψ actually determines the time evolution of $\wp$ via eq. (9.20). Directly from eqs. (9.19) through (9.21) we can see that f satisfies

$$\frac{df}{dt} \equiv \frac{\partial f}{\partial t} + \dot{X} \cdot \nabla f = 0 . \qquad (9.22)$$

Here df/dt is the change along a system trajectory in configuration space. In analogy with the quantity H (essentially the entropy) introduced by Boltzmann, Valentini defines[175]

$$H = \int d\Sigma \, |\Psi|^2 \, f \ln f. \tag{9.23}$$

It follows at once, by virtue of eqs. (9.21) and (9.22), that

$$\frac{dH}{dt} = -\int d\Sigma \, \nabla \cdot (\dot{X} \, |\Psi|^2 \, f \ln f) = 0, \tag{9.24}$$

where the last equality is obtained by using Gauss's theorem relating volume and surface integrals. The *exact* H remains constant (just as in the corresponding classical case).

Let us (again in accord with classical statistical mechanics) coarse-grain over small volumes or cells in configuration space to obtain $\overline{\wp}$, $\overline{|\Psi|^2}$, $\tilde{f}$, and an $\overline{H}$. Then one can demonstrate, Valentini argues, that

$$\frac{d\overline{H}}{dt} \leq 0, \tag{9.25}$$

just as for the famous Boltzmann H-theorem.[176] Valentini's demonstration is then completed with the observation that the necessary and sufficient condition for $\overline{H}$ to have its minimum value is[177]

$$\overline{\wp} = \overline{|\Psi|^2}. \tag{9.26}$$

That is, the coarse-grained probability distribution $\overline{\wp}$ must approach the equilibrium limit of $\overline{|\Psi|^2}$.

Then Valentini establishes a "nesting" property (my terminology, not his) for the quantum equilibrium of subsystems. Once a large system (which contains N particles), with the wave function Ψ, has been driven to (or is simply in) its equilibrium distribution, then the wave function ψ for any individual subsystem drawn from the larger system will also satisfy eq. (9.1) for the distribution of its measured values. It makes sense to speak of the wave function of a subsystem only when, in the notation of appendix 1.2 to chapter 4

$$\Psi(x_1, \ldots, x_N) = \psi_{q_{\text{meas}}}(x) \, \Phi(y) + \Psi^{\perp}(x, y), \tag{9.27}$$

where x refers to the variables of the subsystem (just x_1 in the present case) and y to those of the rest of the entire system (including the measuring apparatus). Let the variables corresponding to the measuring apparatus be $x_2, \ldots, x_M$, with $M < N$. The wave function for the eigenstate of the measured (micro)system is $\psi_{q_{\text{meas}}}(x_1, t)$ and corresponds to some observed value A_q actually read off the measuring instrument (from which we infer the value of q for the observable whose operator is Q). One then defines a coarse-grained probability density $\overline{P}(X_1, t)$ in terms of the $\wp$ for the larger system (which is assumed to be in quantum equilibrium). A direct calculation then leads to the desired result, once it is appreciated that all terms in the resulting expressions must be evaluated at $A_q = A_q^{\text{meas}}$ (just as at the end of appendix 1.2 to chapter 4). That is, this restriction of A confines the values of $(X_2, \ldots, X_M)$ in Y to a domain for which the second term in eq. (9.27) vanishes.

TEN
An Alternative Scenario?

We have now examined the causal quantum theory program and found that there is nothing incoherent or logically inconsistent about it. Since Bohm's theory is observationally equivalent to the standard version of quantum mechanics, its empirical adequacy should not be in question. This alternative formulation is arguably more coherent and understandable than the commonly accepted dogma. That is the situation *today*. I have already indicated, and argue further in the present chapter, that much of the conceptual basis for such an alternative theory already existed, and was relatively familiar, in 1927. My contention has been that historical contingency was a crucial factor in the spread of the Copenhagen hegemony and in the exclusion of a causal theory. I now ask how plausible and sustainable a different choice might have been.

10.1   What if in 1927 . . . ?

A highly "reconstructed" but entirely plausible bit of history could run as follows (all around 1925–1927). Heisenberg's matrix mechanics and Schrödinger's wave mechanics are formulated and shown to be mathematically equivalent. Study of a classical particle subject to Brownian motion (about which Einstein surely knew something) leads to a "classical" understanding of the already discovered "Schrödinger" equation, which is then given a realistic interpretation. A "Nelson" stochastic mechanics underpins this interpretation with a visualizable model of microphenomena.[1] This would have made evident the possibility of a largely classical foundation for that key equation.[2] A realistic ontology would still remain a live option.

Edward Nelson intimated such a possibility in his own papers on stochastic mechanics. In 1966, he "attempt[ed] to show . . . that the radical departure from classical physics produced by the introduction of quantum mechanics forty years ago was unnecessary."[3] In 1985, he wrote that "had the Schrödinger equation been derived [from stochastic mechanics] before the invention of matrix mechanics, the history of the conceptual foundations of modern physics would have been different."[4] Since stochastic mechanics is quite difficult to handle mathematically, people would likely have tackled the less formidable task of exploring the implications of the (equivalent) *linear* Schrödinger equation. The Dirac transformation theory and an operator formalism would still have been available as a *convenience* for further development of the formalism to provide algorithms for calculation.[5]

The entanglement or nonlocality of that formalism would soon become apparent. Einstein did not like such nonlocality and would have rejected any model with this property. Yet the conceptual background existed, I shall argue, even in 1927 to prove a Bell-type theorem. If that had happened at that time, then Einstein and the rest of the quantum physicists would have perceived in sharp relief the choice between determinism and locality in *any* theory. A causal formulation of quantum theory might then have appeared less unpalatable than the Copenhagen version actually chosen by the scientific community.

Einstein might next have made the transition from stochastic mechanics to "Bohmian" mechanics since, as we have seen in the previous chapter, stochastic mechanics turns out to be both indeterministic and nonlocal.[6] It is well known that Einstein was deeply committed to a realistic worldview in which microentities have a continuous, objective, observer-independent existence. The option of keeping (in-principle) determinism with an objective reality and accepting nonlocality as, say, in Bohm's quantum-potential theory might have been pursued. However, it does not seem worth demanding that my counterfactual scenario *had* to go along some highly specific path. What is important is that there were precedents for such moves and that the necessary pieces were already there. If, early on, such a causal quantum-theory program had been pursued, rather than the Copenhagen one, it would have had the internal resources to cope with the generalizations essential for a broad-based empirical adequacy. We could today have arrived at a *very* different worldview of microphenomena. If someone were then to present the (merely) empirically equally adequate Copenhagen version, with all of its own counterintuitive and mind-boggling aspects, who would listen?

There are other, quite different and plausible alternative historical possibilities as well. One suggestion is that Schrödinger's original program for the interpretation of the quantum theory need not have been abandoned so hastily. Theoretical tastes had much to do with this.[7] But for a largely sociological accident, Schrödinger's essentially classical field theory for quantum phenomena could have been successfully pursued.[8] That is a different story and not one I wish to pursue further here.

10.2 A "Bell" theorem

Let me now consider what impact a "Bell" theorem might have had around 1927. Many have wondered why someone did not come up with a Bell-type inequality or theorem long before 1964.[9] Of course, it is one of the hallmarks of a profound insight that it so changes our own way of thinking about a problem that the solution provided by that insight soon comes to appear inevitable and almost "obvious." By 1927, there were precedents for the type of questions that could lead to a confrontation between determinism and locality. For example, Einstein was concerned about the entanglement of separated systems at least as early as 1925, and by 1927 he was aware of such nonseparability for Schrödinger's wave mechanics.[10] Even earlier, he was aware of the lack of independence of systems obeying quantum statistics and by the mid-1920s appreciated that the same was true for interacting massive particles.

> Thus, when we find Einstein objecting both to the failure of separability in quantum mechanics and to the failure of determinism, we must realize that he is lamenting the fact that quantum mechanics forces us to choose to give up one or the other of these desiderata, both of which are built into the foundations of field theories as Einstein conceived them.[11]

Furthermore, in 1924/25, Walther Bothe and Hans Geiger were doing experiments on photon-electron coincidences in Compton scattering that demonstrated that spatially separated quantum systems do not behave as totally independent systems.[12] Finally, (purely *mathematical*) interest in Bell-like inequalities in probability theory have a long history going back well into the nineteenth century, with one of the Bell inequalities (as a *mathematical* formula only) appearing in the literature in the 1930s.[13]

The proof of a Bell-like inequality that I have in mind here would assume both determinism and a version of locality and could be of the form, say, of the arguments used by Henry Stapp and, in an even simpler form, by Asher Peres.[14] Since Einstein was committed to a causal (or deterministic) fundamental theory, there is no need here for me to enter again into the subtleties and technicalities of the distinctions among various types of nonlocality.[15]

10.2.1 The actual historical setting

Already at the Solvay congress of 1927, Einstein was concerned about the possible conflict between the first-signal principle of special relativity and the "collapse" of the wave function upon detection.[16] Figure 10.1 is an

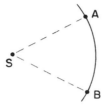

Figure 10.1 Einstein's locality problem

illustration of the type of thought experiment he used. A source S emits (say, one at a time) "particles" in a spherically symmetric pattern and these particles travel toward the (curved) screen AB (e.g., a photographic plate). Quantum mechanics predicts that, prior to detection (e.g., prior to a darkening of *one* spot on the photographic plate), the probability of detecting the particle is uniform over the entire surface of the screen. Once the particle has actually been detected (say, at point A), then the probability is unity there and zero everywhere else. This *instantaneous* collapse (recall eq. [3.6]), Einstein felt, violated (at least the spirit of) relativity and might allow one to circumvent the first-signal principle.

> The interpretation according to which $|\psi|^2$ expresses the probability that *this* particle is situated at a certain place presupposes a very particular mechanism of action at a distance which would prevent the wave continuously distributed in space from acting at *two* places of the screen. In my opinion one can only counter this objection in the way that one does not only describe the process by the Schrödinger wave, but at the same time one localizes the particle during the propagation. I think that de Broglie is right in searching in this direction. If one works exclusively with the Schrödinger waves, [this] interpretation . . . of $|\psi|^2$ in my opinion implies a contradiction with the relativity postulate.[17]

The implication of this last sentence is that the observed results at A and at B are certainly correlated and one might hope to exploit this to transmit information.[18] In his response, Bohr said that he really didn't "understand what precisely is the point which Einstein wants to [make]" and that "the whole foundation for causal space-time description is taken away by quantum theory."[19] In essence, he attempted to make Einstein's question appear ill posed and did not really address the relativity issue. Dirac, in the same general discussion at the congress, made an important distinction between the free choices the experimenters make in setting up an experimental arrangement to interrogate a microsystem and the outcomes of such observations, these latter being chosen by "nature" (his term) and not being subject to any causal evolution.[20] Heisenberg pointed out that these correlations (or the possibility of subsequent interference effects) can persist for any length of time (or, equivalently, over any distance) until one actually makes an observation to obtain a specific result.[21] (Today, we would speak of the entanglement of these quantum states.) These concerns border on the central issue of a no-signaling theorem (i.e., our inability, even in principle, to exploit these long-range influences or correlations to transmit information).[22] This discussion between Einstein and the protagonists of the Copenhagen school was verbal and no quantitative arguments or proofs were given.

In his influential lectures on quantum mechanics, given at the University of Chicago in the spring of 1929, Heisenberg discussed the splitting, by a half-silvered mirror, of the wave packet of a single photon and the subsequent propagation of these packets into far-separated regions of space.[23] If the photon is discovered in the reflected packet, then the other packet must immediately collapse to zero.

> The experiment at the position of the reflected packet thus exerts a kind of action (reduction of the wave packet) at the distant point occupied by the transmitted packet, and one sees that this action is propagated with a velocity greater than that of light. However, it is also obvious that this kind of action can never be utilized for the transmission of signals so that it is not in conflict with the postulates of the theory of relativity.[24]

This is a statement of the content of (but not a proof of) what is today referred to as the no-signaling theorem.

Furthermore, by the mid 1930s at the latest, Schrödinger was aware that no distribution assigning well-defined values to *all* physical variables could give all of the same predictions as quantum mechanics.[25] In a passage from his famous "cat" paper Schrödinger discussed "entangled" states:

> At no moment does there exist an ensemble of classical states of the model that squares with the totality of quantum mechanical statements of this moment. The same can also be said as follows: if I wish to ascribe to the model at each moment a definite (merely not exactly known to me) state, or (which is the same)

to *all* determining parts definite (merely not exactly known to me) numerical values, then there is no supposition as to these numerical values *to be imagined* that would not conflict with some portion of quantum theoretical assertions.[26]

There was already an awareness of the peculiar nature of quantum-mechanical probabilities and correlations.

10.2.2 A counterfactual scenario
The nature of these correlations could have been examined by, say, Einstein early on (as they were later by Einstein, Podolsky, and Rosen in 1935). For example, if apparatuses were placed at A and at B of figure 10.1 to determine correlations for pairs of particles emitted from the source S (but perhaps in the geometry of figure 2.1 for ease of conceptual analysis), then there exists a simple (nonnegative) measurable quantity (given as a linear function of these correlations) that must be bounded from above for *any* local hidden-variables theory.[27] This upper limit of Bell's theorem is violated by the experimentally observed correlations or, equivalently, by those predicted from quantum mechanics.

If such a "Bell" insight had occurred, then Einstein, who did not dispute the empirical adequacy of quantum mechanics, and other quantum physicists could have seen the conflict between determinism and locality in *any* theory. They might have countenanced seriously the possibility that nonlocality may well turn out to be a feature of the *world,* not just a property of some deviant theory or other. Also, given the concerns of Einstein and of Schrödinger and the insights they had (by the late 1920s and early 1930s) into the nature of entangled quantum-mechanical states, a no-signaling theorem could have been established.[28] The statistics garnered at station B by observing some property of particle *b* there carry no information about the type of apparatus chosen or constructed by the experimenter at station A. Not only cannot these quantum correlations be used to signal "faster than light," but they cannot be used to signal *at all.* Such nonlocality is truly, in Abner Shimony's apt phrase, *uncontrollable nonlocality.*[29]

10.3 Einstein and nonlocality

The crucial issue is how, in an observer-independent reality, Einstein would have evaluated or weighted causality (or determinism) versus nonlocality, given that one of these *had* to go.

10.3.1 *Einstein himself*

From an early age, Einstein had a strong commitment to the existence of an objective reality "which exists independently of us human beings and which stands before us like a great, external riddle, at least partially accessible to our inspection and thinking."[30] For him, "physics [was] an attempt conceptually to grasp reality as it is thought independently of its being observed."[31] In these same recollections from 1949, he was quite explicitly aware of the tentative nature of any access to or knowledge about this external world at its most fundamental level: "All our thinking [about the fundamental, unifying concepts we use to organize the phenomena of nature] is of [the] nature of a free play with the concepts; the justification of this play lies in the measure of survey over the experience of the senses which we are able to achieve with its aid."[32] Here Einstein made the Baconian point that the value of an overarching theoretical structure and of the concepts (or conventions) it employs must be judged, ultimately, by its fruits. In his 1933 Herbert Spencer lectures, published as *On the Method of Theoretical Physics,* he expanded on this same theme:

> A complete system of theoretical physics is made up of concepts, fundamental laws which are supposed to be valid for those concepts and conclusions to be reached by logical deduction. It is these conclusions which must correspond with our separate experiences. . . .
>
> The structure of the system is the work of reason; the empirical contents and their mutual relations must find their representation in the conclusions of theory. In the possibility of such a representation lie the sole value and justification of the whole system, and especially of the concepts and fundamental principles which underlie it. Apart from that, these latter are free inventions of the human intellect, which cannot be justified either by the nature of that intellect or in any other fashion *a priori.*[33]

Similarly, in a 1931 essay on Maxwell's concept of reality, he wrote:

> The belief in an external world independent of the perceiving subject is the basis of all natural science. Since, however, sense perception only gives information of this external world or of "physical reality" indirectly, we can only grasp the latter by speculative means. It follows from this that our notions of physical reality can never be final. We must always be ready to change these notions—that is to say, the axiomatic basis of physics—in order to do justice to perceived facts in the most perfect way logically.[34]

This central commitment to judging the concepts in a theoretical framework ultimately on the basis of their usefulness in organizing the phenomena of the world is a question Einstein returned to often.

> A proposition is correct if, within a logical system, it is deduced according to the accepted logical rules. A system has truth-content according to the certainty and completeness of its co-ordination-possibility to the totality of experience. A correct proposition borrows its "truth" from the truth-content of the system to which it belongs.
>
> A remark to the historical development. Hume saw clearly that certain concepts, as for example that of causality, cannot be deduced from the material of experience by logical methods. Kant, thoroughly convinced of the indispensability of certain concepts, took them—just as they are selected—to be the necessary premises of every kind of thinking and differentiated them from concepts of empirical origin. I am convinced, however, that this differentiation is erroneous, i.e., that it does not do justice to the problem in a natural way. All concepts, even those which are closest to experience, are from the point of view of logic freely chosen conventions, just as is the case with the concept of causality, with which this problematic concerned itself in the first instance.[35]

Notice that Einstein distances himself (in 1949) from the acceptance of any particular a priori (Kantian) conditions or concepts for meaningful discourse. The important characteristic here is not that Einstein was not a *positivist* then (even if he ever had been one), but rather that he was, one might say, *pragmatic* enough to countenance whatever conceptual changes might be demanded by the phenomena of nature. That did not require him to be a positivist, since he *did* believe that the central concepts in a theoretical discourse about the nature of physical phenomena are free creations of the human intellect (i.e., conventions), but not capricious conventions—even though our specific choices are not uniquely *demanded* by the phenomena of the world, they must be *consistent* with those phenomena.

Einstein was strongly committed to the field concept as being fundamental.[36] However, even later in life he did acknowledge the important success that a mixture of particles and fields has had in modern physics. He wrote, "There exist two types of conceptual elements, on the one hand, material points with forces at a distance between them, and, on the other hand, the continuous field. It presents an intermediate state in physics without a uniform basis for the entirety, which—although unsatisfactory—is far from having been superseded."[37] Although Einstein might still have found Bohm's theory unacceptable ("too cheap") as a *final* theory (and Bohm himself never suggested it was final theory), he might also not have rejected it out of hand because of the (now benign) nonlocality. Einstein could well have found Bohm's theory less objectionable (once nonlocality had been defanged at the level of phenomena) than standard quantum mechanics.

He took quantum mechanics to be a *statistical* theory only, acknowledged the difficulty of unifying quantum concepts with relativity, and found the loss of any possibility (even in principle) of giving an event-by-event account of microphenomena particularly telling against quantum mechanics.

> I must take a stand with reference to the most successful physical theory of our period, viz., the statistical quantum theory which, about twenty-five years ago, took on a consistent logical form (Schrödinger, Heisenberg, Dirac, Born). This is the only theory at present which permits a unitary grasp of experiences concerning the quantum character of micro-mechanical events. This theory, on the one hand, and the theory of relativity on the other, are both considered correct in a certain sense, although their combination has resisted all efforts up to now.[38]

I shall assume two physicists, A and B, who represent a different conception with reference to the real situation as described by the ψ-function.

A. The individual system (before the measurement) has a definite value of q (i.e., p) for all variables of the system, and more specifically, *that* value which is determined by a measurement of this variable. Proceeding from this conception, he will state: The ψ-function is no exhaustive description of the real situation of the system but an incomplete description; it expresses only what we know on the basis of former measurements concerning the system.

B. The individual system (before the measurement) has no definite value of q (i.e., p). The value of the measurement only arises in cooperation with the unique probability which is given to it in view of the ψ-function only through the act of measurement itself. Proceeding from this conception, he will (or, at least, he may) state: the ψ-function is an exhaustive description of the real situation of the system.[39]

Einstein clearly favored position A and in fact argued (given locality) that the alternative (B) was indefensible. On just this point, though, Bohm offered a completion of quantum mechanics in terms of an objective reality.

10.3.2 Later reconstructions

Arthur Fine, in his *The Shaky Game,* has analyzed extensively Einstein's position on realism and the quantum theory. Fine tells us that the concept of determinism (or causality) was of central importance to Einstein and that he rejected the notion that probability in quantum theory should be accepted as an in-principle and fundamental feature of nature, as opposed to being an interim necessity in the present state of development of the theory.[40] Nonlocality appears to present a prima facie case for a conflict with the first-signal principle of special relativity, but we know that the nonlocality of quantum theory cannot be used for signaling. Today we look upon relativity as a principle theory (to use Einstein's own term). However, the *empirical* basis for this principle is predictive success and a lack of *observational* conflict with its basic hypotheses. Relativistic invariance could turn out to hold only at the observational level, not necessarily at the level of abstract space-time as usually envisioned in special relativity. That is why, had a no-signaling theorem for quantum-mechanical correlations been established, it might reasonably have put to rest Einstein's objections to the nonseparability of the quantum-mechanical formalism.

The point is this. Consider a system S consisting of two subsystems S_1 and S_2 that are spatially separated at some time. Einstein felt that "the real factual situation of the system S_2 is independent of what is done with the system S_1."[41] Einstein worried what it would even mean to do science if this were not the case. But a no-signaling theorem would have shown that relativity could be respected at the practical or observational level and that the nonlocality present in nature was of a "benign" variety. If push had come to shove, it is likely that Einstein would have opted for causality over space-time and, perhaps, to compromise on some of his desiderata for space-time. I quote Fine here:

> Thus, unlike his attitude toward a space/time representation (which he clearly desired but could imagine doing science without) I believe that for Einstein causality was a *sine qua non* for a worthwhile program of realism. I should like to express this by saying that causality and observer-independence were *primary* features of Einstein's realism, whereas a space/time representation was an important but *secondary* feature. . . .[42]

The "round" picture that we now have of Einstein's realism, then, is this. In the center are the linked, primary requirements of observer-independence and causality. Important, but not indispensable, are the secondary requirements of a spatio-temporal representation, which includes separation, and of monism.[43]

However, Don Howard has taken exception to Fine's reading of the relative importance for Einstein, *in the late 1920s,* of the demands of determinism and of locality.[44] Howard characterizes Einstein's primary commitments in terms of determinism (or causality) and realism (or a separable space-time) and claims that "at least near the end of his life [circa 1950], Einstein was willing to compromise on the requirement of determinism."[45] He goes on to state that "during the 1920s, however, Einstein gradually came to realize that, in attempting to explain quantum phenomena, one might be forced to choose between these two commitments, or more specifically, between the independence of interacting systems and the kind of causality embodied in the conservation laws."[46] It is in terms of these characterizations of determinism and realism that Howard argues that Einstein chose realism over determinism.[47]

Relevant, though, for assessing Einstein's position on nonlocality (and hence, for Howard, on realism) in the 1930s is Howard's claim that his own distinction between locality and separability is based upon Einstein's similar (implicit) distinctions as found in essays from the 1930s and 1940s. In his careful reconstruction of Einstein's views, Howard identifies two (logically) distinct principles: the separability principle (two spatially separated systems possess their own separate real states) and the locality principle (all physical effects are propagated with finite, subluminal velocities).[48] When Howard considers the textual evidence for these concepts in Einstein's writings, he first cites a 1935 letter from Einstein to Schrödinger in which a "separation principle" is mentioned according to which the state of one region of space is independent of what happens in another distant region.[49] This, as Howard himself points out, is not *yet* separability, nor is it distinct from (Howard's) locality.[50] To me it seems very much like what I termed Bell locality in section 4.4.1. Howard's somewhat interpretive reading of Einstein leads him to assert that for Einstein separability is more basic than locality. Later passages—from Einstein's "Autobiographical Notes" of 1949 and from his 1948 *Dialectica* article—are also interpreted to support this distinction. I do not deny they *can* be read that way, but I question whether they *must* be.[51] Just as defensible a reading, I claim, is separation = separability = locality.[52] And, in any event, I have already indicated in section 4.4.3 why I do not feel that Howard's distinctions are apposite for Bohm's theory. Separability versus locality may not always be the best framework for a discussion of nonlocality.[53] I take *any* type of instantaneous influence between spatially separated regions to be a form of nonlocality that Einstein would have rejected prior to a no-signaling theorem.

It seems clear that Einstein's primary motivation for realism was that he wanted objectively, actually existing physical objects that were independent of observation. He conceived of these as having to exist in a separable space-time (i.e., one satisfying Howard's separability and locality) because otherwise, it seemed to him, physical objects at a distance from each other would be so "entangled" that there could be no meaningful science.[54] This should have been, ultimately, an *empirical* concern.[55] It is my contention that a "Bell" theorem and a "no-signaling" result would have shown Einstein that, while the physical world is in fact nonlocal, it is nevertheless effectively "separable" enough to allow science as we have known it to function. He could still have had actually existing objects ("realism," as I now use the term) and determinism, once he had modified his views on space-time (without totally abandoning the concept of space-time as physically meaningful). It is defensible to claim that Einstein was most committed to realism (in *my* sense here) and to determinism and was willing to "give" (at least a bit) on locality.[56] So it appears that my counterfactual scenario is not totally unrealistic. For these reasons, I take Einstein's requirements for an acceptable physical theory, in order of importance for him, that it be objectively real, causal, and local.

10.3.3 *An easy choice*

Prior to a Bell result and a no-signaling proof, it would not have been evident that local realism must be a logically untenable position (relative to the observed phenomena). Local realism was a desideratum about which Einstein held strong convictions. What would it have taken for him to modify his stance on this overarching commitment? It is on just this question of the proper role for the speculative element in a theory that Einstein parted ways with Mach.

> The antipathy of these scholars [Ostwald, Mach] towards atomic theory can indubitably be traced back to their positivistic philosophical attitude. This is an interesting example of the fact that even scholars of audacious spirit and fine instinct can be obstructed in the interpretation of facts by philosophical prejudices. The prejudice—which has by no means died out in the meantime—consists in the faith that facts by themselves can and should yield scientific knowledge without free conceptual construction. Such a misconception is possible only because one does

not easily become aware of the free choice of such concepts, which, through verification and long usage, appear to be immediately connected with the empirical material.[57]

I have argued that the disjunction (either deterministic reality or locality) forced by Bell's theorem, plus the no-signaling result to make nonlocality plausibly benign, could reasonably have been an occasion for a man of Einstein's declared views on the conventional nature of basic concepts to reevaluate the status of locality and of his objections to the nonlocal nature of a "de Broglie–Bohm" interpretation of the formalism of quantum mechanics.[58] In 1926 Madelung had the same *equations* Bohm would employ in 1952, but his *interpretation* was very different from Bohm's and did not carry conviction. Given "Bell" and "no-signaling," Bohm's interpretation would certainly have been possible in 1927. The choice would then, early on, have been starkly clear: *either* a realistic, nearly classical worldview based on a theory like Bohm's, with the price of nonlocality, *or* an indeterministic and nonlocal Copenhagen worldview with its truly bizarre ontology and a radical, revolutionary departure from any comprehensible "picture" of physical processes. The causal quantum-theory program could have been off and running.

10.4 Fertility and growth

We have already seen in section 5.3 that "spin" can be accounted for in Bohm's theory. This is quite direct and such a gloss could have been put on spin in 1927. More recent developments in Bohm's theory provide possible new insights into the physics underlying the formalism of quantum mechanics. I now turn to some of these that include a fascinating perspective on quantum statistics and an extension to quantum field theory.

10.4.1 The origin of quantum statistics

It may come as a surprise to some that quantum statistics obtain for *distinguishable* particles provided one assigns to each discrete state an *arbitrary* probability weighting (rather than the usual fixed *equal* probability weighting).[59] Upon averaging over all possible weightings, we obtain for the observed statistical distribution the Bose-Einstein one. If, instead, we limit the state occupation numbers to 0 or 1, then we find the Fermi-Dirac statistics. This procedure of using a random, as opposed to a uniform, distribution is not implausible. "The assumption made here of a uniform random distribution of probability weightings may at first seem arbitrary. However, in the absence of prior knowledge it is arguably more natural (*i.e.*, entails a weaker assumption) to take the probability weights of the states as arbitrary and random than equal."[60] These results have nothing to do with the causal quantum theory program. I mention them only to emphasize that quantum statistics, for which there is much empirical evidence, do not *require* indistinguishable particles, contrary to the received wisdom.

Quantum statistics do, however, follow naturally in a causal interpretation.[61] An immediate connection exists between the many-body quantum potential and the close proximity of particles. If the wave function for a system is antisymmetric, then the "quantum-mechanical" forces will prevent two particles from ever reaching the same point in space, since two or more particles at the same point would require $P = 0$ there (and, by eq. [4.4] or [4.33], $U \to \infty$).[62] If one considers particles that are in principle distinguishable (if only because of their individual past histories along actual trajectories) and that are permanently subject to correlations via the quantum potential, then quantum-potential arguments can be used to *derive* Bose-Einstein (BE) statistics and Fermi-Dirac (FD) statistics.[63] In these causal approaches to quantum mechanics the observed statistics (BE or FD) are *produced* by the quantum potential and these statistics do not themselves constitute a new quantum principle. Arguments show that a random probability weighting is a natural outcome of the causal interpretation via the properties of the quantum potential produced by various symmetries of the wave function.[64]

Treating particle statistics as being produced by interactions is actually very much in the spirit of an observation made in 1925 by Einstein. As has been noted:

> This astonishing result [of obtaining BE statistics for distinguishable particles] can be immediately justified by assuming, following an assumption of Einstein himself [in 1925] that one is not dealing with independent particles . . . but that BE statistics "expresses indirectly a certain hypothesis of a mutual influence, which, for the moment, is of a quite mysterious nature". Of course Einstein himself never developed this suggestion for an obvious reason. For him these "mysterious influences" could not reflect any form of local interaction since these could only lead to MB [Maxwell-Boltzmann] statistics. In his opinion, they would then destroy causality. He did not know then that under certain conditions (satisfied by the many-body quantum potential) actions-at-a-distance do indeed satisfy his own definition of relativistic causality.[65]

Already in 1927, Brillouin appreciated that the assumption of absolute and in-principle indistinguishability was open to question:

188   Chapter Ten

> This identity between the objects is certainly logical in principle; but in practice it corresponds to a process which cannot be physically realized, for all the projectiles are supposed to be thrown at random into the different cells at a given instant. In reality things happen in quite a different way; the atoms of a gas move from one compartment to another as a result of collisions. If they are identical *a priori*, it is easy, nevertheless, to distinguish them by their history.[66]

He then indicated how suitable interactions among distinguishable particles would produce the usual varieties of quantum statistics. Of course, the quantum-potential concept was not available to Brillouin at this time. Nevertheless, this key concept of a causal explanation of quantum statistics had been broached by 1927.

10.4.2   *Relativity and quantum field theory*

A simple illustration of the non-Lorentz invariance of quantum mechanics at the level of individual events is provided by returning to the EPRB experiment of figure 2.1. Let the two electrons be emitted simultaneously from the source and denote their state by Ψ_0.[67] Consider the sequence of events as seen from a Lorentz frame in which the electrons are detected simultaneously at stations A and B.[68] A direct application of the calculational rules of quantum mechanics leads to the joint probabilities of eqs. (2.7) and (2.8).[69] In particular, the joint probability for the detected electrons both having spin up is[70]

$$P_{AB}(++|\theta, \phi) = \frac{1}{2} \sin^2\left(\frac{\theta - \phi}{2}\right). \tag{10.1}$$

Suppose, however, that we choose to observe events from a Lorentz frame in which the detection at A takes place before that at B. Then the single probability for a spin-up detection at A is just[71]

$$P_A(+) = 1/2, \tag{10.2}$$

and the state Ψ_0 collapses to a state Ψ_1 (recall the projection postulate of eq. [3.6]).[72] The conditional probability of then obtaining a spin up at stations B (having already obtained a spin up at station A) is[73]

$$P_{BA}(+|\theta, \phi, +) = \sin^2\left(\frac{\theta - \phi}{2}\right). \tag{10.3}$$

From eqs. (10.2) and (10.3), eq. (10.1) follows identically.[74] If we instead observe these processes from a Lorentz frame in which the electron at B is detected before that at A, then we find

$$P_B(+) = 1/2 \ ,\qquad(10.4)$$

and Ψ_0 collapses to Ψ_2 (which is *different from* the Ψ_1 for the observer in whose Lorentz frame event A came before event B).[75] The conditional probability for a subsequent detection of a spin-up electron at station A is

$$P_{AB}(+|\theta, \phi, +) = \sin^2\left(\frac{\theta - \phi}{2}\right),\qquad(10.5)$$

which again agrees with eq. (10.1).[76]

The point is that, even though the state of the system (Ψ_1 or Ψ_2) between observations as seen by the observers in different Lorentz frames is different, the *predictions* for observations are the same (i.e., Lorentz invariant).[77] At the level of the state vector, all Lorentz frames are *not* equivalent, although all predictions for observables are. If one takes the wave function as being merely a mathematical device for calculation, then he need not see a conflict with Lorentz invariance (i.e., with "relativity," loosely speaking). However, if one takes the wave function as representing a physical reality (as did Bohm, for whom the wave function determines actual particle trajectories through the guidance condition), then all Lorentz frames are not equivalent at the level of individual processes.[78] Such a result is not unexpected when one appreciates that the time evolution of the Schrödinger equation (which is itself not Lorentz covariant) will be different in different Lorentz frames. The dynamics governing the motion of a particle depends nonlocally upon the wave function. If there were instantaneous connections and complete Lorentz covariance (so that there would have to be similar instantaneous connections in other Lorentz frames), then there would exist the possibility of causal anomalies (e.g., killing one's own great-grandfather in the distant past).[79] If there is just *one* preferred frame, then these contradictions do not follow. Although there is a unique (but not experimentally distinguishable) frame in which the nonlocal correlations are instantaneous, the statistical laws are covariant, as are all predictions involving "classical" measuring devices.[80] Since Lorentz covariance is broken only at the level of individual processes, much as instantaneous signaling is possible only at the level of individual particles but not at the level of statistics (cf. section 9.4.1), this invariance may be a symmetry that obtains only in quantum equilibrium.[81]

John Bell has emphasized that whether one takes a "Lorentz" or an "Einstein" view toward the *observational* equivalence of all inertial frames is, finally, a matter of taste.

> Since it is experimentally impossible to say which of two uniformly moving systems is *really* at rest, Einstein declares the no-

tions 'really resting' and 'really moving' as meaningless. For him only the *relative* motion of two or more uniformly moving objects is real. Lorentz, on the other hand, preferred the view that there is indeed a state of *real* rest, defined by the 'aether', even though the laws of physics conspire to prevent us identifying it experimentally. The facts of physics do not oblige us to accept one philosophy rather than the other.[82]

As an aside, I suggest that the existence of a preferred frame could also remove what some see as a problem created by special relativity. "Block-universe" arguments to the effect that relativistic space-time eliminates the possibility of change or becoming have at times been used to conclude that, since quantum mechanics *demands* indeterminism, relativity must be incorrect.[83] One could just as well (or poorly) conclude that this demands a preferred frame and a Bohm-like deterministic theory. Such a move would be somewhat quixotic, since one would be using relativity to argue for determinism and a certain interpretation of quantum mechanics only to conclude, finally, that full Lorentz invariance must be given up.

Let me return to quantum dynamics. Having previously seen the success of the nonrelativistic Bohm theory, both for the Schrödinger equation and for the Pauli equation (to account for "spin"), one naturally asks about its extension to the relativistic domain. A causal interpretation of the relativistic one-particle Klein-Gordon equation is problematic.[84] One reason is that the fourth component of the conserved four-vector current, which one would usually interpret as a probability density, is not positive definite. Also, the four-velocity does not always remain timelike, so causal anomalies can occur because a trajectory can cross the light cone. For these reasons, a direct approach to this one-particle relativistic case may not be workable. As I indicate below, however, boson field theories do have a direct causal formulation and this yields a suitable nonrelativistic limit.[85] A causal interpretation of the Dirac equation does exist and it is free of the difficulties associated with the Klein-Gordon case.[86]

One might claim that, for the standard interpretation of quantum mechanics, turning to the mathematically much more complex area of quantum field theory is unlikely to resolve some of the major foundational problems (e.g., the measurement problem and a coherent classical limit) that already beset the nonrelativistic theory.[87] Even if this is granted (or if one accepts Bohm's interpretation as having resolved those two issues), there do exist new foundational questions that arise, and can only be addressed suitably, in the context of relativistic quantum field theory. Examples of such problems are the relation between quantum theory and relativity and whether a complete and consistent ontology exists for relativistic quantum field theory.[88] This is especially important for the causal quantum theory program.

This causal interpretation can be extended to quantum fields.[89] *All* relativistic quantum field theories, no matter what the interpretation, are beset with mathematical difficulties (such as infinite renormalization "constants" and the like). A cautious attitude would be to treat relativistic quantum field theories as an extremely useful set of rules for making calculations rather than as the basis of a detailed ontology (until a coherent mathematical formulation is available). In any event, causal theories do not suffer from a new class of divergence problems. The modified equations of motion for the quantized fields have their manifest Lorentz invariance broken by the appearance of a super quantum potential term.[90] In the causal interpretation it appeared as though there could be an essential difference between fermions and bosons because a relativistic field-theory extension indicated to some that fermions should be regarded as particles, while bosons must be treated as fields where boson "particles" turn out to be quantized excitations of the field.[91] Already in his 1952 paper Bohm had pointed out that photons cannot be treated as particles.[92] There is no "photon" as a localized object, but the nonlocal structure of the super quantum potential of causal quantum field theory localizes electromagnetic energy upon measurement.[93] While the relativistic boson case goes through nicely enough in the causal interpretation, the corresponding fermion case remained for some time undecided.[94] Recent work indicates, however, that a field ontology may be suitable for both bosons and fermions.[95] In spite of the lack of manifest Lorentz invariance, the *observable* consequences of these theories do remain Lorentz invariant (as is the case for Bohm's causal program extended to the relativistic domain). It has long been known, for example, that the statistical predictions (i.e., the results of measurements) for the quantized electromagnetic field are the same in the causal formulation (with its super quantum potential) as for the standard formulation of quantum electrodynamics.[96] Since the predictions of the latter are known to be covariant, those of the former must necessarily be so too.

10.4.3 *Truly, an alternative*
One could continue a counterfactual line of argument by pointing out that von Neumann's celebrated impossibility "proof" would never have come to be an obstacle to hidden-variables theories since an early version of a Bohm theory would already have contradicted such a claim. It would also have provided an explicit counterexample to the Copenhagen interpretation's claim to completeness and finality. Most important, it would have been an antidote to that dogma's establishing its hegemony. This

"story" is neither ad hoc (in the sense of these causal models having as their sole justification an origin in successful results of a rival program) nor mere fancy, since all of these developments exist in the physics literature.

A decade ago John Bell described his own reaction when in 1952 he read David Bohm's paper and saw the impossible done. After pointing out that de Broglie had already advanced the basic idea in 1927, Bell asked rhetorically:

> But why then had Born not told me of the 'pilot wave'? If only to point out what was wrong with it? Why did von Neumann not consider it? More extraordinarily, why did people go on producing 'impossibility' proofs, after 1952, and as recently as 1978? When even Pauli, Rosenfeld, and Heisenberg, could produce no more devastating criticism of Bohm's version than to brand it as 'metaphysical' and 'ideological'? Why is the pilot wave picture ignored in text books? Should it not be taught, not as the only way, but as an antidote to the prevailing complacency? To show that vagueness, subjectivity, and indeterminism, are not forced on us by experimental facts, but by deliberate theoretical choice?[97]

However, Copenhagen got to the top of the hill first and, to most practicing scientists, there seems to be no point in dislodging it. A pertinent observation in the same vein as Bell's is that of Valentini: "The fact that [Bohm's theory] is not widely accepted as an alternative formulation, as is Feynman's path-integral formulation, appears to be purely a matter of historical accident."[98]

Bohm's perspective may also provide a useful matrix within which to address problems in quantum theory extended to general relativity and cosmology.[99] When standard quantum mechanics is applied to the universe, the status of the wave function becomes problematic. Who or what "observes" the universe to produce the reduction from the superposition of many possibilities initially present in the actual universe in which we do find ourselves? The homogeneity problem for the universe may find a natural explanation in terms of instantaneous interactions occurring prior to the onset of quantum equilibrium.[100] Useful insights into quantum chaos could also follow. A basic difficulty for standard quantum theory is specifying just what quantum chaos *is* because the typical hallmark of chaos is the highly irregular behavior of the *trajectories* of a system. Even in the (nonrelativistic) quantum domain, trajectories still exist for Bohm, but not for Copenhagen. While these are at present no more than tantalizing possibilities for Bohm's program, they should not be held *against* it.

Had some of the realizations sketched above occurred early on, is it likely that Copenhagen would have been the theory of choice? It has *not* been my purpose here to argue in favor of Bohm's interpretation over Bohr's, but rather to question the supposed necessity of the Copenhagen interpretation.[101] After all, *any* candidate for an interpretation of quantum mechanics has problematic aspects.[102] The main unsettling concern that I have raised is this: but for a plausible temporal reordering of certain historically contingent events, our worldview of fundamental microprocesses might well be one of determinism rather than indeterminism—a worldview requiring a less radical departure from then already-ensconced classical principles. This is a particular instance of a larger issue. I claim that asking what might have happened at certain critical junctures of theory construction and selection, and why it did not, is more than just idle speculation fit only for a free Saturday afternoon.[103] This raises epistemological and general philosophical issues about whether our most successful scientific theories are even effectively unique and about the reliability of the knowledge science gives us concerning the structure of our world at the most fundamental level.[104] These issues are the subject of my final chapter.

Appendix 1 A simple derivation of Bell's theorem

The relevant figure for this discussion is figure 2.1 of section 2.3.3. The spin correlation for the outcomes r_A and r_B at each station is given by the *formalism* of quantum mechanics as (in admittedly modern notation)[105]

$$\langle (r_A r_B(\theta, \phi)) \rangle = -\cos(\theta - \phi). \qquad (10.6)$$

The basic definition of the correlation in terms of the actual measurement results $r_{A_k}(\theta, \phi)$ and $r_{B_k}(\theta, \phi)$ at stations A and B, respectively, is

$$\langle r_A r_B(\theta, \phi) \rangle \equiv \frac{1}{N} \sum_{k=1}^{N} r_{A_k}(\theta, \phi) \, r_{B_k}(\theta, \phi). \qquad (10.7)$$

Here k labels the run (or repetition) of the experiment so that $k = 1, 2, 3, \ldots, N$. If we accept determinism (as I have argued that Einstein would have) and assume (in the spirit of relativity) no superluminal influences ("locality"), then we can write[106]

$$r_{A_k}(\theta, \phi) = r_{A_k}(\theta), \, r_{B_k}(\theta, \phi) = r_{B_k}(\phi). \qquad (10.8)$$

Let each experimenter (one at A and the other at B) be able to make two choices (θ_1 and θ_2 at A and ϕ_1 and ϕ_2 at B) for apparatus settings for which the system can respond with $r_A = +1$ or $r_A = -1$ and $r_B = +1$ or $r_B = -1$.[107]

For any *actual* outcome $r_{A_k}(\theta_1)\, r_{B_k}(\phi_1)$ on, say, the kth run, we could contemplate another *possible* experiment in which the choice ϕ_2 (rather than ϕ_1) had been made by experimenter B (while experimenter A still made choice θ_1). The outcome (on that run) would then have been $r_{A_k}(\theta_1)\, r_{B_k}(\phi_2)$, where we would *expect* (because of locality and determinism) that $r_{A_k}(\theta_1)$ would still have the *same* value in this (possible) alternative experiment that it has in the actual one. We do not know, however, the value of $r_{B_k}(\phi_2)$ (i.e., whether it is a $+1$ or -1). Similarly, we could consider other possible experiments with corresponding results (on a run-by-run basis) $r_{A_k}(\theta_2)\, r_{B_k}(\phi_1)$ and $r_{A_k}(\theta_2)\, r_{B_k}(\phi_2)$. We can then ask whether or not there is *any* conceivable set (or collection) of outcomes $\{r_{A_k}(\theta_1)\, r_{B_k}(\phi_1)\}$, $\{r_{A_k}(\theta_1)\, r_{B_k}(\phi_2)\}$, $\{r_{A_k}(\theta_2)\, r_{B_k}(\phi_1)\}$, and $\{r_{A_k}(\theta_2)\, r_{B_k}(\phi_2)\}$ such that the experimental correlations of eq. (10.1) would be satisfied. Realize that *each* $r_{A_k}(\theta)$ and $r_{B_k}(\phi)$ is $+1$ or -1. Now, because of this, observe that with the definition

$$P_k \equiv r_{A_k}(\theta_1)[r_{B_k}(\phi_1) + r_{B_k}(\phi_2)] + r_{A_k}(\theta_2)[r_{B_k}(\phi_1) - r_{B_k}(\phi_2)], \quad (10.9)$$

it is evident, for *any* possible set of the r_{A_k}, r_{B_k}, that

$$|P_k| = 2. \quad (10.10)$$

Therefore, the measurable quantity

$$R \equiv |\langle r_A r_B(\theta_1, \phi_1)\rangle + \langle r_A r_B(\theta_1, \phi_2)\rangle + \langle r_A r_B(\theta_2, \phi_1)\rangle - \langle r_A r_B(\theta_2, \phi_2)\rangle| \quad (10.11)$$

can, with the use of eqs. (10.9) through (10.11), be bounded as

$$R \equiv \left|\frac{1}{N}\sum_{k=1}^{N} P_k\right| \leq \frac{1}{N}\sum_{k=1}^{N}|P_k| = 2. \quad (10.12)$$

That is, *any* deterministic, local theory must predict for R a value such that

$$R \leq 2. \quad (10.13)$$

However, if we accept (and Einstein indicated no desire to do otherwise) the empirical adequacy of the predictions of the *formalism* of quantum mechanics, then we can employ the result of eq. (10.6) for the various correlations $\langle r_A r_B(\theta, \phi) \rangle$ entering into the definition of R. If we now make the choices

$$\theta_1 = \phi_1 = 0°, \theta_2 = -\phi_2 = 60°, \tag{10.14}$$

then eq. (10.11) becomes

$$R_{QM} = |-\cos(0°) - \cos(60°) - \cos(60°) + \cos(120°)| = 2.5. \tag{10.15}$$

Since this contradicts eq. (10.13), no empirically adequate theory can be *both* deterministic *and* local.

Appendix 2   The no-signaling theorem

I can illustrate the no-signaling theorem in terms of the two-particle EPRB arrangement of figure 2.1 of section 2.3.3 and of the state-vector reduction of eq. (3.6) that was discussed in section 3.5.1. In the notation of figure 2.1, I denote by $\mathcal{A}$ an apparatus or measuring device at the station A on the left and by $\{\psi_j\}$ a (complete, orthonormal) set of eigenvectors for "particle" a that travels to the left (i.e., to station A). Similarly, $\{\chi_j\}$ is a set of eigenvectors for "particle" b that travels to station B and O_b is an operator representing some observable to be detected at station B (i.e., on "particle" b). Let the entangled state of the system (after, say, the production and emission of the particle pair from the source in the center of figure 2.1) be given as

$$\Psi = \sum_j c_j \chi_j \psi_j. \tag{10.16}$$

Then an observation at station A on a, yielding a value corresponding to the eigenvalue of, say ψ_k, would (with probability $|c_k|^2$) result in a state (via reduction as in eq. [3.6])[108]

$$\Psi = \sum_j c_j \chi_j \psi_j \rightarrow \chi_k \psi_k. \tag{10.17}$$

The outcome at A has "forced" particle b (*instantaneously,* no matter how great the distance between stations A and B) into the state χ_k, so that one might suspect that this nonlocality could be used for signaling between A and B (and in arbitrarily short time).

This is the type of situation that Einstein might have had in mind in his comments at the 1927 Solvay congress (cf. figure 10.1 in section 10.2.1 above). The hope would be to learn about the choice or type of apparatus $\mathcal{A}$ at station A by making a series of observations (via O_b) on particle b at station B (*i.e.*, to use the statistics gathered at station B to learn about the apparatus $\mathcal{A}$ at A). Just as in the discussion of the measurement process in the appendix to chapter 3, I take H_1 to be an interaction Hamiltonian that couples apparatus $\mathscr{A}$ and particle a during the interaction that produces a measurement at that station. Notice that H_1 does *not* interact in any way with particle b at any time. If $\{\phi_j\}$ is the set of eigenvectors for the apparatus $\mathscr{A}$ and ϕ_0 the state of the apparatus before the interaction begins, then the state Ψ_0 of the entire system (apparatus $\mathscr{A}$, particle a, and particle b) before the interaction is (cf. eq. [3.17])

$$\Psi_0 = \phi_0 \sum_j c_j \chi_j \psi_j , \qquad (10.18)$$

and this becomes, after the interaction between $\mathscr{A}$ and a (cf. eq. [3.27])[109]

$$\Psi_0 \rightarrow \Psi = \sum_j c_j \phi_j \chi_j \psi_j . \qquad (10.19)$$

Now the quantum-mechanical expectation value of O_b is simply (cf. [iv] of section 3.2)

$$\langle O_b \rangle_\Psi \equiv \langle \Psi | O_b | \Psi \rangle = \sum_j |c_j|^2 \langle \chi_j | O_b | \chi_j \rangle . \qquad (10.20)$$

However, this result (or quantum-mechanical prediction) has *no* dependence at all upon $\mathscr{A}$. This means that the statistics observed at station B cannot give us any information about choice of apparatus settings made at station A.

Appendix 3   Scalar causal quantum field theory[110]

As an illustration of how one develops a causal version of quantum field theory, I sketch such a program for the scalar field.[111] The usual Lagrangian formulation for a scalar field $\phi(x, t)$ begins with the Lagrangian L and its related Lagrangian density $\mathscr{L}$[112]

$$L = \int \frac{1}{2}\left[\left(\frac{\partial \phi}{\partial t}\right)^2 - (\nabla \phi)^2\right] dV =$$
$$\int \frac{1}{2}\left[(\dot{\phi})^2 - (\nabla \phi)^2\right] dV \equiv \int \mathscr{L} dV. \qquad (10.21)$$

The definition of the canonical momentum and the Euler-Lagrange equation then yield, respectively[113]

$$\pi(x, t) = \frac{\delta L}{\delta \phi} = \frac{\partial \mathscr{L}}{\partial \dot{\phi}} = \frac{\partial \phi}{\partial t} \qquad (10.22)$$

and

$$\frac{\partial^2 \phi}{\partial t^2} = \nabla^2 \phi . \qquad (10.23)$$

We have simply obtained the usual wave equation for the scalar field $\phi(x, t)$.

In standard quantum field theory, second quantization is implemented by treating the field $\phi(x, t)$ and its canonical momentum $\pi(x, t)$ as operators satisfying canonical commutation relations. In the causal interpretation of quantum field theory, one equivalently accomplishes quantization by introducing a super wave function $\psi(\ldots \phi(x, t) \ldots) = Re^{iS}$ that depends on ϕ at every space-time point.[114] Then $P = |\psi|^2$ is the probability of finding a given field configuration. The momentum operator becomes the functional derivative

$$\pi_{op}(x, t) = -i\frac{\delta}{\delta\phi(x, t)} \tag{10.24}$$

that acts on ψ. The basic ontology is a wave one and, as usual, particles are the excitations of the field. A Schrödinger equation governs ψ as

$$i\frac{\partial \psi}{\partial t} = H\psi, \tag{10.25}$$

where the Hamiltonian operator H is

$$H = \frac{1}{2}\int_{space}\left[-\frac{\delta^2}{(\delta\phi(x,t))^2} + (\nabla\phi(x,t))^2\right]dV. \tag{10.26}$$

The integration is over all of *three*-space only so that $H = H[\phi, t]$. The usual separation of the Schrödinger equation into real and imaginary parts produces two equations,

$$0 = \frac{\partial S}{\partial t} + \frac{1}{2}\int\left[\left(\frac{\delta S}{\delta\phi}\right)^2 + (\nabla\phi)^2\right]dV + Q$$
$$\equiv \frac{\partial S}{\partial t} + H_0 + Q \equiv \frac{\partial S}{\partial t} + \int \mathcal{H}_0 dV + Q \tag{10.27}$$

and

$$\frac{\partial P}{\partial t} + \int \frac{\delta}{\delta\phi}\left[P\frac{\delta S}{\delta\phi}\right]dV = 0, \tag{10.28}$$

where the super quantum potential is

$$Q = -\frac{1}{2}\int\frac{[(\delta^2/\delta\phi^2)R]dV}{R}. \tag{10.29}$$

Equation (10.27) is just the Hamilton-Jacobi equation. The momentum conjugate to the field ϕ is given as (recall eq. [5.49a])

$$\pi(x, t) = \frac{\delta S}{\delta \phi}. \tag{10.30}$$

Hamilton's equations of motion give (cf. eqs. [5.45a])

$$\dot{\phi} = \frac{\delta H}{\delta \pi} = \frac{\partial \mathcal{H}_0}{\partial \pi} = \pi,$$

$$-\dot{\pi} = \frac{\delta H_0}{\delta \phi} + \frac{\delta Q}{\delta \phi} = \frac{\partial \mathcal{H}_0}{\partial \phi} - \nabla \cdot \left(\frac{\partial \mathcal{H}_0}{\partial \nabla \phi}\right) + \frac{\delta Q}{\delta \phi}$$

$$= -\nabla^2 \phi + \frac{\delta Q}{\delta \phi} \tag{10.31b}$$

From these two equations follows the modified wave equation for $\phi(x, t)$

$$\frac{\partial^2 \phi}{\partial t^2} = \nabla^2 \phi - \frac{\delta Q}{\delta \phi}. \tag{10.32}$$

The term $\delta Q/\delta t$ is a functional of ϕ and a function of t (but *not* of *x*). Therefore, eq. (10.32) is both nonlinear in ϕ and noncovariant.

Just as the quantum potential U for the case of particle mechanics is responsible for producing quantum effects (cf. the discussion in section 4.1.1), so the super quantum potential Q of eq. (10.29) effects quantum properties of the field ϕ by modifying the wave equation from eq. (10.23) to eq. (10.32).

ELEVEN
Lessons

Neither internal factors (such as logical consistency and empirical adequacy) *alone* nor external ones (such as metaphysical or psychological commitments and sociological pressures) *alone* have been sufficient to account for the wide acceptance of the Copenhagen worldview in place of a causal one. Nature provides (often tight) constraints, but there still remains latitude in theory choice—in the present case, an interpretation or a worldview. The actual route taken to closure was a complex one with many overlapping factors playing important roles. Science, even in its products or laws, remains historical or contingent in an essential manner. Developments might have gone a very different way at certain critical junctures. Why they did not may be as important as the reasons for the "right" choices that science has made. We are not particularly uncomfortable with a lack of inevitability in other areas of history, but for science such a possibility may strike many as bordering on the sacrilegious, or at least as teetering on the brink of an abyss of rationality skepticism. However, a lack of absolute certainty or inevitability need not result in total arbitrariness in our theories.

11.1 The underdetermination thesis

I begin with some general background on the nature of this element of arbitrariness in our scientific theories and then apply these considerations to the case study I have developed in this book.

11.1.1 A little "ancient" history

The origin of the so-called Duhem-Quine thesis is usually located in Pierre Duhem's *The Aim and Structure of Physical Theory*.[1] Duhem was quite explicit there about what he took to be *the* basis for judging whether or not a given physical theory is acceptable: "The sole purpose of physical theory is to provide a representation and classification of experimental laws; the only test permitting us to judge a physical theory and pronounce it good or bad is the comparison between the consequences of this theory and the experimental laws it has to represent and classify."[2]

When one attempts to move beyond (mere) successful prediction of and accounting for empirical regularities to the level of warranting hypothesis by means of a "crucial experiment" or a decisive disjunction between or among a putatively exhaustive set of alternatives, Duhem argued that we are necessarily doomed to failure since "the physicist is never sure he has exhausted all the imaginable assumptions."[3] If a theory is contradicted by an experimental result, Duhem tells us, it is a conjunction of hypotheses that is refuted. Different scientists may choose to modify different hypotheses. All are equally justified logically, as long as they all save the phenomena.[4]

Not even Duhem claimed that we are left hopelessly adrift in uncertainty. He observed that we use "good sense" to decide which of two or more approaches to theory modification is preferable or acceptable. In the current jargon, I might say that nonevidential criteria are used to eliminate underdetermination in specific cases. While we can, and in fact do, make such moves, there is a problem here because "these reasons of good sense do not impose themselves with the same implacable rigor that the prescriptions of logic do."[5] They can be vague and often there is no common agreement on their value. This turns out to be *the* key issue in the underdetermination debate and I return to it later.

While Duhem took the appropriate unit of appraisal to be (at least) a scientific theory, Willard Quine held that "the unit of empirical significance is the whole of science."[6] Any scientific theory must meet the hard boundary conditions of physical reality, but "there is much latitude of choice as to what statements to reëvaluate in the light of any single contrary experience."[7] As for the role of common sense, or "germaneness" in his terminology, Quine saw "nothing more than loose association reflecting the relative likelihood, in practice, of our choosing one statement rather than another for revision in the event of recalcitrant experience."[8] This is consonant with his picture "of the conceptual scheme of science as a tool, ultimately, for predicting future experience in the light of past experience."[9] For him, not only microentities, but physical objects generally, are introduced into a theory as a conceptual or pragmatic convenience:

> But in point of epistemological footing the physical objects and the gods [of Homer] differ only in degree and not in kind. Both sorts of entities enter our conception only as cultural posits. The myth of physical objects is epistemologically superior to most in that it has proved more efficacious than other myths as

a device for working a manageable structure into the flux of experience. . . . [10]

Each man is given a scientific heritage plus a continuing barrage of sensory stimulation; and the considerations which guide him in warping his scientific heritage to fit his continuing sensory promptings are, where rational, pragmatic.[11]

Two forms of such an underdetermination thesis can be distinguished.[12] Assume that a given hypothesis and a set of auxiliary assumptions together entail some specific prediction. The strong thesis holds that, should the prediction not accord with observation (i.e., with reality), then it is *always* possible to find a new set of auxiliaries that, together with the *same* hypothesis, will successfully account for the observations. The weak thesis claims that, unless there is a *proof* that no saving auxiliaries exist, then disagreement with observation does not conclusively refute the hypothesis under test. With these as alternatives, it seems that Duhem held only the weak version, namely that refutation of a *chosen* hypothesis is not possible.[13] Quine's thesis is essentially the strong one. Duhem believed that one could never be certain of refuting a specified hypothesis, while Quine presumes that saving hypotheses always exist.[14] There would seem to be some ambiguity, or at least some latitude, in just what we take the Duhem-Quine thesis to be. It is fairly clear that the more general claim—some version of the strong thesis—is what philosophers are typically interested in.

So far, my discussion has been about an underdetermination of possible successful modifications of a theory when it encounters a refuting instance. This is not the same as the claim that, if there is one scientific theory that successfully accounts for a body of empirical evidence, then there must *in principle* be another.[15] The two types of underdetermination are, however, closely linked, as Duhem himself indicated.[16] In the context of this case study, I do not differentiate between the Duhem-Quine thesis and theory underdetermination (or simply underdetermination).[17] If one could make a case for radical and universal underdetermination, then there would arguably exist a serious problem for the ability of science to select a theory that would give us a reliable representation of the physical world in terms of hypothesized microentities. However, even if underdetermination could be shown *not* to hold for *every* conceivable theory, it would not necessarily follow that *no* theory would be underdetermined or that such possible cases of individual underdetermination might not have wide-ranging implications for the reliability of science as an enterprise. One might, as some scientists do, take a strongly reductionist view of scientific theories—attempting to explain successful scientific theories in all areas of the physical sciences in terms of a single underlying foundational theory, such as quantum mechanics. Then an essential underdetermination in that foundational theory could raise questions as well for those theories built on it. Even if underdetermination were not guaranteed, it could turn out to hold contingently for our best scientific theory.

11.2.2 The Laudan-Leplin thesis

I can focus on an aspect of the underdetermination thesis that is relevant to my case study by summarizing a recent argument against universal underdetermination. Larry Laudan and Jarrett Leplin separate the question of empirical equivalence from that of underdetermination.[18] They attempt to break the inference from empirical equivalence to underdetermination by allowing other criteria of theory choice than those of pure logic and of empirical evidence. I am less concerned here with assessing the validity of their arguments against the thesis of *general* (i.e., *universal*) empirical equivalence and of *general* underdetermination of *all* scientific theories than with finding space, even if their *general* arguments were convincing, for singular but important cases of empirical equivalence and of underdetermination. In this spirit, just *two* equivalent theories will do nicely.

Laudan and Leplin appeal to their "three familiar theses": (i) variability of the range of observables, (ii) the need for auxiliaries in prediction, and (iii) the instability of auxiliary assumptions. This finally results in the claim that one cannot be certain that empirical equivalence, if it once existed, would be maintained indefinitely. Even if cogent, this line of defense would not imply that such equivalence might not obtain in a specific case. On which side of the question does the burden of evidence and of argument lie?[19] Any appeal to some principle that different ontologies *must* eventually issue in different empirical consequences would seem to be more an act of faith than an argument. In the example already treated at length in this book, we have two theories with the same mathematical structure and rules for calculation. The formalism of a theory may eventually change (i.e., fail)—as has always been and will probably always remain the fate of theories—so that that particular case of underdetermination will fail. However, what guarantee, or even good reason to believe, do we have that the successor theory will not also be similarly underdetermined? Why should we grant Laudan and Leplin's (subsequent) "atemporal thesis" that empirical equivalence must be permanent?[20] And even if this equivalence should be permanent between the two theories I have considered, those theories themselves will eventually fail and be rejected (together). The discussion must then start anew. Even if it is granted that empirical equivalence is contextual, it does not necessarily follow, as Laudan and Leplin claim, that it need be defeasible.[21]

For argument's sake, though, Laudan and Leplin are willing to countenance empirical equivalence but then attempt to block the inference from that to underdetermination. Their line of argument is that, in actual scientific practice, evidential results are often taken as supporting a theory even when those results are not consequences of the theory.[22] That is a statement about what scientists at times do, not an argument for its validity. Such practice, which is based on certain pragmatic choices or on prior commitments, need not give a firm basis for a refutation of the *possibility* of underdetermination—except by fiat that it *shall not* occur. So, assuming that this type of actual underdetermination *may* obtain, how are we now to assess Bohm-Copenhagen as a possible example of this?

11.2 Contingency and scientific theories

Let me begin by addressing what some may feel has been a misappropriation of a term on my part. In the discussion of Copenhagen vis-á-vis Bohm, my use of 'empirical equivalence' has been essentially synonymous with 'observational equivalence.' Philosophers may find it useful to distinguish between these two terms, but I have not, and basically for the following reason.[23] The *Random House Dictionary* defines *empirical* as "derived from or guided by experience or experiment"; "depending upon experience or observation alone . . . "; and "provable or verifiable by experience or experiment." Even though a dictionary of American English is hardly the court of final appeal in a philosophical discussion, use of the term 'empirical', prima facie, might not seem indefensible for what I have had in mind. I really have been interested in observational equivalence. I suggest cashing out the finer distinction 'empirical' may have over 'observational' in terms of different *ontologies,* rather than of different empirical contents, for Copenhagen versus Bohm. Each interpretation makes different *ontological* statements about the world—for instance, the actual location of a pointer before observation—but both make the same *observational* claim. Is this an *empirical* difference? There is an in-principle ontological difference, but de facto practical equivalence.[24] Perhaps this *is* an issue for philosophers to decide, but I do not address it here.

So, we may have two *actual* (not just fancifully concocted for argument's sake) empirically indistinguishable scientific theories that have diametrically opposed ontologies (indeterministic/deterministic laws and nonexistence/existence of particle positions and trajectories).[25] The formalism of quantum mechanics may change (as *all* previous formalisms have eventually), but, should that be required, Copenhagen and Bohm may both have to go. In evaluating the import of my case study, much turns on how much significance one attaches to an *actual*, long-standing case of underdetermination. If the equivalence of these two theories is taken seriously, what are some of the implications that may follow?

11.2.1 *The relevance for scientific realism*

Such underdetermination could represent, in a sense, a double threat to the scientific realist. To begin with, the almost universally accepted Copenhagen interpretation has traditionally been a serious (arguably an insurmountable) challenge to a realistic construal of quantum mechanics.[26] The core of the difficulty is the measurement problem, one entailment of which is usually taken to be that a physical system cannot (*in principle*) possess even a definite (but merely unknown to us) value of the physically observable attribute of position. Taking such a theory seriously as an actual representation of the physical world (at the level of *individual* microentities) requires that we accept a rather bizarre ontology, one that may not even be conceptually coherent. The measurement problem has been around now for nearly seventy years and has resolutely defied any successful, generally accepted solution. This has provided effective ammunition for the antirealist who begins at the level of microphenomena, accepts Copenhagen quantum mechanics as *the* fundamental and exact theory of all physical processes, and then throws down the challenge to the realist to construct a coherent (*realistic*) ontology consistent with the demands of quantum theory.[27] Such an antirealist begins his or her argument in the microrealm, extrapolates to the macrorealm of everyday experience and leaves the ensuing conundrum at the doorstep of the realist. On the other hand, the realist has relatively easy going in the domain of macrophenomena (everyday objects, bacteria, dinosaurs, etc.), but then encounters difficulties in carrying these explanatory resources down to the domain of microphenomena.[28]

At first sight, the Bohm interpretation of quantum mechanics would seem to offer consolation and a potentially powerful means of rebuttal (of antirealism) to the realist. This interpretation, which in its nonrelativistic version represents a microentity as a particle guided by a quantum potential (not as a wave *or* a particle as does Copenhagen), lends itself readily to a realist construal of even fundamental physical processes that develop completely deterministically in a continuous space-time background. There are some highly nonlocal, nonclassical effects present, but this is so for the Copenhagen interpretation too. While this Bohm interpretation is consonant with and even conducive to a realist position, it is empirically indistinguishable from an ontologically incompatible Copenhagen interpretation. So does the realist have grounds for requiring a realistic interpretation other than predilection or fiat? Realism is in double jeopardy here: Copenhagen can be seen as anathema to realism, while Bohm, which provides a consistent realistic interpretation, presents an underdetermination dilemma and thus blocks the realist from achieving the desired goal. That is, *if* one can erect mutually incompatible ontologies on a given formalism, then that may pose a genuine problem for the realist. A proponent of realism would have to prove or argue strongly for a claim that once genuinely different ontologies are proposed it will be possible to extend the formalism along different lines, *because distinct ontologies involve distinct physical magnitudes*. This *may* happen, but *need* it? Once again we have a move that seems more a declaration of belief than an argument for a position.

There is, though, one obvious option open to the realist at this point. He can claim that any two theories that are empirically indistinguishable can (by *definition*) differ only in inessentials. To set the mood, let me recall that Erwin Schrödinger, in his inaugural address at the University of Zurich in late 1922, expressed the belief that it is impossible to decide on the basis of observation whether the world is basically deterministic or indeterministic.[29]

> The demand for an absolute law in the background of the statistical law—a demand which at the present day almost everybody considers imperative—*goes beyond the reach of experience*. Such a dual foundation for the orderly course of events in Nature is in itself improbable. *The burden of proof falls on those who champion absolute causality, and not those who question it*. For a doubtful attitude in this respect is to-day by far the more natural.[30]

In another inaugural address—this time in 1929 before the Prussian Academy of Sciences—Schrödinger claimed:

> In my opinion this question [about the possibility and advisability of an acausal concept of nature] does not involve a decision as to what the real character of a natural happening is, but rather as to whether the one or the other predisposition of mind be the more useful and convenient one with which to approach nature. Henri Poincaré explained that we are free to apply . . . any kind of . . . geometry we like to real space. . . . The same statement probably applies to the postulate of rigid causality. We can hardly imagine any experimental facts which would finally decide whether Nature is absolutely determined or is partially indetermined. The most that can be decided is whether the one

or the other concept leads to the simpler and clearer survey of all the observed facts.[31]

Which description one uses is a purely pragmatic, or possibly in some cases an accidental, matter, dictated by convenience or by contingency. Of course, the scientific realist *could* write off the difference between indeterminism and determinism in the ontology of the world as an insignificance, but that would indeed be strange for one concerned with a reliable and meaningfully complete picture of the world. This radical conceptual difference between inherent indeterminism and absolute determinism also makes it virtually impossible to conceive of a "dictionary" that would map the language of one of these theories onto the other (i.e., map a concept onto its negation).

11.2.2 Actual underdetermination

It is not yet absolute endgame for the realist who wants to have *some* theory that will yield *the* correct and actual ontology of our world. We have seen that empirical adequacy and logical consistency together do not alone provide sufficient criteria to choose between the two theories discussed in this book.[32] One can enlarge these criteria to include factors such as fertility, beauty, coherence, naturalness, and the like. Partly on the basis of such criteria, the scientific community quite early on did make a decisive selection in favor of the Copenhagen interpretation. Much as Einstein claimed in the quotation in section 1.2.2, *one* theory was in fact chosen. An examination of the actual historical record has shown that some key motivating factors, for certain crucial assumptions about the features that an "acceptable" theory must have, were based upon the philosophical predilections of the creators of the Copenhagen version of quantum mechanics and upon highly contingent historical circumstances that could easily have been otherwise.

Either of these two theories passes a test for fertility, in the sense of possessing the internal resources to cope with anomaly and new empirical developments that actually occurred, as well as for suggesting new avenues for research and generalization. My historically counterfactual scenario of chapter 10 has indicated that Bohm's theory was not an ad hoc, stillborn creation that could have matched Copenhagen only for the simplest cases (say, circa the late 1920s and early 1930s), but then would not have been able to grow and mature with its own internal resources and without recourse to ad hoc moves at each turn. I do *not* claim that Bohm's theory and extensions thereof have yet been generated that have matched *all* of the successes of the standard approach or that there are *no* internal inconsistencies yet to be discovered.[33] *Few* people, after all, have worked on this alternative program. Nor am I claiming that more people *should* work on it.[34] Had there been a different majority view, there is little reason to suppose that we would necessarily have arrived at the same set of questions and experiments we are at now.[35]

If the historical record indicates that a rearrangement of highly contingent factors could plausibly have led to a radically different scientific theory and worldview being accepted as correctly and uniquely representing the physical world, then one can reasonably question the value of a philosopher's rational reconstruction to pass judgment on a scientific theory. Philosophers typically study the successful theory already accepted by the scientific community. They doubtless have the ability to reconstruct rationally, and hence to legitimate, *any* theory that has already survived the scrutiny of the scientific community. But for historical contingency, though, they might find themselves doing just as well reconstructing and justifying an *essentially* different, equally successful, and widely accepted theory. What, then, is the value of the exercise, except as a check for noncontradiction? Each of these reconstructions could be equally *rational*, but there would not necessarily be anything rational to choose between them.

One purpose of this book has been to consider an underdetermination thesis, not just as some in-principle or abstract logical possibility, but as a real and practical problem that should be faced in our most successful scientific theory to date—quantum mechanics. If we do not "see" this feature as an obvious fact, it is only because we are not sufficiently sensitive to the essential role played by highly contingent historical factors at certain crucial junctures of theory construction and selection. With a different but equally plausible set of contingent factors, our rationally reconstructed and justified view of the physical world could be in several important features diametrically opposed to what it now is. One is then left with a rather counterintuitive result that I now discuss.

11.3 Determinism and indeterminism—equivalent?

Suppose one grants that either an inherently indeterministic (Copenhagen) theory or an in-principle completely deterministic one (Bohm) is equally adequate empirically, so that only other types of criteria can be decisive in making a choice between the two. We have seen that Schrödinger in 1922 and again in 1929 expressed the opinion that there is no empirical way to decide whether the world is at base deterministic or indeterministic and that which type of theory one finally chooses is a matter of convenience and efficiency of application.[36] A useful and favored theory may be either deterministic or indeterministic in its *structure*, but there is no solid warrant for transferring that fundamental property to the structure of the world itself. Can so fundamental a property of the external world be undecidable by empirical test and can the choice be observationally irrelevant?

11.3.1 Some precedents
This question is not a new one. We can find its roots already (but by no means for the *first* time) in early Newtonian mechanics. The mechanical determinism of the seventeenth century reinforced views of *divine* foreknowledge. An in-principle absolute determinism in the equations of mathematical physics was consonant with the generally accepted belief in an omnipresent, omniscient God. Such a belief *preceded* Newton's formulation of his mathematical laws of motion. Although Newton certainly believed in the existence and action of such a God, he was less certain that the laws of mechanics could of themselves represent the deterministic evolution of a stable universe. In Newton's opinion, as expressed in the queries to his *Optics,* the mechanical universe required the active intervention of God, not just to create and order it, but also to maintain it.

> To what end are Comets, and whence is it that Planets move all one and the same way in Orbs concentrick, while Comets move all manner of ways in Orbs very excentrick; and *what hinders the fix'd Stars from falling upon one another?* . . . [37]
>
> For it became him also who created [all material things] to set them in order. And if he did so, it's unphilosophical to seek for any other Origin of the World, or to pretend that it might arise out of a Chaos by the mere Laws of Nature; though being once form'd, it may continue by the Laws for many Ages. For while Comets move in very excentrick Orbs in all manner of Positions, blind Fate could never make all the Planets move one and the same way in Orbs concentrick, some inconsiderable Irregularities excepted, which may have risen from the mutual Actions of Comets and Planets upon one another, and which will be apt to increase, *till this System wants a Reformation.*[38]

Here, in the words I have set off in italics, Newton touches upon the *stability* of the solar system.

By the latter part of the eighteenth century, the absolute determinism and self-sufficiency of the mechanical universe had become an accepted article of faith for many. This classical determinism was boldly stated by Pierre-Simon Laplace (1749–1827), the great mathematician and theoretical astronomer who had perfected perturbation-theory calculations and used them to argue for the stability of the solar system. In his *Philosophical Essays on Probabilities,* Laplace enunciated an absolute determinism for *all* events in the physical universe, even those insignificant ones of whose true causes we remain ignorant.

> Present events are connected with preceding ones by a tie based upon the evident principle that a thing cannot occur without a cause which produces it. This axiom, known by the name of *the principle of sufficient reason,* extends even to actions which are considered indifferent. . . .
>
> We ought then to regard the present state of the universe as the effect of its anterior state and as the cause of the one which is to follow.[39]

In the introduction to the 1814 edition of his *Analytic Theory of Probability,* he continued with an often-quoted, important, passage:

> If an intelligence, for one given instant, recognizes all the forces which animate Nature, and the respective positions of the things which compose it, and if that intelligence is also sufficiently vast to subject these data to analysis, it will comprehend in one formula the movements of the largest bodies of the universe as well as those of the minutest atom: nothing will be uncertain to it, and the future as well as the past will be present to its vision. The human mind offers in the perfection which it has been able to give to astronomy, a modest example of such an intelligence.[40]

As is well known today, Laplace's confidence in the absolute predictability of the future behavior of a system, given the initial position and velocity, was an overstatement, even within the context of classical mechanics. Familiar and simple mechanical systems (such as a pendulum subject to *small* periodic inputs, or "kicks") can exhibit chaotic behavior. Suppose the initial positions and velocities of each part of such a system have been specified quite accurately and the interaction forces, that are continuous and finite, have been given. Then this system develops according to Newton's laws, but in what is for all practical purposes a nonpredictable fashion. Arbitrarily small variations in the initial conditions of even so simple a system will produce huge (in fact, extremely rapidly increasing) variations in the behavior of the system as time passes. Such solutions had not been expected on general analytical grounds and have been discovered through numerical solutions to the equations of motion. These numerical solutions have been made possible by the advent of high-speed electronic computers. I mention this only to remind the reader that classical mechanics ought not naively to be equated with a determinism that necessarily allows complete and meaningful predictability.[41] The relevant observation is that a *belief* in determinism for mechanical systems was founded upon our ability to predict accurately the precise future behavior of a large class of systems.

Nevertheless, we tend to take classical mechanics as *the* paradigm warrant for a belief in a completely deterministic, clockwork universe. I use as a rough-and-ready definition of 'determinism' the requirement that the present state of the universe (or system) plus the laws of mechanics uniquely determine the future state of the universe. Let me ask what the basis is for this belief in determinism. Prior to Newton's time, there was already a *theological* underpinning in terms of a God seen as a lawgiver. If one accepts a God who runs the universe in an orderly and lawlike fashion, then it makes sense to seek to discover these laws as represented in the workings of his creation. A belief in or an inclination toward the acceptance of a lawlike evolution of the universe predated any specific set of analytical or mathematical laws of physics. Newton was a transitional figure in that, while he definitely did believe in the existence of a God who was responsible for the orderly evolution of the physical universe, he did not believe that the mathematical laws as represented in his *Principia* were in themselves sufficient to explain or to predict the long-term stability and future evolution of the physical universe. It was only after Newton that the determinism and complete predictive accuracy of his laws of mechanics became accepted. A general cultural background had been conducive to a deterministic gloss on the laws of mechanics.

11.3.2 Deterministic chaos[42]

To help settle the question of predictability, one might first ask whether classical systems governed by Newton's second law of motion and his law of universal gravitation evolve deterministically and whether they are stable against small perturbations.[43] This is a *very* difficult mathematical problem since it takes the form of a differential equation

$$\ddot{x} \equiv \frac{d^2x}{dt^2} = f(x) \tag{11.1}$$

for the function $x = x(t)$. Since $f(x)$ can be a *nonlinear* function of x, relatively little can be said without rather stringent restrictions being placed on $f(x)$.

It is well known, however, that nature was kind (or, perhaps, given the ultimate "deception" played on us, cruel). For a $1/r^2$ force this nonlinear equation can be converted into a linear one (by a simple change of dependent variable) and solved exactly to yield conic sections. This much was known to Newton, but it was unclear to him that a many-body system, such as the solar system, would remain stable under the mutual perturbations caused by the planets on each other's orbits about the sun. Laplace is generally credited with having taken care of the perturbations, resolving the issue in favor of the long-term stability of the solar system. The actual historical record is more complicated. Of necessity, Laplace employed perturbation (i.e., approximate) calculations in his analysis and truncated his equations in low orders of small expansion parameters. His first memoir on the subject was published in 1766, and by 1773 he had shown that all terms in his solutions were periodic (i.e., long-term recurrent, as opposed to secular).[44] Lagrange in 1776 and Poisson in 1809 extended these results. By 1876 Newcomb had demonstrated that, for the planetary perturbations, it is possible to represent the solutions by purely *periodic* functions, *provided these series converge*. It appeared that the stability of the solar system was within epsilon of being proved (a mere "technicality"). However, in 1892 Poincaré showed that these series are in general *divergent*.[45] Such results were, as we know today, the seeds of modern deterministic chaos theory.

By the term 'chaos' I mean the complicated, unpredictable, seemingly random behavior of a system. The weather is a paradigm case of this type of time evolution. Such behavior is often contrasted with the orderly, highly predictable motion of the planets and comets in the heavens. And there are common systems that, under different conditions, exhibit now one, then the other, of these two patterns of evolution. A fluid flowing at moderate rates shows a smooth, orderly behavior (termed laminar flow). When the fluid moves rapidly, the flow becomes highly disordered or turbulent. A single billiard ball moving on a smooth square table and suffering only elastic collisions with the rigid sides of the table has a simply predictable behavior. On the other hand, one billiard ball moving among many others (even initially at rest, as in a typical game of billiards) is unstably susceptible to arbitrarily small external influences (such as the variation in the gravitational force due to displacing an *electron* light-years away—just to make my point). After a relatively short time, we have *no* effective predictive power about its whereabouts and velocity on the table, even if we had had *perfect* initial information about its location and velocity.

The traditional view had been that systems consisting of only a few parts (e.g., the planets moving about the sun) were amenable to the precise calculations necessary for meaningful prediction, whereas complex systems, such as a large collection of gas molecules (or the atmosphere) were simply beyond our calculational abilities. It was assumed that, regarding determinism, there was no in-principle difference between these two types of systems. One was just too complicated to handle computationally—a mere practical limitation. This view, or intuition, was based on an examination of the (relatively few) physical problems that we could solve analytically (as in the case of the two-body central-force problem). Such integrable systems are, essentially by definition, the ones that can be treated by the methods of (classical) mathematical analysis. All of the exactly soluble problems of classical mechanics turn out to be *separable*, which means, roughly, that the equations describing their behavior separate into sets of individual one-body problems. This should give us some appreciation of just how *special* and, not surprisingly, *atypical* these cases are. Insights, formed on the basis of these atypical systems, were assumed to be typical of *all* physical systems. (Of course, what else could one reasonably do but form a picture of the physical universe based on a theory that appeared to be enormously successful?) Since such integrable systems turned out to be highly atypical, our intuition, or general picture of the world, was based on a poor induction from too narrow a range of systems. For nearly 300 years we thought we understood classical mechanics, but we didn't. There are many *simple* mechanical systems that exhibit chaotic behavior and for which we have *no* effective predictive power (even though these systems are governed by completely deterministic laws).

We are brought full circle back to the question: What is the warrant for our belief in the completely deterministic evolution of classical systems? Historically, an important element supporting that belief has been the predictive accuracy of the deterministic laws of classical physics. Suppose that the situation had been different and that a proposed set of deterministic laws of classical physics had, in the *overwhelming* majority of the cases to which they were applied, proved incapable of yielding reliable predictions. A defender of "in-principle" determinism could still claim that, at the most fundamental level, the universe is governed by deterministic laws, but that we are simply unable to specify initial data precisely enough to do the requisite calculations that would allow us to "see" this determinism reveal itself. Who would be impressed by such an argument or theory, since its effective "predictions" would be empirically indistinguishable from those that would obtain in a universe that was at base completely indeterministic? Today we do know that "most" classical physical systems do exhibit chaotic behavior. Is the world fundamentally deterministic and is this determinism masked in a few atypical situations (the traditional view), or is it fundamentally indeterministic but exhibiting apparent or effective deterministic behavior in a few atypical situations? Is there any way to decide this issue?

There is both an *epistemological* and an *ontological* dimension to this question. The former refers, in the present context, to what we know about the world and the latter to actual existence claims. Determinism is both a property of the theories or equations of physics and a property of the actually existing world itself. We begin with the observed phenomena of the world (say, astronomical data as embodied in Kepler's three empirical laws) and then construct a theoretical framework, or a system of equations (here, Newton's laws of motion and of gravitation), to account for these phenomena and to make new quantitative predictions. It is in these laws or equations, which *represent* our world, that we discover and explore the property of determinism. Once we have satisfied ourselves of the accuracy and reliability of these mathematical laws, we then are willing to transfer (perhaps unconsciously) the generic features of these laws (or *representations* of our world) to the physical world itself. We go beyond a mere instrumentalist view of the laws of science (in which we would demand *only* empirical adequacy of our laws) and take them realistically as literally true representations of the world. What is our warrant for this transfer?

11.3.3 *The choice—an act of faith*

The subject of deterministic chaos has become so important in recent years because it is now apparent that chaotic classical dynamical systems occur in so many different fields of science. Examples of chaotic behavior can be illustrated in disparate types of physical systems. The generic source of this classical dynamical chaos is the *exponential* separation (in time) of system trajectories (in phase space) so that there is extreme sensitivity to initial conditions, leading to loss of effective predictive ability for the long-term behavior of the system.[46] (This is sometimes referred to as the "butterfly effect," since, fancifully, a butterfly flapping its wings in South America may, eventually, set off a tornado in Indiana.) Such systems exhibit extreme sensitivity to initial conditions so that once-neighboring trajectories in phase space diverge exponentially in time, rendering prediction of future behavior effectively impossible. There are precise, technical mathematical criteria for chaotic behavior, but I do not need these here.

Deterministic chaotic systems can be as irregular (and unpredictable) as a truly random system. For a random sequence, the length of a set of instructions (e.g., a computer program) to compute the sequence is roughly as long as the sequence itself. There is no simple rule that will allow one to compute or predict the nth term in the sequence short of writing out the entire sequence itself. No *finite* algorithm exists for computing a chaotic orbit. That is why the amount of time (or computational effort) increases without limit when we attempt to calculate the future values of the state variables (say, the position and velocity). For a chaotic system there is no faster way to find out how such a system will evolve than to watch the system itself actually evolve—there is no usable "rule" for passing directly from the initial conditions or state to the final one. Still, even though an *individual* system's behavior cannot be predicted, the *average* behavior of a large set of them can often be (with probabilities).

It is for these reasons that belief in determinism as being more fundamental than probabilities in the actual physical universe is an act of *faith,* rather than a position demanded, or even particularly well warranted, by the laws of (even classical) physics and by the observed behavior of physical systems.[47] Usable determinism reigns almost nowhere and chaos nearly everywhere. Newtonian (Laplacian) determinism may remain only a theorist's unattainable dream.[48]

In classical mechanics unwarrantedly general conclusions about a universal determinism were read out of (or, perhaps better, into) the formalism of the theory. In this book, we have encountered the randomness-determinateness conundrum in quantum mechanics. Once again, it seems, science made an overgeneralization. This time it was about the *necessity* of indeterminism and acausality in the Copenhagen version of quantum mechanics. We have seen that Bohm's quantum-potential formulation is completely deterministic and exhibits essentially chaotic behavior. But it and Copenhagen are empirically equivalent. This can be taken as yet another illustration of Schrödinger's claim of the undecidability between determinism or indeterminism at the most fundamental level of physical phenomena.

11.4 So, finally . . .

Relative to the background of classical physics, some largely "commonsense" possibilities for interpretations of quantum mechanics were rejected or simply dismissed as even *conceivably* correct. Important factors in this were some of the (broadly speaking) philosophical commitments of several of the major protagonists in the quantum-theory debates. Still, the positions of Heisenberg, Pauli, Bohr, and Born, say, did not simply *make* the Copenhagen program a logically and empirically viable one. There were logical and empirical grounds for rejecting certain attempts at hidden-variables theories (such as Einstein's in 1927—those are the *easy* cases) but such considerations did not uniquely constrain the choice.

In the end, we may be left with an essential underdetermination in our most fundamental physical theory and this issues in an observational equivalence between indeterminism and determinism in the basic structure of our world. Whether this is of any particular interest or importance is, of course, up to the reader to decide. One possible conclusion is that, as a pragmatic matter, we can simply choose, from among the consistent, empirically adequate theories on offer at any time, that one which allows us best to "understand" the phenomena of nature, while not confusing this practical virtue with any argument for the "truth" or faithfulness of representation of the story thus chosen.[49] Successful theories can prove to be poor guides in providing deep ontological lessons about the nature of physical reality. Such an enterprise would appear to be consistent with Quine's dictum that the world intrudes first as a surface irritation and remains thereafter as a constraint on our imaginations (in constructing scientific theories).[50]

It is unfortunate that David Bohm, whose incisive thinking demonstrated the possibility of so dramatically different a view of the physical world, is no longer with us to savor the renewed interest in his work.

NOTES

Preface

1. Much of the necessary history that I use in this book is already in the published literature of the history and philosophy of science and I simply give the necessary citations at appropriate places in the text. Even though historians of science will find here little that is new to them, I nevertheless hope to avoid, in the words of an infamous review, mere "opportunistic and impertinent compilation" and "so many dustbins for irrelevant detail." I shall have to wait and see.

2. A notable exception to this is David Albert's book (1992), which has (in chapter 7) an evenhanded and clear discussion of the general conceptual features of Bohm's theory.

3. I thank J. B. Kennedy for suggesting this emphasis.

4. I am not attempting to "refute" the claim made by Laudan and Leplin (1991, 449–450) that "there is no general *guarantee* of the possibility of empirically equivalent rivals to a given theory" (my emphasis). Rather, I offer *one* important candidate for such underdetermination.

5. Gillispie 1960, 117; Moritz 1914, 167; Moulton 1906, 199. Michael Crowe provided me with the specifics of the Moritz reference.

6. Gould 1989.

7. van Fraassen 1989.

8. Bohm himself (1987, 39) conjectured that things might have gone very differently had de Broglie carried the day at the 1927 Solvay congress.

9. Feyerabend (1989) has recently commented on the historically contingent nature of our (even scientific) view of the world.

10. Holland 1993a; Bohm and Hiley 1993; Valentini 1992.

11. Cushing 1990b.

Chapter One

1. See, for example, Laudan (1984).

2. Cushing 1990b, especially chapter 10.

3. This is the thesis defended by Laudan (1984).

4. Rorty (1988, 60) has pointed out that the use of hypotheses is not peculiar to science and has claimed that there is no *special* scientific method. He has also (1988, 70) suggested a possible psychological basis for the need (in some quarters) to give science a special status.

5. I discuss this at length in my 1990a.

6. Modern science may often appear quantitatively different from common-sense reasoning in being better scrutinized and from, say, some forms of philosophical or theological reasoning in being more immediately tied to "reality," but it is notoriously difficult to cash this out precisely.

7. Watkins 1984, 130.

8. For example, in a series of nearly twenty case studies, Donovan et al. (1988) document change in science, including changes in "metalevel" methodological ground rules themselves. Laudan (1981) has shown how the method of hypotheses (the hypothetico-deductive method) was effectively forced upon the scientific community by its success at the level of practice. See also Cushing 1990b, section 10.4.

9. Forman 1971; Pinch 1977; Laudan 1984; Pickering 1984; Shapin and Schaffer 1985; Donovan et al. 1988; Cushing 1990b.

10. In section 4.4 I return to nonlocality/nonseparability.

11. Bacon 1620, Aphorism XIX.

12. Whewell 1857, vol. 1, 146.

13. Galison 1987.

14. The terms *Copenhagen* and *Bohm* as used in this book often refer to schools or types of attempts, rather than necessarily just to the work of some specific individual or small group.

15. This is what Pickering (1984, 7–9) refers to as the "scientist's account."

16. Post 1971, 217, 228.

17. Here the terms *Bohr* and *Bohm* are used to stand for entire casts of actual historical characters belonging to two schools, of which one (Bohr) formulated the now almost universally accepted Copenhagen version of quantum mechanics and the other the now largely ignored causal version.

18. Post 1971, 218.

19. Bohm 1952a, 168–169. Post (1971, 219–220) refers only to the Copenhagen version of quantum mechanics and does not even acknowledge the existence of viable alternatives.

20. Post 1971, 233–234. I return to this in section 3.5.3.

21. Harding 1976.

22. Ben-Menahem 1990, 267.

23. Einstein 1954b, 221–222.

24. In conversation (8 October 1991) Don Howard pointed out to me that the standard translation (Einstein 1954b) of this 1918 address by Einstein leaves out an important caveat actually contained in the original paper. Just following the passage I have quoted in the text, Einstein (1918, 31) goes on to say (my loose translation): "Furthermore, this system of ideas, unequivocally coordinated with the world of experience, is deducible from a few fundamental laws, from which the entire system can be developed logically. The investigator sees here in each new important advance his expectations exceeded, because every fundamental law is more and more simplified under the pressure of experience. With wonder, he sees the apparent chaos united into a sublime order, which issues, not from

an individual intelligence, but, on the contrary, from the nature of the world of experience." This caveat does somewhat lessen the sharp contrast between the in-principle multiplicity of systems of theoretical physics and the one uniquely determined by the phenomena of nature. However, even this extended quotation really only indicates how, in Einstein's opinion, a successful program becomes ever more sharply focused in the hands of its creators and users as they continually mold it to fit the boundary conditions dictated by physical reality. It does not preclude other such possibilities. I thank Professor Howard for having brought this additional material to my attention. Howard (1990b) has also discussed the influence that Duhem's 1906 *The Aim and Structure of Physical Theory* (1974) had on Einstein, after 1910 or so, with regard to the theme of conventionalism in science. In fact, Howard (1990b, 379–380) has even located in Duhem (1974, 255–256) a passage that expresses sentiments similar to those stated here by Einstein.

25. Pickering 1984; Cushing 1984a.
26. Heisenberg 1971, 76.

Chapter Two

1. Of course the correspondence rules between the mathematical symbols that appear in a theory (e.g., the momentum operator $-i\hbar\nabla$ in quantum mechanics) and the physical observables in the world (the momentum p in my example) constitute an interpretation in a sense. (The symbol $\hbar$ represents Planck's constant divided by 2π.) However, it is not these correspondences (which I bracket with the formalism) that I am concerned with in discussing various interpretations of the formalism of quantum mechanics.

2. von Weizsäcker 1987, 279; Jammer 1974, chapter 1. MacKinnon (1982) has presented his own extensive views on scientific explanation in the context of the development of atomic physics. There, as here, one sees that what counts as an explanation is an evolving and context-dependent one, as opposed to there being fixed canons of rationality (cf. Cushing 1984b, 1990a).

3. A more detailed discussion of my views on this can be found in my 1991a.

4. A contemporary reassessment of inference and explanation is the subject of Earman's (1992) collection of essays.

5. This is also the position taken by Friedman (1974).

6. This is consistent with Salmon's (1984, 1985) view that scientific understanding is produced by suitable explanations.

7. Such a position is exemplified by Duhem (1974). I am not claiming any absolute observational/theoretical distinction, but am emphasizing differences among phenomenological rules, formal mathematical frameworks, and physical, event-by-event causal glosses on a formalism.

8. I am *not* arguing for a realist interpretation of theories or claiming that understanding can be had only within a realist tradition. My point here is that, within the ground rules of modern science (in which, for example, final causes are not really acceptable explanations), those explanations that produce scientific understanding are typically susceptible to (but do not require) a realist interpreta-

tion. Understanding is not tied by any type of *necessity* to a realist interpretation of a theory.

9. On the pragmatic aspects of understanding, see Achinstein (1985).

10. Toulmin 1961.

11. Meyerson 1908; Bachelard 1934.

12. Salmon (1984, 1985) classifies three basic conceptions of explanation as the epistemic (covering-law model), the ontic (causality), and the modal (necessity). He (1984, 263) sees causal explanation as the key to producing understanding.

13. Whewell 1857, books V, VII, and IX.

14. Einstein 1954d, 223.

15. This is admittedly a highly "reconstructed" form of the empirical data summarized by Kepler's first law. However, I believe this is a fair enough simplification to make my point. Also, the specific form of the conic section given in eq. (2.1) is *as it stands* valid only for an ellipse (or a circle), but not for a parabola or hyperbola. The necessary modification for open orbits is trivial and would contribute nothing to the discussion here. In eq. (2.1), a is the semimajor axis and b the semiminor axis of the ellipse, ε is the eccentricity, and θ is measured from the semimajor axis.

16. As a related example, one might claim to understand why objects of different masses free fall at the same speed at or near the surface of the earth *because* he knew Newton's laws of motion. In a sense, that would beg the question (although it does rather effectively shove it into the background) since what is really at issue is the origin of the gravitational force.

17. Whewell 1857, vol. II, 155.

18. The field concept *alone* (either in classical gravitational theory or in classical electromagnetic theory) did not mandate a causal mechanism propagated at *finite* speed and thereby eliminate the possibility of *instantaneous* action-at-a-distance theories. For classical electrodynamics the theory was refined beyond mere instantaneous (electro*static*) Coulomb-force situations to account for phenomena in terms of electromagnetic waves that propagate at the speed of light. But no such refinement ensued for classical (Newtonian) gravitational theory.

19. I do invert the actual historical sequence here since a simple kinetic-theory model of an ideal gas is older (e.g., Bernoulli) than the formalism of statistical mechanics (e.g., Maxwell, Boltzmann, Gibbs). However, the logical point I wish to make is independent of this sequence of events. Also, the global aspects of statistical mechanics (like those of classical mechanics as encapsulated in eqs. [2.2] and [2.3] above) are important since many other empirically testable results besides eq. (2.4) can be obtained.

20. The details of such a calculation can be found, for instance, in Huang (1963, 143–153). For more discussion of this example, see Cushing (1991a, 343–344).

21. Details and relevant caveats for this example can be found in Cushing (1991a, 344–345).

22. Bohm 1951, sections 22.15–22.18; Einstein, Podolsky, and Rosen, 1935.

23. I have labeled each emitted particle an "electron," but they could just as well be, say, photons, as they often are in the actual experiments used to test the Bell inequalities.

24. As indicated in the preface, calculations and other technical details are confined to appendices at the end of the appropriate chapter.

25. Bell 1964; 1966. A recent version of such an experiment has been carried out by Aspect, Dalibard, and Roger (1982). Such comparisons with experiment are possible, though, only when auxiliary assumptions (such as fair sample or no enhancement) are made to compensate for detector inefficiencies and the like (Clauser and Shimony 1978). Here 'determinate' means that results are fixed prior to measurement and 'local' implies (for now) no interactions that are propagated faster than light.

26. See, for example, Jarrett (1984).

27. Fine 1982. As Fine (1989 182 n. 3) himself points out: "A common-cause explanation . . . is what the foundational literature calls a factorizable stochastic hidden-variables model." What I mean by an event z being the common cause of two other events x and y is the following. Let neither of the events x or y be the (direct) cause of the other and let x and y *not* be statistically independent, so that the joint probability $P(x, y)$ does *not* factor: $P(x, y) \neq P(x) P(y)$. Then an event z (located in the common past of x and y) is the common cause of x and y if, once x and y are conditioned on z, statistical independence obtains as $P(x, y) = P(x|z) P(y|z)$. This factorizability is the condition for a common cause (cf. Butterfield 1989, 121; Cartwright 1989, 19).

28. Jarrett 1984; Shimony 1984.

29. Here the term 'realistic' refers to an observer-independent reality in which objects have (intrinsic) possessed values that are context independent. While it would be more precise to speak of a determinate, local theory, 'local realism' has become current in the literature.

30. In the present section I discuss this question in general conceptual terms only. In section 3.5.1 I consider the same issue at a more technical level.

31. Rohrlich 1986, 1987. Shimony sees an ontological coherence problem for Bohr since that program involves "the renunciation of any ontological framework in which all types of events—physical and mental, microscopic and macroscopic—can be located" (1963, 755). Shimony further claims that "any ontology whatever is alien to his [Bohr's] thought" (1993, 183) and observes that "one can wonder whether such an artifice [Bohr's renunciation of metaphysics] will not lead to more obscurity than illumination" (1993, 313).

32. Stapp 1985.

33. Bell 1987b; 1990.

34. Ghirardi, Rimini, and Weber 1986; Ghirardi and Rimini 1990; Pearle 1990.

35. Turning to quantum-field theory does not help with such problems, but only introduces further technical complications, as I have argued elsewhere (Cushing 1988).

36. Teller 1986; 1989.

37. Teller 1986, 73.

38. It seems to me that Bohm's theory could warrant such relational holism, but only if one were to accept nonlocality rather than just what is termed nonseparability in the Bell literature (cf. section 4.4).

39. Fine 1989. A similar position that the EPRB correlations do not need an explanation has been developed by van Fraassen (1985). There he considers various types of explanation and effectively rules out all but causal ones, which are excluded as a possibility by Bell's theorem (and by subsequent work on its implications).

40. What constitutes a satisfying stopping point in an explanation can certainly be influenced by our cultural heritage. I ask whether there is anything beyond or more fundamental than that.

41. In fact, he attacked it (McMullin 1978, chapter 3).

42. This statement (perhaps too sweepingly) greatly simplifies a complex historical story in that several outstanding physicists did challenge the doctrine of absolute action at a distance (van Lunteren 1991), but no successful, generally accepted alternative to instantaneous action at a distance emerged. I thank F. A. Muller for bringing van Lunteren's work to my attention.

43. I thank J. B. Kennedy for pointing this out to me. See also Howard (1993) on the historical background to the conception of space as a means of individuation for physical systems.

44. Edwards 1967, vol. 6, 47–51; vol. 8, 369–379.

45. Kant 1952, 24–26; see also Krieger 1992, 3.

46. Meyerson 1908. I return to Meyerson's views in the next subsection.

47. de Broglie 1948, 330.

48. McMullin 1989.

49. McMullin 1989, 301–302.

50. Frans van Lunteren (1991), in his extensive recent study of conceptions of gravity in the eighteenth and nineteenth centuries, demonstrates that a quest for intelligibility was uppermost for several major physicists who attempted ether-type explanations for action at a distance.

51. Maxwell 1890, 480–490.

52. Bell 1987b. For discussions of the locality/separability issue from several different perspectives see the collection of essays in Cushing and McMullin (1989).

53. In fact, Bohm and Hiley's (1989) stochastic version of causal quantum theory and Vigier's (1982) covariant ether quantum theory attempt to reinstate *local* causality with influences transmitted at superluminal (but finite) speeds. A similar move has recently been suggested for Bohm's 1952 theory as well (Squires 1993). See also Eberhard (1994).

54. Schweber 1986.

55. Cushing 1982, 39.

56. MacKinnon (1982, chapter 2) has emphasized the importance, for Kant, of a distinction between purely deductive inference (my 'explanation') and contextual or content-dependent explanations based on our modes of organizing our perceptions of the world.

57. Meyerson 1908.

58. In a similar vein, Poincaré held that euclidean geometry was a concept that conferred on early man an advantage for survival (Miller 1984, 25).

59. By an object being identical with itself, Meyerson does not mean some logical tautology, but rather a statement about an object's maintaining its identity over *time*. "Thus the principle of causality is none other than the principle of identity applied to the existence of objects in time" (Meyerson 1908, 43).

60. Meyerson 1908, 384 (emphasis in original).

61. Meyerson 1908, 395.

62. Faye 1991, 158. In chapter 6 I argue that even Bohr did not rid his theory of causality, but bent its meaning to fit his interpretation.

63. Bachelard 1934.

64. Bachelard 1934, 3.

65. Bachelard 1934, 54, 175.

66. On this point, David Bohm (1987, 39) observed: "To have some kind of intuitive model was better, in my view, than to have none at all."

67. Bohm and Hiley 1993, section 7.7.

68. Bohm, Hiley, and Kaleyerou 1987, 331 (final emphasis mine).

69. Margenau 1954, 7.

70. This can also be seen by writing the $|\Psi_0\rangle$ out in terms of the $\psi_\pm(\theta)\otimes\psi_\pm(\phi)$ basis as

$$|\Psi_0\rangle = \frac{1}{\sqrt{2}}\Bigl[-i\sin\left(\frac{\theta-\phi}{2}\right)\psi_+(\theta)\otimes\psi_+(\phi) + \cos\left(\frac{\theta-\phi}{2}\right)\psi_+(\theta)\otimes\psi_-(\phi)$$
$$-\cos\left(\frac{\theta-\phi}{2}\right)\psi_-(\theta)\otimes\psi_+(\phi) + i\sin\left(\frac{\theta-\phi}{2}\right)\psi_-(\theta)\otimes\psi_-(\phi)\Bigr].$$

Chapter Three

1. Gell-Mann 1981, 169–170 (emphasis in original). Similarly, Richard Feynman (1965, 129) was of the opinion: "I think I can safely say that nobody understands quantum mechanics."

2. Bohr 1913.

3. Folse 1993, 9–10. Here 'separability' means, rather loosely, independence of events in spatially distant regions. A technical and more complete discussion of this concept is given in section 4.4. Bohm (1985, 689) has characterized the fundamental difference between Einstein and Bohr as one over an intuitive and unambiguous visualizability versus an ambiguous and severely limited visualizability.

4. For more details on Einstein's views and the relative strength of his commitment to various requirements on reality, see section 10.3.

5. On consistency versus uniqueness claims in Bohr's work, see also Bohm (1957, 93–95) and Vigier (1985a, 653).

6. Folse 1993, 11 (emphasis in original).

7. Folse 1985, 20. I quote Folse here, not Bohr.

8. Laurikainen 1988c, 53–57.

9. Folse 1985, 22–23, 204.

10. Folse 1985, 257.
11. Folse 1989, 268.
12. Einstein 1949b, 674.
13. See, for example, von Neumann (1955); Messiah (1965); d'Espagnat (1976); and Wigner (1976). No claim is made that these postulates are complete, independent, or the most general ones possible. They are intended only as an *illustration* of a formal structure when a state vector ψ can be used to represent a specific physical situation.
14. A specific and detailed application of these rules can be found in appendix 1 to chapter 5.
15. Even though I have made a distinction between the terms 'theory' and 'interpretation', I shall often speak of the Copenhagen *interpretation* and the Bohm *interpretation* when, strictly, I should refer to the Copenhagen *theory* or to the Bohm *theory*. I do this because 'Copenhagen interpretation' has become a commonplace expression for standard quantum theory. They are, nevertheless, different *theories*.
16. Stapp 1972.
17. Stapp 1972, 1100.
18. Ballentine 1970; Home and Whitaker 1992.
19. Ballentine 1970, 374.
20. Bohr 1934, 48–51.
21. Bohr 1934, 53.
22. Bohr 1934, 54–56.
23. Bohr 1934, 108. This same exercise of redefining terms to suit their own purposes, yet often trading on wider, common connotations in subsequent discussions, had begun in earnest already in Bohr's (1927) Como lecture and in Heisenberg's (1927) "uncertainty" paper.
24. Bohr 1949, 223.
25. Bohr 1961, 90.
26. Stapp 1972.
27. Heisenberg 1958, 44.
28. Heisenberg 1958, 46.
29. Heisenberg 1958, 48.
30. 'Epistemology' and 'ontology' are the terms used by Heisenberg himself (Heisenberg 1958, 48).
31. Heisenberg 1958, 49.
32. Heisenberg 1958, 50–51.
33. Heisenberg 1958, 52, 54.
34. Heisenberg 1958, 129–130. In later years Heisenberg supported a more Platonic view of reality (Heisenberg 1976).
35. Born 1951, 155 (emphases in original).
36. Born 1951, 157 (emphases in original).
37. Born 1951, 163–164. A similar view on the statistical interpretation is found in a more technical work of his from the same period (Born 1936, 130–135).

38. Born 1936, 84–85.
39. Born 1951, 162.
40. Dirac's (1958), von Neumann's (1955), and Pauli's (1980) treatises (the first editions of which all appeared in the early 1930s) are formal codifications of these principles and need not be considered separately here.
41. Of course, Bohr (1935, 699) specifically denied the possibility of "a mechanical disturbance of the system" and opted for *"an influence on the very conditions which define the possible types of predictions regarding the future behavior of the system"* (emphasis in original). However, as I argue at length in section 4.4, even the latter should be counted as nonlocality.
42. Bohr's oft-quoted dictum that the language of classical physics must always be used to describe the results of a measurement or observation is not equivalent to claiming that the ontology of the world is just that of classical physics.
43. On the recent project of attempting to salvage Bohr from incoherence, see Cushing 1993b.
44. Beller 1992.
45. Beller 1992, 149.
46. Beller 1992, 176.
47. Folse 1985, 10.
48. Quoted in Folse 1985, 9.
49. Folse 1985, 18.
50. I do not mean that physicists had not, ever since Einstein's 1905 photon hypothesis, been puzzled by *some* type of dualism between the wave and particle aspects of basic entities. However, a precisely stated position came only much later.
51. Beller 1992, 176 (cf. Pauli 1980, 1–7).
52. Faye 1991, 185.
53. Honner (1987, 36) claims that Bohr's concept of complementarity grew out of the earlier idea of wave-particle duality.
54. Heisenberg 1958, 43.
55. Pauli, 1980, 7 (emphasis in original). This same statement can be found in his 1933 *Handbuch* article on quantum mechanics (Pauli 1933).
56. Bohr 1934, 78.
57. Bohr 1949, 210 (emphasis in original).
58. Bohr 1949, 210.
59. Bohr 1949, 223.
60. Bohr 1934, 10 (emphasis in original).
61. Don Howard strongly disagrees here. For him, 'complementarity' refers to the relationship between observables for conjugate variables, while 'wave-particle duality' denotes the existence or nonexistence, respectively, of quantum correlations (or interference effects). Pan Kaloyerou has suggested to me that one should simply distinguish between complementarity concepts that follow directly from the mathematical formalism (such as those associated with conjugate variables) and those that do not (such as mutually exclusive classical concepts). The former are an intrinsic part of quantum theory, the latter not.

62. Folse (1985, 22) does claim that there is more to complementarity than wave-particle dualism and the use of complementary "pictures." But that does not invalidate such duality as a useful illustration.

63. Bohr 1934, 10.

64. Home 1992.

65. Scully and Walther 1989; Scully, Englert, and Walther 1991. As these authors put it, "in all standard examples, including Heisenberg's famous microscope, complementarity is enforced with the aid of Heisenberg's position-momentum uncertainty relation." (Scully, Englert and Walther 1991, 112). They discuss other experiments in which complementarity is "enforced" by means other than this uncertainty relation. They do, however, take complementarity to be a *given* for quantum mechanics and do not claim to have a way to circumvent it. When "which-path" information is obtained, interference effects disappear. Hauschildt (1990) offers a general proof that particle path information and interference are impossible to have simultaneously.

66. Ghose and Home 1992; Home 1992.

67. Ghose, Home, and Agarwal 1991, 1992. The quotation is from Home (1992).

68. Dewdney et al. 1992, 1263–1264.

69. For a presentation of several of the (ultimately unsuccessful) approaches to the measurement problem, see Lahti and Mittelstaedt (1991). Albert (1992) discusses, and raises serious objections against, all of the traditional, and even some of the nonstandard, programs to resolve that central dilemma. Questions of internal consistency remain for the "preferred-basis" solution proposed by Kochen (1985), by Dieks (1989), and by Healey (1989) (cf. Albert 1992, 191–197).

70. Bohm 1951, chapter 22.

71. Pais 1991, 354; Wheeler and Zurek 1983, 699–700.

72. See eq. (4.40) for an H_1 to be used in eq. (3.19).

73. For simplicity I consider here only the case of a *pure* state (i.e., one in which the quantum-mechanical state of a system can be represented by a state vector or wave function). Mixed states require a density matrix and I do not enter into the formalities of such states. A simple example of a mixed state is an ensemble of unpolarized electrons (i.e., one for which $\langle \sigma \rangle = 0$) whose appropriate density matrix is $\rho = (1/2)\mathbf{1}$, where $\mathbf{1}$ is the 2×2 identity matrix. This situation is *different* from one in which the ensemble has a definite (nonzero) polarization (as in eq. [3.4]) whose axis simply does not happen to be along the z-axis. For the ψ_0 of eq. (3.4) we find $\langle \sigma \rangle = \langle \psi_0 | \sigma | \psi_0 \rangle = \hat{n} = (\sin \theta \cos \varphi, \sin \theta \sin \varphi, \cos \theta)$ where φ is the relative phase between α and β^* (i.e., $\beta \alpha^* = |\alpha| |\beta| e^{i\varphi}$) and $\sin \theta = 2|\alpha| |\beta|$.

74. Here 'linearity' means that a superposition of two solutions to the Schrödinger equation is also a solution (cf. eq. [3.26]).

75. This caveat of physical acceptability refers to the constraint that any Hamiltonian must be a hermitian operator so that the time evolution of the state vector will be unitary.

76. See, for example, Ghirardi, Rimini, and Weber (1986) for one such attempt and Albert (1992, chapter 5) for a criticism of this "solution."

77. Daneri, Loinger, and Prosperi 1962. Albert (1992, 91) points out that this approach is not a *solution* to the collapse problem, but rather an argument that (at present) we cannot in practice tell at what stage collapse occurs—*given* that it does occur. Furthermore, one can see Daneri, Loinger, and Prosperi's proposed solution as, ultimately, a restriction on the types of measurements we are, in practice, able to perform. If one is restricted to their *classical* (i.e., mutually commuting) observables, then there is no way to distinguish between a superposition $|\psi\rangle = \Sigma c_i |\psi_i\rangle$ and a mixed state $\rho = \Sigma |c_i|^2 |\psi_i\rangle\langle\psi_i|$. (Here $\{|\psi_i\rangle\}$ is the set of common eigenvectors of these mutually commuting operators.) However, such a move is an avoidance of the measurement problem, not a solution to it.

78. d'Espagnat 1976, 193. This issue, as regards the practical versus the in-principle possibility of constructing observables to detect differences between the actual pure state of the compound system and an *effective* mixture, is actively debated (Daneri, Loinger, and Prosperi 1966). Even more important, though, Daneri, Loinger, and Prosperi's (1966) extensive discussion of the amplification aspect of measurement sidesteps the heart of the measurement problem, since, in principle, a "measurement" could be performed by a single molecular structure (Bub 1968; Ballentine 1970, 371). A currently popular approach to dynamical "reduction" mechanisms is *decoherence* (Omnès 1992; Zeh 1993), in which, because of interaction with the environment, different collective states rapidly become orthogonal. At present, however, decoherence remains more a program than a completed theory. It also would not solve the puzzle of how the long-range EPRB correlations are produced. Decoherence is compatible with Bohm.

79. On these two types of limits, see, respectively, Hepp (1972) and Bub (1988).

80. Bell 1975.

81. Bell 1975 [98].

82. Schrödinger 1935. There is, admittedly, a bit of "theater" in the rhetoric of this paradox. One could claim that the "reduction" of the state vector takes place long before the physical chain of events reaches any system as complex as a cat. However, the reduction process still requires an explanation and the dilemma Schrödinger presents us with remains.

83. Goldstein 1950, section 9.8.

84. The question of any type of conceptual match is even more problematic since the ontology of classical physics is arguably incompatible with any consistent candidate for a "quantum ontology." But this is just what it means to find an interpretation of the standard version of quantum mechanics.

85. If one insists upon letting $\hbar \to 0$, then this limit is nonanalytic and one does *not* simply recover the results of classical mechanics (Berry 1989).

86. Messiah 1965, 216–222.

87. Things are really not even this good since, in general, $\langle F(r)\rangle \neq F(\langle r\rangle)$.

88. Messiah 1965, 222–228.

89. It is a simple technical matter to exhibit explicitly specific interaction Hamiltonians H_I (e.g., one like that of eq. [4.40]) that will produce the results stated in eqs. (3.21) and (3.22) (for example, see Bohm 1951, chapter 22). However, nothing essential relative to my discussion here would be gained by writing these out.

90. Here I neglect the relative motion of the two subsystems since it is not relevant for illustrating the measurement problem.

Chapter Four

1. There is, incidentally, a middle-of-the-road interpretation—the so-called sum-over-histories one—in which one has actual physical paths, but no wave function (Sinha and Sorkin 1991). But that's yet more underdetermination. I do not pursue this topic here since one viable alternative interpretation—Bohm—will suffice.

2. Bohm 1952a. Although these are *two* papers published together, I refer to them as Bohm's (1952) *paper* for convenience.

3. Bohm 1952a. Here and in the rest of this book I shall be concerned almost exclusively with Bohm's 1952 interpretation of quantum mechanics, as opposed to his later attempts (beginning roughly with Bohm [1957]) to underpin this interpretation with an all-encompassing worldview.

4. In the Hamilton-Jacobi formulation of mechanics, the energy of the system is given as $E = -\partial S/\partial t$ (Goldstein 1950, 278). From eq. (4.23) we then see that $E = K + V + U$.

5. Of course, the quantum world *is* a strange place. For example, in a highly regarded theory of fundamental interactions, namely quantum chromodynamics, the force between two quarks is assumed to *increase* with distance and this increase continues no matter how great the separation between the two quarks. That is surely unexpected from any intuition we have about the behavior of known classical forces.

6. Bohm 1952a, 170.

7. Since eq. (4.23) in appendix 1.1

$$\frac{\partial S}{\partial t} + \frac{1}{2m}(\nabla S)^2 + V + U = 0$$

is just the Hamilton-Jacobi equation for Hamilton's principal function $S(x, t)$, it follows that if $p_0 = \nabla S(x_0, t_0)$ initially, then $x = x(t)$, such that $x_0 = x(t_0)$, will be the trajectory that is orthogonal to the surfaces $S = const.$ (Goldstein 1950, section 9.8). In fact, Bohm (1987, 35) tells us that the classical Hamilton-Jacobi theory, which relates wave and particle theories, and the WKB (Wentzel-Kramers-Brillouin) approximation for Schrödinger's equation started him thinking about what would happen if one did *not* make any approximations in establishing this equivalence. This produced the additional quantum-potential term.

8. Although Bohm's original papers were written in 1952, well before the advent and popularity of modern chaos theory, his general approach and several of

his insights are forerunners of, and certainly consonant with, this field of current activity.

9. That is, from eqs. (4.2) and (4.6) we have

$$\langle\psi|p_{op}|\psi\rangle_{QM} = \int \psi^* p_{op} \psi dV = \int R\, e^{-iS/\hbar}(-i\hbar\nabla)R\, e^{iS/\hbar} dV$$
$$= \int \left[-\frac{i\hbar}{2}\nabla(R^2) + R^2\nabla S\right] dV.$$

The first term in the final integral is zero (by a version of Gauss's theorem for a gradient integrated over all space) and the remaining term is just $\int P\nabla S\, dV$, as claimed. However, $\langle\psi|p_{op}^2|\psi\rangle_{QM}$ does *not* equal $\langle\psi|p^2|\psi\rangle$.

10. Bohm 1952a, 171.

11. A nontechnical discussion of the overall features of Bohm's theory is given in Albert 1992, chapter 7.

12. Bohm 1952a, 171. Keller (1953) has argued that if (iii) would have to be assumed as a separate postulate, rather than being derivable from the dynamics, then Bohm's interpretation would not be an ordinary statistical mechanics of a deterministic theory (as is classical statistical mechanics).

13. Here I am commenting on the one-body system in some given environment. When the theory is extended to a larger system, such as a (closed) universe, the wave function *includes* the environment. Recent work by Aharonov, Anadan, and Vaidman (1993), in which they show how the wave function of a *single* particle can, in principle, be measured experimentally, adds support to Bohm's view that $\psi(x)$ is an objectively real physical entity and not simply a probability amplitude.

14. Bohm 1953b. Of course, from a purely logical point of view, one could simply demand (iii) by fiat, as is essentially done for the Copenhagen interpretation. Bohm's argument for $P \to |\psi|^2$ is outlined in appendix 1.4 to this chapter. In keeping with later terminology, I often refer to the condition $P = |\psi|^2$ as *quantum equilibrium*.

15. Bohmian mechanics *does* lead to the usual probabilities for outcomes of other (*façon de parler*) observables besides position, as illustrated in the example treated in appendix 1.3 to this chapter. Daumer et al. (1994) show that the entire operator formalism for predicting the outcomes of measurements, such as $\langle A\rangle = \langle\psi|A|\psi\rangle$, emerges from Bohmian mechanics. Interestingly enough, from a historical perspective, Vigier (1990, 148) pointed out that the mapping of canonical transformations in phase space onto unitary transformations in Hilbert space was demonstrated long ago by Koopman (1931). Sudarshan (1992, 151) has also discussed such a mapping and has claimed that Koopman used the wrong vector space for his proof.

16. Bohm's ψ-wave can certainly be given a coherent representation in *configuration* space, while such a representation is more problematic in ordinary three-dimensional space (cf. note 33 in chapter 8). For a single point particle, configuration space has just three dimensions and then this distinction is merely formal.

Bohm (1953a) also discussed the apparently insuperable difficulties of attempting to formulate a causal interpretation in momentum space. He concluded (1953a, 320): "It would appear, therefore, that a causal interpretation of the quantum theory can be obtained only if we use the space-time representation of the wave function as a basis."

17. In this picture it is consistent for the measuring apparatus as well to be governed by the laws of Bohmian mechanics (Nelson 1985, 113). Perhaps it is worth mentioning that a criticism of the GRW collapse mechanism by Albert and Vaidman (1989) based on a measuring process not involving macroscopic motions of apparatus constituents does not apply to Bohm's theory. The reason is that in GRW these macroscopic changes are necessary to effect the collapse of the wave function in coordinate space while in Bohm there is no collapse that must be produced—the microsystem must simply be *registered* at one or another of several macroscopically separated instrument positions.

18. This is the elementary quantum-mechanical problem of an infinite square well.

19. This just means that the spatial part of the wave function has a *constant* phase.

20. For the nth energy level, the wave function is $\psi_n = A_n \sin(2\pi nx/L)$, where L is the width of the well. The corresponding energy is $E_n = (1/2m)(nh/L)^2 = p_n^2/2m$. Strictly speaking, the correct ψ_n for this problem is nonvanishing only on $(0, L)$ and vanishes everywhere outside this interval. I thank Dr. C. R. Leavens for pointing this out to me. The general point being made here is not affected by this fact, however.

21. Not only does Bohm discuss this in his original paper (Bohm 1952a, 184), but he also returns to this in an essay (Bohm 1953c) addressed specifically to Einstein's (1953) unease.

22. Bohm 1952a, 184.

23. For any given wave function for the *universe*, the positions of particles *always* have determinate values at any instant. If we consider a subsystem, and represent the environment by a wave function that we can change by making a "decision" of choosing an apparatus setting (e.g., setting the direction of the magnetic field in a Stern-Gerlach device), then we can countenance situations in which the resulting *instantaneous* change of the wave function causes a particle to "jump" discontinuously from one trajectory to another. However, this absence of a definite position at that instant seems to be more an artifice of our inconsistently mixing the determinism of Bohm's theory (for the "universe") with our alleged ability to modify "freely" the wave function representing the environment.

24. In section 4.4 I discuss my reasons for not distinguishing, in the context of Bohm's theory, between nonlocality and nonseparability.

25. Bell 1964; 1981 [132]; Cushing and McMullin 1989.

26. Bohm 1952a.

27. Since I do not discuss how spin is incorporated into Bohm's theory until chapter 5, I use for illustration the determination of the angular momentum of an atom, rather than of the spin of an electron. This is the same case treated by Bohm

and Hiley (1984). Mathematical details of the necessary calculations can be found in appendix 1.2 to this chapter.

28. Figures 3.1 and 4.1 are different schematic representations of the same basic physical process. Strictly speaking, I am considering an ensemble of particles, all having the same wave function Ψ and having their initial positions distributed according to $|\Psi|^2$. For simplicity, though, I often speak of a beam of particles and focus on one particle in the beam.

29. Both f and the ψ_n of eq. (4.7) are normalized to unity. The internal coordinates of the atom are represented by ξ. Here and in the following, $f(x) \equiv f(x, t = 0)$.

30. This approximation for the transit time is thoroughly discussed by Bohm (1951, 595–596).

31. The *support* of a function is the region within which the function is not identically zero.

32. This is discussed further in appendix 1.2 to this chapter.

33. See Bohm and Hiley 1984.

34. Recall that, for Bohm, *all* measurements are position measurements.

35. These claims become really clear and convincing only when one goes through some of the mathematics in appendix 1.2. That the wave function does not vanish *identically* between the shells does not matter, since it is vanishingly small so that P is effectively zero there. Once P (or R) has flattened out in a region, the quantum potential also (effectively) vanishes there.

36. Recall that, because of eq. (4.5), once x_0 is known, so is p_0.

37. Here the term 'chaotic' refers to the often erratic, unpredictable nature of the motion.

38. If one takes ψ simply as a mathematical tool for encoding the influence of the environment on the microsystem, then there is no conceptual difficulty in "throwing away" those pieces of ψ that have become irrelevant for the future dynamical evolution of the system.

39. Bohm 1952a, 178.

40. Bohm and Hiley 1984, 262–269.

41. Remember that the wave function for the microsystem plus the apparatus is defined on a configuration space of $3n$ dimensions (where n is, roughly, the number of particles in the *entire* system). Once a sufficient number of these variables become different (as in a registration process) in two (or more) apparatuses, the corresponding *total* wave functions have practically *no* overlap and interference effects are not possible.

42. Bohm and Hiley 1984, 266.

43. Holland 1993a, chapter 6.

44. Bohm and Hiley 1985.

45. Bohm 1952a, 175.

46. Bohm 1952a. This argument is given in appendix 1.3 to this chapter.

47. Bohm 1952a, 180.

48. There is a fairly extensive literature on the constraints quantum mechanics imposes on probability distributions of canonically conjugate variables in terms

of information entropy (Bialynicki-Birula and Mycielski 1975; Deutsch 1983; Partovi 1983; Bialynicki-Birula 1984; Bialynicki-Birula and Madajczyk 1985). Some of these bounds are, in fact, stronger than the Heisenberg uncertainty relations.

49. See appendix 1.3 to this chapter.
50. See note 23 above.
51. Bohm 1952a, 187.
52. In this section I assume that quantum equilibrium ($P = |\psi|^2$) obtains. In section 9.4, I return to the implications of possible deviations from such equilibrium.
53. Bohm 1952a; Bohm, Hiley, and Kaloyerou 1987; Bohm and Hiley 1993. Albert (1992, 134) makes the interesting claim that no other theory can have exactly the same empirical content as standard quantum mechanics because standard quantum theory is at present unfinished (i.e., it gives no coherent account of wave function "collapse") and so has no well-defined empirical content.
54. Home 1986, 49. I thank Dr. Home for bringing this interview to my attention.
55. Bell 1980 [111].
56. Pauli's *Handbuch* article (1933, 140; 1980, 63) already gave an argument that time cannot be an operator T conjugate to the Hamiltonian H (energy operator) satisfying the commutation relation $[T, H] = i\hbar$. A standard argument shows that the existence of such a time operator would imply that the energy spectrum of H would always be continuous and could never be bounded from below. But that is contrary to observation. See also Jammer (1974, 141) and Ballentine (1990, 239).
57. For further discussion see Hauge and Støvneng (1989), Fertig (1990, 1993), Leavens (1990a, 1990b, 1991, 1993a, 1993b, 1993c), Leavens and Aers (1990, 1991, 1993), Muga, Brouard, and Sala (1992), Olkhovsky and Recami (1992), and Holland (1993a, section 5.5.1). This type of tunneling through a barrier is quite different from the decay of a metastable state (as represented by the tunneling of a nucleon through the potential barrier of a nucleus), which is an old and quantitatively well understood quantum-mechanical process.
58. Specifically, a consistent set of dwell, transmission and reflection times must be real (as opposed complex), positive quantities and satisfy the constraint of eq. (4.88) in appendix 2 to this chapter. Sokolovski and Connor (1993) have recently argued that a proper application of the quantum-mechanical rules for quantizing classical time parameters allows a path-integral formulation of *observable* tunneling times (and keeps the *observed* times from being complex, a difficulty that had beset previous attempts). It is unclear that their distinction actually resolves the tunneling-time problem in conventional quantum theory.
59. Hauge and Støvneng 1989, 917. They discuss this question of various tunneling times only within the framework of the standard interpretation of quantum mechanics and do not consider Bohm's interpretation.
60. Leavens 1990a, 1990b, 1991, 1993a, 1993b, 1993c; Leavens and Aers 1990, 1991.

61. An important theoretical matter is whether or not the tunneling times are such sensitive functions of the *shape* of the initial incoming packet that there might be no way to have quantitatively meaningful predictions for them without having to be able to violate, say, the uncertainty relations. Furthermore, the time scale for a tunneling process may depend sensitively upon the means used to measure it. Such issues are unresolved at present.

62. The relevant expressions are eqs. (4.97) through (4.99) in appendix 2 to this chapter.

63. There already is some indication of such a move in the technical literature with regard to the expression of eq. (4.96) for τ_D.

64. In other words, I am suggesting that, for the purposes of this book, we *ignore* all of the distinctions I now discuss and refer simply to locality/nonlocality. However, some readers will want me to *justify* ignoring those distinctions that are now common in the foundations literature. That is the purpose of section 4.4. The reader who has no qualms about using only the undifferentiated term 'nonlocality' need not be concerned with this section.

65. Much in my summary here can be found in Cushing (1989).

66. The assumption that each experimenter can make choices that are *free* (i.e., effectively decoupled from the physical environment) is crucial for Bell's argument. An obvious way out is "superdeterminism," in which these choices are, in fact, constrained to follow the proper statistics (i.e., they are themselves predetermined).

67. I do not display explicitly the station indices A and B [as, for instance, $P^{AB}(x, y \mid i, j)$] because that information is already contained in the (x, i) for A and the (y, j) for B; such superfluous indices can cause confusion for marginal and conditional probabilities when those probabilities become independent of outcomes or settings at a distant station.

68. Bell 1964.

69. Fine 1986, 59. Notice that this is framed, as it was for Bell himself, in terms of *outcomes,* not in terms of *probabilities* of outcomes.

70. Bell's actual argument was framed in terms of observables and their expectation values, rather than in terms of probabilities. I shall return to this difference shortly. However, general philosophical discussions typically work with the probabilities and I follow this line of presentation in order to get directly to the distinctions I am interested in.

71. This term is due to Fine (1981a). It is just the condition of stochastic independence.

72. The existence of Fine's prism models demonstrates this (Fine 1981a).

73. Jarrett 1984; Shimony 1986; Howard 1985.

74. The quantum-mechanical joint probabilities for the EPRB arrangement are given in eqs. (2.7) and (2.8). In that case, each marginal is just 1/2. This may be overkill, but eq. (4.15) *is* satisfied there. There λ is just the state vector.

75. I address only being able to signal via the quantum correlations here since one could, for example, in a *nonrelativistic* theory, fire a "bullet" faster than light to establish a superluminal cause-effect link. The relevant question is being able

to signal with the correlations (actually, at *any* speed here). Furthermore, even if one cannot signal with the $P(x \mid i)$ themselves (i.e., given locality), it does not follow from this alone that the string of results $\{x\}$ does not contain structure that would allow signaling (e.g., 010101010101 ... → 0.5000, as does 001100110011 ...). I thank Mark Adkins for pointing this out to me (letter of 6 January 1993). This observation applies equally to standard quantum mechanics as to Bohm's theory. The *formal* requirement of "locality" in eq. (4.15) may be too restrictive to capture the full sense of physical locality. Of course, if standard quantum mechanics is complete, then there is *no* observable at one station whose expectation value can depend upon a choice made at the other (spacelike-separated) station (Shimony 1984, 228).

76. Jarrett 1984, 573, 587 note 15. Shimony (1984, 227) makes a similar point.

77. Valentini 1991b, 4. Again, this claim is made only for the *possibility* signaling with the *correlations* (i.e., with the *statistics* one can gather).

78. Valentini 1991b, 5.

79. My use of 'uncontrollable' here is very different from the technical sense in which Shimony has made that term popular.

80. This terminology is consistent with the definitions of eqs. (4.12) and (4.13), since $\sum_y P_\lambda(y \mid i, j) = 1$. Notice, further, that separability *alone* yields only $P_\lambda(x, y \mid i, j) = P_\lambda(x \mid i, j) \, P_\lambda(y \mid i, j)$, and this is *not* sufficient to prove the Bell inequality.

81. Fine (1982) has shown that the factorizability of eq. (4.14) is the necessary and sufficient condition for the existence of a deterministic hidden-variables model that will produce the joint distributions for a correlation experiment (of the EPRB type). The existence of such a complete set of state variables λ is equivalent to a common-cause explanation (in the common past of the parts of the system to be observed).

82. For example, the quantum-mechanical correlations of eqs. (2.7) and (2.8) yield for the conditionals $P(+|\theta, \phi, +) = \sin^2\left(\frac{\theta - \phi}{2}\right)$, $P(+|\theta, \phi, -) = \cos^2\left(\frac{\theta - \phi}{2}\right)$ and these are not equal.

83. Shimony 1984; Ghirardi, Rimini, and Weber 1980. See appendix 2 of chapter 10 as well. Of course, one must distinguish between a no-signaling theorem established within the framework of quantum mechanics and a more general claim that *no* type of violation of eq. (4.16) could *ever* allow signaling. Jarrett's (1984, 580–581) original argument against the possibility of signaling with a violation of eq. (4.16) amounts essentially to the observation that a deterministic hidden-variables theory would necessarily be complete so that a violation of completeness would require a nondeterministic theory. However, so the argument goes, in such a theory the outcomes are not under the control of the experimenter and therefore cannot be used to send a signal. In fact, Shimony (1986, 192) discusses a possibility of signaling via a violation of eq. (4.16), but not superluminally. Jones and Clifton (1993) go further and argue that a violation of eq. (4.16)

could be used for *superluminal* signaling, but in their model a violation of eq. (4.16) actually implies a violation of eq. (4.15). In that case, one might view this as just another case of the possibility of signaling with nonlocality.

84. I appreciate that there *are* still further fine distinctions to be made that are not spelled out in this list. Those readers who know these details don't need this list, but others may find it useful for orientation. See Howard (1989) and Jarrett (1989) for fuller discussions of these distinctions.

85. The $P(x \mid i, j)$ and the $P(x, y \mid i, j)$ are, respectively, just the $P_\lambda(x \mid i, j)$ and the $P_\lambda(x, y \mid i, j)$ integrated over the $\rho(\lambda)$. But, the $P(x \mid i, j, y)$ is *not* simply $P_\lambda(x \mid i, j, y)$ so integrated, since $P(x \mid i, j, y) \equiv P(x, y \mid i, j)/P(y \mid i, j) \neq \int [P_\lambda(x, y \mid i, j)/P_\lambda(y \mid i, j)]\, \rho(\lambda)\, d\lambda$. The point is that a theory *local* on the fine-grained (i.e., λ) level *must* be *local* at the observable (i.e., integrated) level, but a theory *separable* at the fine-grained level *need* not be *separable* at the observable level. These observations are relevant to Bohm's theory.

86. I do not say this to diminish in any way the importance of these distinctions of his in *general* philosophical analyses of hidden-variables issues. However, it remains an open question whether or not his separability is a *physically* meaningful concept (apart from locality). This question is related to Teller's (1986, 1989) concept of relational holism (see section 2.4).

87. Again, this simply reflects the point that $P_\lambda(x \mid i, j, y) = P_\lambda(x \mid i, j)$ (i.e., separability at the fine-grained or λ level) does *not* necessarily imply that $P(x \mid i, j, y) = P(x \mid i, j)$ (i.e., separability at the observable level), even though locality at the λ level *does* imply locality at the observable level. Similarly, nonlocality (in Jarrett's sense) at the fine-grained level *need not* imply nonlocality at the observable level. For Bohm's theory, of course, $\rho = \rho(\lambda; i, j)$.

88. Recall that a wave function Ψ for a system consisting of two (or more) subsystems (say 1 and 2) is *separable* if it factors into the simple one-term product $\Psi = \psi\phi$, where ψ is the wave function for system 1 and ϕ that for system 2. In general, however, Ψ will be a *sum* of such terms, as $\Psi = \sum \psi_j \phi_j$, and is then termed an *entangled* (or nonseparable) state.

89. Bell's own notation would be $A(a, b, \lambda)$ where a and b are the analyzer axes (i.e., the parameters i and j). I have used the notation $A_\lambda(i, j)$ to facilitate comparison with the Ps above.

90. Again, Bell (1964) writes $P(a, b)$ for my $\langle AB(i, j)\rangle$.

91. That Bell chose to work directly with the As and Bs may be related to the fact that he had been familiar with Bohm's theory and much impressed with it as a counterexample to Copenhagen dogma (Bell 1964, 1966, 1987a). Since Bohm's theory is deterministic, separability in terms of the A_λs and B_λs would not make much sense.

92. The interested reader should consult Bohm (1952a) for more details. I use a direct transcription from Bohm's 1952 paper both because his own presentation is so clear and because I wish to emphasize that these results and arguments were readily accessible in the physics literature as long ago as 1952. They are *not* an artifice of a retrospective reconstruction or rewriting of history.

93. The identification of eq. (4.27) is *consistent* with an interpretation of eq.

(4.26) as a continuity equation. Equation (4.23) is just the Hamilton-Jacobi form of Newton's second law of motion. This is spelled out in more detail in appendix 2 to chapter 5, eqs. (5.44) through (5.51).

94. In eq. (4.30) I have taken what is sometimes termed the "convective derivative" $df/dt = v \cdot \nabla f + \partial f/\partial t$ where $f = f(x, t)$ and $x = x(t)$ [i.e., df/dt is computed along a flow- or stream-line $x = x(t)$].

95. Typically, a Gaussian form is used for $g(k)$ since that also produces a Gaussian for $f(x)$.

96. There is also an overall phase factor $e^{-i\varepsilon_1 t/\hbar}$, where ε_1 is the energy corresponding to the $n = 1$ level of the atom. However, only elastic scattering takes place in this example and this phase factor plays no role in the actual calculation. For simplicity I have dropped it.

97. The atomic wave functions are given as $\psi_n(\xi) = h(r)Y_1^n(\theta, \phi)$, where $\ell_z Y_1^n(\theta, \phi) = n\hbar\, Y_1^n(\theta, \phi)$ (for $n = 0, \pm 1$) and $\xi = (r, \theta, \phi)$. The $Y_1^n(\theta, \phi)$ are the usual spherical harmonics.

98. Bohm (1951, chapter 22) has discussed the general validity of this "impulsive" approximation to a measurement.

99. Mott and Massey 1933, chapter VIII; Messiah 1965, 372–380; Cushing 1975, 509–511; 1990b, 292–294.

100. I have been somewhat elliptical in writing eq. (4.49) since the integration variables in the exponential function should actually be $\{i\, k\, r - i[E(k)\, t/\hbar]\}$ with $E(k) = [(\hbar k)^2/2m] + E_0 - E_n$. It is only *after* the stationary-phase approximation that $k \approx k_0$ so that $E \approx E_n$, and k_n is then given by eq. (4.50). See Messiah (1965, 376–378) for the details of the type of argument required to obtain, finally, the results implied in eqs. (4.49) through (4.51).

101. In the integral of eq. (4.49), the integration variable k is given by the k_n of eq. (4.50). The actual form of $g_n(\hat{y}, k)$ (aside from normalization factors) is the matrix element $\langle k_n|V|\psi_k\rangle$ where V is the *classical* potential energy (*not* including the quantum potential U) and ψ_k is the time-independent scattering state for incident particle momentum k.

102. The separation of the packets is important not so that the particle may *have* a definite trajectory, but so that we may *know* in which packet it is (and in which one it will remain *after* we have ascertained its trajectory). That the separation is never *absolutely* complete is not an issue since once the wave function is essentially zero between the packets, then the probability P of finding a particle there is effectively zero.

103. That is, according to the probability interpretation (i.e., rule [v] of section 3.2), the probability of ψ_n obtaining after the scattering process has been completed should be just the square of the coefficient of ψ_n in eq. (4.49). Once we integrate over the support (a "shell") of the outgoing spherical wave, we obtain eq. (4.52), as claimed.

104. Here I do not display explicitly the time variable t because my arguments now depend only upon the spatial variables denoted collectively by x and y.

105. That is, the guidance condition (eq. [4.2]) and the quantum potential (eq. [4.4]) are invariant under the substitution $\psi \to \alpha\psi$, where α is any constant.

106. Here I use the notation of Bohm's 1952 paper, not that of the previous section, where lowercase referred to "free" variables and uppercase to actual particle positions. In this section, q and r are microsystem observables, represented by operators Q and R, x is the spatial variable for the microsystem, and y and z are spatial coordinates for two (different) apparatuses. The ("momentum") operators P_y and P_z are conjugate, respectively, to y and z.

107. There are many discussions of an interaction Hamiltonian that will result in a "measurement" (see, for example, Bohm [1951, chapter 22]; von Neumann [1955, section VI.3]).

108. If we let $r = y - aqt$, $s = y + aqt$ and chain-rule differentiate, we find that eq. (4.67) reduces to $\dfrac{\partial f_q}{\partial s} \equiv 0$, so that f_q can depend only upon $r = y - aqt$.

109. Strictly speaking, δq is the *smallest* separation of the (*discrete*) qs and eq. (4.75) then becomes $\delta y \geq at\delta q$.

110. A concise summary of Bohm's proof is given in Belinfante (1973, 186).

111. Bohm 1953b.

112. Messiah 1965, 119–121.

113. The basic reason for this uniqueness result follows from the observation that the solution to $\partial f/\partial x + \partial f/\partial t = 0$ is just $f(x, t) = f(x - t, 0)$. This is easily seen by transforming the differential equation to the new variables $r = x + t$, $s = x - t$.

114. Messiah 1965, chapter 3.

115. Hauge and Støvneng 1989, 934.

116. The constants A and B in eq. (4.90d) are determined by the continuity of the logarithmic derivative of ψ at $x = 0$ and at $x = d$.

117. Hauge and Støvneng (1989) refer to these as the *extrapolated* phase times because they hold for $x_1 = 0$, $x_2 = d$.

118. That is, while the incident wave packet is still coming into the barrier, part of it is already being reflected back and there is interference between the two.

119. The problematic point here is that there is no particle position (or trajectory) in the standard interpretation.

120. Hauge and Støvneng 1989, 934–935.

121. This claim can be checked directly from eqs. (4.88) through (4.96). Hauge and Støvneng (1989, 920–921) show that, for an opaque (i.e., $T \ll 1$) rectangular barrier, $\tau_T^\phi = \tau_R^\phi \approx 2m/(\hbar k \kappa)$ and (with $x_1 = 0$, $x_2 = d$) $\tau_D \approx 2mk/(\hbar \kappa k_0^2)$, in clear violation of eq. (4.88).

122. Hauge and Støvneng 1989.

123. Leavens (1990a) has shown that Bohm's theory leads directly to eq. (4.96) and that the corresponding expressions for τ_T and τ_R satisfy eq. (4.88).

124. Dewdney et al. (1992, 1225) show graphically the time evolution of $x(t)$ for a collection of incident particles. Leavens (1990a, 1990b) has done numerical calculations for the various τs.

125. There is some indication that such extreme sensitivity to the shape of the wave packet may obtain (Leavens 1990b, 261).

Chapter Five

1. In making this statement I bracket for now the possibility (but not yet a certainty) discussed in section 4.3 of a class of phenomena that may allow Bohm's theory to make a testable calculation when standard quantum mechanics cannot.
2. Rosenfeld 1961, 384.
3. Rosenfeld 1961, 388. Rosenfeld took the old classical worldview to be one in which a fixed, observer-independent reality was assumed, while a new vision required the observer to play an ineliminable role in the very existence of reality. He wanted this reality to transcend any *particular* observer.
4. Causal interpretations of quantum mechanics have been discussed before (and usually dismissed) by philosophers of science. See, for example, Krips (1987), Albert and Loewer (1989), and Healey (1989).
5. Bohm 1952a, 166.
6. Bohm 1952a, 168.
7. Bohm 1952a, 169.
8. The interferometer is cut from a *single* crystal in order to ensure that the diffraction planes are absolutely parallel (on the scale of atomic dimensions).
9. In this section I often speak of the "spin" of the neutron and of this "spin" being "up" or "down" along one axis or another, even though I stress later that the Bohm interpretation need *not* endow a particle, such as a neutron or an electron, with such an intrinsic property. I employ here the usual language of the Copenhagen interpretation, both because the reader is probably familiar with it and because it is convenient. It will turn out, even in Bohm's interpretation, that certain point particles behave "as though" they possessed the usual quantum-mechanical spin. The actual *calculation* itself does not depend upon any literal commitment to a picture of a "spinning" electron.
10. In figure 5.1 two beams (labeled O and H) are shown emerging (on the far right) from the interferometer. The intensity of neither of these beams separately will remain constant as the interference conditions are varied. No *total* flux is lost, but the intensity of either beam will vary. I am concerned only with the O beam here.
11. Badurek, Rauch, and Tuppinger 1986a, 1986b.
12. Classically one might expect that the superposition of two beams of equal intensity, one spin-up and the other spin-down, should produce a state of zero polarization. Instead, the average spin direction is perpendicular to the initial direction of polarization (cf. eq. [5.15]).
13. There is a specific resonant frequency ω_r at which these coils must be operated (cf. eqs. [5.23] and [5.24]).
14. The locus for a classic discussion of the standard double-slit experiment is Feynman (1965).
15. There is a vast literature on these experiments and on the characteristics and operation of these neutron interferometers. The reader wishing to pursue this in more detail can consult the following references: Rabi (1937); Bloch and Siegert (1940); Drabkin and Zhitnikov (1960); Abragam (1961); Rauch, Treimer, and Bonse (1974); Rauch and Suda (1974); Badurek et al. (1976); Eder and Zeilinger (1976); Bonse and Rauch (1979); Alefeld, Badurek, and Rauch (1981); Summ-

hammer et al. (1983); Badurek, Rauch, and Summhammer (1983); Badurek, Rauch, and Tuppinger (1986a, 1986b); Weinfurter et al. (1988). The main reference for the numbers given in the text is Badurek, Rauch, and Tuppinger (1986a) and references cited therein.

16. Greenberger 1983, 877.

17. Badurek, Rauch, and Tuppinger 1986b, 135. The source of the neutrons is a nuclear reactor.

18. Here χ is the phase shift introduced during the first leg of path II.

19. Dewdney et al., 1984; Badurek, Rauch, and Tuppinger 1986a, 2605.

20. Badurek, Rauch, and Tuppinger 1986a, 2601.

21. Badurek, Rauch, and Tuppinger 1986a, 2605.

22. For a general discussion, set in some historical and philosophical background, of wave-particle duality and the relevance of neutron interference experiments, see Combourieu and Rauch (1992).

23. Vigier 1987; Dewdney et al., 1988; Holland 1993a, chapters 9 and 10.

24. Unnerstall (1990) has shown in detail that these neutron interferometry experiments do not *force* one to accept Einstein's view over Bohr's on an interpretation of quantum mechanics. Either view is logically consistent (and compatible with experimental results).

25. Badurek, Rauch, and Tuppinger 1986a, 2601.

26. The Pauli equation (cf. eqs. [5.54] and [5.55]) is a nonrelativistic wave equation that accommodates the spin degrees of freedom of an "electron" with a two-component wave function.

27. The body's internal structure, such as a moment of inertia, is not relevant then. This is termed the "minimalist approach" by Holland (1993a, 420). It has been advocated by Bell (1966; 1981 [131]) and by Bohm and Hiley (1989). One takes the current density j that satisfies the continuity equation, $\nabla \cdot j + \partial \rho/\partial t = 0$, and *defines* the guidance-condition velocity v as $v = j/\rho$ with $\rho = \Psi^{\dagger}\Psi$. (Realize that for the Pauli equation Ψ will be a two-component spinor.) See eq. (5.57) for an explicit expression for this v. A similar assignment of v also works for the Dirac equation (Bohm and Hiley 1989, 114–118).

28. Bell (1981 [127–133]) discusses this in some detail. The argument is similar to that given in appendix 1.2 to chapter 4 for a beam of particles passing through a Stern-Gerlach apparatus.

29. Vigier 1986; 1987; 1988; Dewdney et al., 1988. Some details of this theory can be found in appendix 2 to this chapter.

30. One can, as Holland does (1993a, chapter 10), advocate instead actually quantizing the equations of motion for a rigid rotator. The technical difference between the treatment of the Pauli equation outlined in the text and that advocated by Holland is that in the former case the configuration space is $R^3 \times SU(2)$, while in the latter it is $R^3 \times SO(3)$.

31. The angles φ and θ give the orientation of the spin vector, while ξ is simply a rotation about the spin axis. For my discussion, the only relevant angles will be φ (which lies in the xy-plane of figure 5.1 and is measured down from the y-axis) and θ (which is measured down from the z-axis of that figure).

32. Dewdney et al., 1988.

33. Vigier 1987, 9.

34. Here, as previously, χ is the phase introduced by the phase shifter in the interferometer. It is *not* an Euler angle.

35. Vigier 1987, 13. See eq. (5.36).

36. For a detailed treatment of the EPRB experiment in the framework of the causal interpretation of quantum field theory, see Kaloyerou (1992).

37. Dewdney et al., 1988; Holland 1993a, chapter 11.

38. This initial state is of the form $\psi_0(x_1, x_2) = f(x_1) f(x_2) \Psi_0$ where Ψ_0 is given in eq. (2.9).

39. The Pauli equation now becomes a two-body version of eqs. (5.54) and (5.55) and the guidance condition is essentially eq. (5.57).

40. The spin is defined as the expectation value of σ as in eq. (5.53).

41. The Pauli equation is solved in impulsive approximation, much as was done for eq. (4.39) to obtain eq. (4.44).

42. The detailed process of angular momentum transfer is much like that discussed in section 5.3.1 and appendix 2 to this chapter.

43. See Holland 1993a, 468–469 for explicit expressions showing this behavior.

44. Valentini 1991b; Holland 1993a, section 11.3.

45. Bohm and Hiley 1989, 114.

46. However, for a very different point of view on this, see Holland (1993a, especially chapter 10), where the spin degrees of freedom are described by continuously variable Euler angles. In particular, he asks how, if individual spin vectors are denied, it is possible to recover the classical theory of spinning bodies in a suitable limit. The essential point to be made here is just this. If one considers *all* particles as point particles and position as the *only* objective, possessed property, then there is no difficulty (either mathematical or conceptual) in generalizing Bohm's approach from the single-particle to the many-body case. If, however, one attempts to generate an even more visualizable model of, say, an extended spinning particle, then one encounters difficulties such as those of intrinsic properties becoming relational. So a coherent causal interpretation simply eschews any extension beyond point particles. Bohm and Hiley (1993, section 10.4) discuss in detail how, in Dirac theory, the magnetic moment of the electron is not an intrinsic localized property associated with "spin," but arises rather from the circulating movement of a *point* particle.

47. Bohm and Hiley (1989, 111) point out that an extended particle would have its periphery moving at speeds in excess of that of light in order to produce the observed value of $\hbar/2$ for the spin of an electron.

48. That position is a possessed property does *not* imply that all position measurements are necessarily faithful (i.e., that the observed value is what the value was before the measurement). See also the comment in note 23 of chapter 4.

49. From eq. (5.3) we see that the *magnitude* of s remains constant so that only its *direction* changes with time (i.e., s precesses about the direction of $\hat{n}$).

50. Equations (5.11) and (5.12) are standard fare and can be found in many references. (See, for example, Cushing 1975, 586–589.)

Notes to Pages 86–88   241

51. In the general case with $\mathbf{B}_1 = B_1 \hat{n} = B_{1_x}\hat{i} + B_{1_y}\hat{j}$ and $\hat{n} = \cos \gamma \, \hat{i} + \sin \gamma \, \hat{j}$, one finds that

$$\exp\left(-i\frac{\pi}{2}\boldsymbol{\sigma}\cdot\hat{n}\right)|+z\rangle = -i\boldsymbol{\sigma}\cdot\hat{n}|+z\rangle = -e^{i(\gamma+\pi/2)}|-z\rangle$$

52. Here and elsewhere in these discussions I often neglect overall phase factors and multiplicative constants in writing down state vectors.

53. Abragam 1961, 22–24, 57; Badurek, Rauch, and Tuppinger 1986a, 2601–2603.

54. A straightforward (if somewhat lengthy) way to verify the result of eq. (5.18) is to begin with the Schrödinger equation

$$i\hbar\frac{\partial\psi}{\partial t} = H\psi = -\boldsymbol{\mu}\cdot\mathbf{B}\psi = \mu\mathbf{B}\cdot\boldsymbol{\sigma}\,\psi$$

and construct a unitary operator $U(t)$ so that $\psi(t) = U(t)\psi(0)$. Define an auxiliary function $\phi(t)$ such that $\phi(t) = e^{i\sigma_z\omega t/2}\psi(t)$. It then follows from direct calculation that the "Schrödinger" equation for $\phi(t)$ becomes

$$i\hbar\frac{\partial\phi}{\partial t} = -\sigma_z\frac{\hbar\omega}{2}\phi + e^{i\sigma_z\omega t/2}\mu\mathbf{B}\cdot\boldsymbol{\sigma}\,e^{-i\sigma_z\omega t/2}\phi \,,$$

which, after more algebra, reduces to

$$i\frac{\partial\phi}{\partial t} = \left[\frac{\omega_1}{2}\sigma_x - \frac{1}{2}(\omega-\omega_L)\sigma_z\right]\phi \,,$$

where $\omega_1 \equiv 2\mu B_1/\hbar$ and $\omega_L \equiv 2\mu B_0/\hbar$. This first-order equation has the solution

$$\phi(t) = \exp\left\{-i\left[\frac{\omega_1}{2}\sigma_x t - \frac{(\omega-\omega_L)}{2}\sigma_z t\right]\right\}\phi(0) \equiv e^{-i\boldsymbol{\beta}\cdot\boldsymbol{\sigma}(t/2)}\phi(0)$$

$$= \left[\cos\left(\frac{\beta t}{2}\right) - i\sin\left(\frac{\beta t}{2}\right)\hat{\boldsymbol{\beta}}\cdot\boldsymbol{\sigma}\right]\phi(0) \,,$$

such that $\phi(t=0) = \phi(0) = \psi(t=0)$. The vector $\boldsymbol{\beta}$ is just that defined in eq. (5.19).

55. The result of eq. (5.22) is obtained by projecting the $\psi_+(t) = U(t)|+z\rangle$ onto $|-z\rangle$ and squaring the modules (in the notation of the previous note and with the use of eq. [5.18]) as

$$|\langle -z|\psi_+(t)\rangle|^2 = |\langle -z|e^{-i\boldsymbol{\beta}\cdot\boldsymbol{\sigma}(t/2)}|+z\rangle|^2$$

$$= \left|\left\langle -z\left|\exp\left\{-i\left[\frac{\omega_1}{2}\sigma_x - \frac{(\omega-\omega_L)}{2}\sigma_z\right](\Delta t)\right\}\right|+z\right\rangle\right|^2$$

$$= \left|\left\langle -z\left|\left[\cos\left(\frac{\beta\Delta t}{2}\right) - i\sin\left(\frac{\beta\Delta t}{2}\right)(\hat{\beta}_x\sigma_x + \hat{\beta}_z\sigma_z)\right]\right|+z\right\rangle\right|^2$$

$$= \sin^2\left(\frac{\beta\Delta t}{2}\right)\hat{\beta}_x^2 = \left[\sin^2\left(\frac{\beta\Delta t}{2}\right)\right]\bigg/\left(1 + \frac{\beta_z^2}{\beta_x^2}\right).$$

56. We have already obtained solutions ψ_1 and ψ_2 to the problems

$$i\hbar\frac{\partial \psi_1}{\partial t} = \left[\frac{1}{2}H_0 + \frac{1}{2}H(\omega t)\right]\psi_1(\omega t), \quad i\hbar\frac{\partial \psi_2}{\partial t} = \left[\frac{1}{2}H_0 + \frac{1}{2}H(\pi - \omega t)\right]\psi_2(\omega t)$$

and we want the solution to

$$i\hbar\frac{\partial \psi}{\partial t} = \left[H_0 + \frac{1}{2}H(\omega t) + \frac{1}{2}H(\pi - \omega t)\right]\psi_2(\omega t)$$

We can readily see that the superposition $\psi = \psi_1 + \psi_2$ does *not* provide a solution since

$$\left[H_0 + \frac{1}{2}H(\omega t) + \frac{1}{2}H(\pi - \omega t)\right](\psi_1 + \psi_2) = i\hbar\frac{\partial \psi_1}{\partial t} + \frac{1}{2}H(\pi - \omega t)\psi_1 + i\hbar\frac{\partial \psi_2}{\partial t}$$
$$+ \frac{1}{2}H(\omega t)\psi_2 = i\hbar\frac{\partial}{\partial t}(\psi_1 + \psi_2) + \frac{1}{2}[H(\pi - \omega t)\psi_1 + H(\omega t)\psi_2] \neq i\hbar\frac{\partial}{\partial t}(\psi_1 + \psi_2) \, .$$

57. There are some fine, but essential, points involving small frequency shifts due to the Bloch-Siegert effect (Bloch and Siegert 1940). Although these are taken into account in a proper analysis of the experiments under discussion here, I neglect them in order to simplify matters. The conclusions I draw are not altered by these small corrections. In particular, the resonance frequency ω_r for 100% efficiency of the spin flipper is modified from the result stated in eq. (5.23) to

$$\omega_r = \omega_L\left(1 + \frac{B_1^2}{16B_0^2}\right) .$$

58. The *total* Hamiltonian for the neutron is $H = H_0 + H_B(t)$, where H_0 represents (essentially) the kinetic energy E and $H_B(t) = -\boldsymbol{\mu} \cdot \boldsymbol{B}(t)$ represents the influence of the time-dependent spin-flip field (which is on only for the time of transit Δt).

59. Eder and Zeilinger 1976.

60. Badurek, Rauch, and Tuppinger 1986a, 2606.

61. To simplify the form of eq. (5.42), I have written as the *constant* phase factor α what is actually the sum of a nuclear phase shift χ (produced by the "χ" slab of figure 5.1) and a phase difference Δ between the two rf generators driving the two flip coils. However, all that is important for my discussion is that $I_0(t)$ of eq. (5.42) has a time dependence [i.e., the $(\Delta\omega)t$ term in the argument of the cosine].

62. It was, incidentally, through the Hamilton-Jacobi formalism that Bohm (1952a) first formulated his theory.

63. Goldstein 1950, chapter 9.

64. Goldstein 1950, sections 4.4–4.5. What I really need here is a means to orient a (body-fixed) set of axes. It is a bit more picturesque simply to speak of a "rigid body."

65. Here I am merely summarizing (essentially by direct transcription, although at times with some license) the types of calculations presented by Vigier (1987) and his coworkers (Dewdney, Holland, and Kyprianidis 1987; Dewdney et al., 1988). See Holland (1993a, chapter 9) for more details. The basic techniques for handling such problems go back to Bohm, Schiller, and Tiomno (1955), who use what are basically the Cayley-Klein parameters of classical rigid-body dynamics to describe the motion of the spinning particle (cf. Goldstein 1950, section 4.5). Bohm, Schiller, and Tiomno (1955) use the Pauli equation as the equation of motion governing the time evolution of these Cayley-Klein parameters.

66. For convenience, in this section I set $\hbar = c = 1$. Of course, the spin *vector* s in eq. (5.53) is *not* the same as the spin *operator* s in eq. (5.1).

67. Bohm, Schiller, and Tiomno 1955, 50. Here I have written the equations for a "spinning" electron (rather than for a neutron with its "anomalous" magnetic moment μ).

68. One can verify directly eq. (5.56) by use of the definition for v in eq. (5.57) and the dynamical equation, eq. (5.54), with the Hamiltonian of eq. (5.55).

69. If, in succession, $\Psi^\dagger$ and then $\Psi^\dagger\boldsymbol{\sigma}$ are contracted on eq. (5.54) and the real and imaginary parts are separated, one (eventually) obtains four equations that can be combined to yield eqs. (5.56), (5.58), and (5.61). The gradient of eq. (5.58) leads to eq. (5.62) (cf. Holland, 1993a, chapter 9). It turns out (Bohm, Schiller, and Tiomno 1955, 62) that (up to signs) ρ and $\xi/2$ are canonically conjugate variables, as are $\rho \cos \theta$ and $\phi/2$. When U and H_s can be neglected compared to the other energies, eq. (5.58) reduces to the equation for a classical spinning particle (Dewdney et al., 1988).

70. Bohm, Schiller, and Tiomno 1955, 64.

71. Bohm, Schiller, and Tiomno 1955, 65–66. If the electron is in an inhomogeneous B-field, the right side of eq. (5.62) has an additional term $\dfrac{e}{m}\left(\dfrac{\partial B}{\partial x_j}\right) \cdot s$.

72. Bohm, Schiller, and Tiomno 1955, 65; Holland 1993a, 391.

73. Dewdney et al., 1988.

74. Vigier 1987, 9.

Chapter Six

1. Brush 1980.
2. Sprinkle 1933; Jammer 1989, section 4.2; Faye 1991.
3. While the last two issues raised here do not necessarily impact upon determinism as such, they do question the then-common acceptance of determinism.
4. Boutroux 1920.
5. Edwards 1967, vol. 1, pp. 355–356, and vol. 7, pp. 180–182.
6. Jammer 1989, 178; Feuer 1974, 205–206. I thank Martin Eger for reminding me of the relevance of Feuer's book regarding the influence of Poincaré and of Bergson on the young de Broglie.
7. Faye 1991. Favrholdt (1992) claims that, in his opinion, virtually all of the

major commentators (e.g., Jammer [1966], Folse [1985], Murdoch [1987], Faye [1991]) have simply been misguided in their representations of the relevance of Høffding's ideas to Bohr's physics. Even Favrholdt, though, does not deny at least the *early* influence of these ideas. De Regt (1993) basically reiterates Favrholdt's position.

8. Jammer 1989, 364. Again, Favrholdt (1992) and de Regt (1993) marginalize these acknowledgments made on public occasions as being vague and motivated by courtesy.

9. I thank Professor Linda Wessels for some helpful information on this topic, although I do not mean to imply that she agrees with my presentation.

10. Forman, 1971.

11. I cite here the corresponding passages from the later (1989) edition of Jammer's book (the first edition of which was published in 1966).

12. Jammer 1989, 174.

13. Jammer 1989, 185.

14. Forman 1971, 3.

15. Forman 1979, 14.

16. Forman 1979, 11.

17. Forman 1971, 7–8.

18. Forman 1984.

19. In 1924–1925, causality was taken as including a space-time description.

20. Heisenberg 1927; Forman 1984, 338.

21. Forman 1984, 342.

22. Forman 1984, 343–344.

23. Forman 1967, iv.

24. Forman 1971, 39.

25. Forman 1971, 62.

26. Forman 1971, 110.

27. Forman 1971, 3.

28. Kaiser (1994) has recently analyzed the personalities of Einstein and of Bohr and claimed that their divergent views on fundamental issues in physics were rooted in their fundamentally different personalities. This difference in personalities, rather than broad cultural forces that were common to both, provides a key, Kaiser argues, to an understanding of the positions of each in their debate over quantum mechanics.

29. Faye 1991. See also Feuer (1974, 112–126, 131–146) on the impression made on Bohr by Høffding and, indirectly through him, by Kierkegaard with their concepts of "qualitative leaps" of ideas from one stage to another and of "renunciation" in the ethical and religious sphere. Bohr, Feuer argues, transferred these concepts to the arena of physical phenomena. Again, Favrholdt (1992) minimizes, but does not totally discount, these factors.

30. Brush 1980.

31. Brush 1980, 393.

32. Wise 1987.

33. See, for example, Hendry (1980) and Kraft and Kroes (1984).

34. Cushing 1990b, 6–7, 216–217.
35. Degen 1989.
36. Frank 1947, 42–43.
37. Edwards 1967, vol. 6, 360–363; Einstein 1949a, 21.
38. Faye 1991.
39. Jammer 1966; Jammer 1989, 173–186.
40. Petersen 1963. Favrholdt (1992) takes a similar position.
41. Pais 1991, 420–425.
42. Pais 1991, 420, 424, respectively.
43. Faye 1991, xi. Hendry (1993), citing Favrholdt (1992), takes exception to a few of Faye's stronger claims, but does grant that some of Bohr's ideas may have been shaped under Høffding's influence. In any event, it is hardly necessary to make a case that all, or even most, of Bohr's ideas about the interpretation of quantum mechanics originated with Høffding. A significant influence is what is relevant.
44. Faye 1991, xi.
45. Faye 1991, xx.
46. Faye 1991, 75. I am quoting Faye's conclusion, which he based on an examination of an essay published by Høffding in 1930.
47. Faye 1991, 146, 158.
48. Murdoch 1987, 72.
49. Wise 1987, 395.
50. A discussion of some of the methodological issues relevant to the shaping of quantum mechanics can be found in Cushing (1982).
51. Einstein 1909a, 189; Klein 1964, 8, 11.
52. Einstein 1909b, 817; Klein 1964, 5.
53. Einstein 1917.
54. Debye 1923.
55. Compton 1922; 1923.
56. Einstein and Ehrenfest 1922, 31.
57. Jammer 1989, 131–132.
58. Einstein 1917; van der Waerden 1968, 76 (my emphasis).
59. de Broglie 1962, vii–viii.
60. Feuer 1974, 206.
61. de Broglie 1973, 12.
62. Meyerson 1936, vi–xiv.
63. Edwards 1967, vol. 5, 307–308.
64. Feuer 1974, 208–212.
65. Mehra and Rechenberg 1982, vol. I, part 2, 578–581.
66. Quoted in Kubli (1970, 55).
67. Abragam 1988, 27.
68. Jammer 1989, 247.
69. de Broglie 1923a; 1923b; 1923c. According to Howard (1990a, 75), as early as 1909 Einstein had begun to think along lines that would lead to his own concept of "ghost" or "guiding" fields.

70. Quoted in Jammer (1989, 258).
71. Heitler 1961, 222.
72. Schrödinger 1926d; 1928, 46; Klein 1964, 4. While both de Broglie and Schrödinger took waves as basic, the former had waves in physical space and the latter in configuration space. This distinction would again be important decades later for Bohm's theory versus de Broglie's revised version of these early ideas.
73. Wessels 1979, 313.
74. MacKinnon 1980, 16, 19.
75. Wessels 1980, 62–63.
76. Mehra 1987, 44.
77. MacKinnon 1982, 246.
78. Schrödinger 1926a, 95; quoted in Klein (1964, 43), from which this translation is taken. Wessels (1975, 2) supports Klein on the importance of this work on gas theory for Schrödinger's development of wave mechanics.
79. Schrödinger 1926a, 95; quoted in Klein (1964, 43), from which this translation is taken. Actually, de Broglie's earlier theory had both a wave and a particle for each microentity, rather than just a wave with the particle represented by a crest on it. I develop this in more detail in chapter 8.
80. Schrödinger 1926b; 1926c; Jammer 1989, 261.
81. Wessels 1975, 28, 58. Kragh (1982), while agreeing with Wessels on the essentials of her account of Schrödinger's route to wave mechanics, does give more weight to the Hamilton-Jacobi analogy for that process.
82. Jammer 1989, 257, 264, note 264. In his second paper on wave mechanics (1926c) Schrödinger cited Courant and Hilbert's book. Wessels (1975, 59) points out that their book is referenced sixteen times in his five wave-mechanics papers.
83. Jammer 1989, 258.
84. Schrödinger 1926d; 1928, 46. 'Perspicuity' is the rendering Schrödinger gave in translation to his original term '*Anschaulichkeit*'. I have already indicated in section 6.2.1 the importance this concept would assume in the interpretation of quantum mechanics.
85. If de Broglie had succeeded with this theory, he might have satisfied Schrödinger's wish for a theory based on waves only, rather than on waves and particles.
86. Hendry 1984, 7. This is Hendry's own opinion, not a quotation of Einstein, de Broglie, or Schrödinger.
87. Beller 1985, 342.
88. Mehra 1975, 146.
89. Jammer 1989, 19.
90. Bohr 1913.
91. Peierls 1959, 176.
92. Beller 1983a, 155ff; Ben-Menahem 1989; Hendry 1984, 132.
93. Bohr 1949, 202.
94. Bohr 1949, 203.
95. Bohr 1949, 206 (my emphasis).
96. Bohr 1949, 222 (my emphasis).
97. Bohr 1949, 232.

98. Bohr 1949, 235 (emphasis in original).

99. Bell's theorem provides an instructive contrast to this situation. Prior to, say, 1964 most physicists *believed* that a "hidden-variables" completion of quantum mechanics was impossible. After Bell's work it was *proven* that a local, deterministic theory, agreeing in all of its predictions with quantum mechanics, is impossible.

100. Rosenfeld 1970, 239.

101. Hendry 1984, 13–14.

102. Pauli 1921, 4.

103. Pauli 1921, 206.

104. Vladimir Jankovic (1991) has made a useful distinction between 'operationalism' as a prescription for being able (at least in principle) to measure or observe a physical entity and 'operationalism' as a way to define the essence or the very meaning of a concept. He argues that the founders of matrix mechanics were, in fact, operationalists only in the former (misnomer) sense of the term.

105. Hendry 1984, 22.

106. Hendry 1984, 30–50; Cassidy 1990, 396. Darrigol (1992, chapter 8) discusses in great detail "the catastrophe of helium" encountered by Born, Pauli, and Heisenberg in the early 1920s with the "old" quantum theory and their conclusion (Darrigol 1992, 177–178) that the laws of (classical) mechanics had failed beyond saving.

107. Peierls 1959, 177.

108. Hendry 1984, 36.

109. Hendry 1984, 64. This "classically nondescribable two-valuedness" was Pauli's own expression (Pauli 1964).

110. Peierls 1959, 178.

111. Hendry 1984, 36.

112. Dresden 1987, 212.

113. AHQP (1962–1964), (first) interview with Slater, p. 33.

114. AHQP (1962–1964), (first) interview with Slater, p. 32.

115. AHQP (1962–1964), (first) interview with Slater, p. 32. That Slater had lost respect for Bohr did not mean, of course, that he would not work with the Copenhagen version of quantum mechanics, which was, so to speak, the only ball game in town.

116. AHQP (1962–1964), (first) interview with Slater, p. 32.

117. Heisenberg 1925, 261.

118. Born 1926c, 68.

119. Beller 1988, 147.

120. Heisenberg 1971, 62–69.

121. Heisenberg 1971, 63.

122. Quoted in Feuer (1974, 83–84).

123. Kragh 1990, 2, 32.

124. Dirac 1977, 118.

125. Kragh 1990, 5.

126. Forman 1971, 58–63.

127. Kalckar 1985, xix.
128. Beller 1985, 349.
129. Hendry 1984, 90.
130. Hendry 1984, 91.
131. Heilbron 1988, 203–204.
132. Heilbron 1988.
133. Quoted in Heilbron (1988, 211).
134. Beller 1992. See also Ben-Menahem (1989) on various uses of the term 'causality' in these early debates.
135. Heilbron 1988, 219.
136. Heilbron 1985, 391.
137. Other indications of the role such factors played can be found in Beller (1983a; 1983b).

Chapter Seven
1. The theme of this chapter is similar to Feyerabend's (1975, chapters 6–12) argument that the diffusion and eventual acceptance of Galileo's theory of motion owed as much to propaganda ploys as to accepted "scientific" styles of reasoning. I thank Peter Holland for pointing this out to me.
2. de Broglie 1925.
3. Raman and Forman 1969, 295.
4. Mehra and Rechenberg 1982, vol. I, part 2, 580.
5. Again, I here use the term 'interpretation' in the same sense that I specified in section 2.1.
6. von Weizsäcker 1977. Of course, this recollection by von Weizsäcker, like Heisenberg's own quoted in the interview below, is a reconstruction long after the events of 1927. Von Weizsäcker (1989, 185) stated about the period from 1930 to 1932, during which he was working with Heisenberg in Leipzig, that "Heisenberg already possessed at that time the contours of his conception of the historical development of theoretical physics, which he formulated more than 15 years later in his *Dialectica* article."
7. Heisenberg 1948.
8. Beller (1985, 349) has also emphasized Heisenberg's belief in the completeness of the mathematical scheme of quantum mechanics. (See quotation from Beller [1985, 349] toward the end of section 6.5.) For Heisenberg, the formalism was more basic, determining the interpretation, and any disagreement between nature and this interpretation (fixed by the formalism) required a change in the formalism. Don Howard informs me that this methodological principle, that a formalism should determine its interpretation uniquely, was a widely respected one in the 1920s (cf. Howard 1992).
9. I thank Professor Thomas S. Kuhn for permission on behalf of the archive and Dr. Helmut Rechenberg for permission on behalf of the executor of Heisenberg's papers to quote this material.
10. AHQP (1962–1964), interview number 7 (of 12), pp. 1–2.
11. AHQP (1962–1964), interview number 7 (of 12), pp. 2–3. Heisenberg

went on to cite his "fluctuations" paper (1926) to support his belief that the formalism fixes the interpretation.

12. AHQP (1962–1964), interview number 7 (of 12), p. 6.
13. AHQP (1962–1964), interview number 7 (of 12), p. 6.
14. AHQP (1962–1964), interview number 7 (of 12), p. 30.
15. Schrödinger 1926d. In a letter to Jordan on April 12, 1926, Pauli had already established the connection between matrix and wave mechanics (prior to the publication of Schrödinger's "equivalence" paper) (van der Waerden 1973, 277). The question of the complete mathematical equivalence of these two versions of the formalism is not as clear-cut as is usually presumed (cf. van der Waerden 1973).
16. Beller 1983b; Cassidy 1992, chapter 11.
17. Pauli 1926.
18. Beller 1983a.
19. Quoted in Cassidy (1992, 225).
20. Heilbron 1988, 204–209.
21. Beller (1992) argues that a careful reading of Bohr's famous Como lecture (delivered in September of 1927, prior to the Solvay congress) shows that Bohr was actually disposed favorably toward the de Broglie–Schrödinger ideas. This is paradoxical since Como is usually taken as the first public formulation of the complementarity principle, which is itself often seen as a cornerstone of the Copenhagen interpretation. There was, at this time, serious disagreement between Bohr and Heisenberg (along with other younger members of the Copenhagen group) over the central issue of continuity/discontinuity. This makes all the more remarkable the ability of this group to reach sufficient consensus to present a unified front on essentials.
22. Schrödinger's *techniques* for calculation, but not his *conceptual framework*, were exploited by the Copenhagen school (e.g., in Born's probability interpretation).
23. AHQP (1962–1964), interview with Kronig, pp. 19–20.
24. Robertson 1979, 156–159. This is not to imply that there was a conspiracy by Bohr and his followers to keep the French from visiting Copenhagen. In large measure the French excluded themselves (even if unconsciously) by their choice of the areas of theoretical physics in which they worked. Quantum physics simply was not a field of great activity in France at this time.
25. Hendry 1984, 107; Darrigol 1992, 287–288.
26. Jammer 1989, 325.
27. It is not my intention here to create the impression that the main topic of discussion at this historic conference was de Broglie's theory and reaction against it. The Bohr-Einstein confrontation is what is most often recalled about this meeting. There is no shortage of commentary on that debate. I emphasize de Broglie's work here because it is less well known and because it is relevant to my case study.
28. de Broglie 1926; 1927a; 1927c.
29. de Broglie 1960, 175; de Broglie 1928a.
30. Pauli 1928, 280–282. Details of Pauli's objection are given in the appendix to this chapter.

250 Notes to Pages 118–125

31. de Broglie 1960, 181; Bohm and Hiley 1982, 1003; Jammer 1974, 113–114.
32. Bohr 1985, 102.
33. de Broglie 1973, 16.
34. Slater 1973, 23.
35. de Broglie 1930, 7.
36. de Broglie 1930, 119–121, 130–132; de Broglie 1960, 183–184.
37. von Neumann 1955, 313–328; de Broglie 1962, 99.
38. Fermi 1926.
39. See eq. (7.7) in the appendix to this chapter.
40. That is, if $f = g + h$ and we write $f = |f|e^{i\theta}$, $g = |g|e^{i\varphi}$, $h = |h|e^{i\chi}$, then θ is a complicated function of $|g|$, $|h|$, φ, and χ.
41. de Broglie 1928b, 282.
42. de Broglie 1928a, 122.
43. de Broglie 1928b, 282; de Broglie 1960, 181. The basic point de Broglie is making here is that a true (*monochromatic*) plane wave would have *infinite* extent and there would be no way to keep such diffracted waves (not all traveling in a common direction) from overlapping (even if each wave were of finite extent along its direction of propagation).
44. Enz and v. Meyenn 1988, 242.
45. de Broglie 1960, 175.
46. de Broglie 1955b, 163.
47. de Broglie 1955b, 163.
48. AHQP (1962–1964), interview with Kronig, p. 24.
49. Vigier 1985a, 653–655.
50. Heilbron 1988, 219. Heisenberg's 1929 lectures have been republished (Heisenberg 1949).
51. Sopka 1988, 102–104.
52. Sopka 1988, 162.
53. Sopka 1988, 166–167.
54. Sopka 1988, 196–201.
55. de Broglie 1960, 175–180; Jammer 1974, 111–113. Pauli's original criticism of de Broglie's model is reprinted in Enz and v. Meyenn (1988, 240–242).

Chapter Eight
1. Jammer 1974, 24–33.
2. Jammer 1974, 32–33.
3. Born 1926a, 1926b; Wessels 1981, 194; Beller 1990.
4. Born 1927, 356.
5. Jordan 1927, 567. On the other hand, Beller (1990, 581) sees Jordan's appraisal of this question as "the most eloquent" in recognizing that indeterminism had not yet been *proven*.
6. Heisenberg 1927; Beller 1990.
7. Madelung 1926; Jammer 1974, 33–36.
8. Madelung 1926, 325; Jammer 1974, 36.
9. Temple 1934, chapter 3. However, all reference to Madelung's work disap-

peared from later versions (e.g., 1951) of Temple's popular little treatise on quantum theory, which, in one edition or another, was published from 1934 at least through 1951.

10. Temple 1934, 47.
11. de Broglie 1955b, 157.
12. de Broglie 1955b, 164–165.
13. de Broglie 1927c.
14. de Broglie 1960, 3.
15. de Broglie 1965, 246–247.
16. de Broglie 1960, 99.
17. That is, eq. (8.8) or (8.33).
18. Notice that de Broglie's phase ϕ is identical (in my notation) with Madelung's S (cf. eqs. [8.7] and [8.8]), while de Broglie's A_0 is Madelung's R (cf. eqs. [8.2] and [8.21]).
19. de Broglie 1960, 100–102; Jammer 1974, 48–49.
20. de Broglie 1960, 103.
21. de Broglie 1927c, 241; de Broglie 1960, 174–175; de Broglie and Brillouin 1928, 135.
22. See eq. (8.8) of appendix 1 to this chapter.
23. de Broglie 1960, 220ff. Mathematical solutions for singularity waves exhibiting the type of behavior de Broglie postulated have *recently* been found (Barut 1990a; 1990b; 1991; Vigier 1991). Barut (1990a and references therein) has essentially completed de Broglie's double-solution program.
24. de Broglie 1955b, 159–160.
25. Fine 1986, 98; Kirsten and Treder 1979, 134–135 and plate facing 129; Howard 1990a, 89–91. I thank Arthur Fine for calling my attention to this document, David Cassidy for confirming its relevance to my interests, Jürgen Renn and Robert Schulmann for providing me a photocopy of this handwritten manuscript (in German) (#2-100-1-5) in April of 1990, and Dr. Ze'ev Rosenkranz for permission granted by the Albert Einstein Archives, the Hebrew University of Jerusalem, Israel, to refer in print to the contents of this manuscript. See appendix 3 to this chapter for some of the details of Einstein's theory.
26. Born 1971, 96.
27. My use of the term "entanglement" here refers (in a colloquial sense) to an influence of one subsystem on the motion of another subsystem. This is not the same as the (technical) use of "entanglement" today for a wave function.
28. Born 1971, 96.
29. Einstein was already aware of such nonlocality by 1927 (Jammer 1974, 117).
30. Fine 1981b, 148. Howard (1990a) has also discussed Einstein's early intimations (at the very least by the mid 1920s) of quantum-mechanical long-range correlations and nonlocal interactions.
31. Kennard 1928, 876.
32. His equation was just my eq. (8.14), which is, of course, the same as Bohm's (my eq. [4.3]).
33. Kennard 1928, 878 (emphases in original). Kennard considers a system of

$3n$ degrees of freedom (i.e., n point particles). His basic claim (by no means obvious, even if true) in this last paragraph is that *one* function of $3n$ variables in configuration space can be *replaced by* (not that it is *equal to*) n functions in physical (three-dimensional) space. That is, he may have had in mind something like

$$\psi(x^{(1)}, x^{(2)}, x^{(3)}, \ldots, x^{(n)}) \to \psi_1(x, x_2, x_3, \ldots, x_n) + \psi_2(x_1, x, x_3, \ldots, x_n) + \cdots + \psi_n(x_1, x_2, x_3, \ldots, x)$$

where each of the ψ_j are actual waves in three-space. For computing the velocity or the quantum potential for the jth particle, one would take gradients, etc., of S or of R with respect to that *free* variable x and then set $x = x_j$, the coordinate of the jth particle. Even such a gloss is not without its problems. The only reason for such an attempt appears to be a desire to have only *real* (i.e., actual, physical) waves in three-dimensional space, rather than a fictitious wave in configuration space. According to Wessels (1979, 333), in the early days of wave mechanics, there were many attempts to "show how the ψ in $3n$-dimensional space [might determine] n waves in 3-dimensional space" and MacKinnon (1980, 32) discusses a series of papers by Schrödinger in 1927 (e.g., Schrödinger 1927) to replace his own ψ-wave with three-dimensional de Broglie waves.

34. Kennard 1928, 879.
35. Rosen 1945, 67.
36. Rosen 1945, 70–71.
37. Schrödinger 1931; Métadier 1931; Jammer 1974, 418–419.
38. Fürth 1933; Schrödinger 1932.
39. Fürth 1933, 157.
40. Nelson 1966.
41. Fürth 1933, 158.
42. Beller 1988.
43. Bopp 1947; Bass 1948; Fényes 1952; Weizel 1953; Nelson 1966.
44. Pinch 1976, 1977, 1979. I thank Robert Clifton for bringing Pinch's work to my attention. Although von Neumann's famous proof is found in his *Mathematische Grundlagen der Quantenmechanik* published in 1932, the bulk of these formal developments had already been published by him in 1927 (von Neumann 1927a, 1927b, 1927c).
45. de Broglie 1962, 99.
46. Pauli 1985, 409–418.
47. Jordan 1936a, 91–92; the quotation is taken from the English translation of this work (Jordan 1944, 109).
48. Jordan 1936b, 283–286.
49. Bohr 1939, 16.
50. Bohr 1939, 38. See also pp. 30–32 there for a summary of von Neumann's remarks on hidden variables.
51. Born 1949, 108–109.
52. Pauli 1948, 309.
53. van Hove 1958, 98.

54. That is, he cited explicitly von Neumann's only caveat (von Neumann 1955, 210).

55. An excellent, complete, and technically very competent discussion of "no-hidden-variables" theorems and of hidden-variables theories in general can be found in Belinfante (1973).

56. von Neumann 1955, section IV.2; Jammer 1974, 265–270.

57. d'Espagnat 1976, 62, 103–104.

58. von Neumann 1955, 305.

59. von Neumann 1955, 323. See appendix 4 to this chapter for a discussion of some of the technicalities of the proof.

60. von Neumann 1955, 325.

61. Bell 1966. See appendix 4 to this chapter for specific examples of the violation of this sum rule.

62. A somewhat formal way to see this is that *if* (for $A = A^\dagger$ and $\|\psi\| = 1$) $\langle A \rangle \equiv \langle \psi | A | \psi \rangle$, then $\Delta A = 0$ requires $\langle A^2 \rangle = \langle \psi | A^2 | \psi \rangle = \|A\psi\|^2 = \langle A \rangle^2 = |\langle \psi | A | \psi \rangle|^2$. The Schwarz inequality (here, an *equality*) then demands that $A\psi = a\psi$. That is, a dispersion-free state for A must be an eigenstate of A, and this would have to be true for *all* of the allowed eigenvalues. But by a well-known result for operators on a Hilbert space, two noncommuting operators do not have a common complete set of eigenvectors.

63. Jammer 1974, 272–277; Pinch 1977, 186.

64. Pinch 1977; 1979.

65. Macrae 1992.

66. Bell 1966. In a dissertation published in Paris in 1964, Mioara Mugur-Schächter had given, prior to Bell's paper, a discussion of the invalidity of von Neumann's proof.

67. See Redhead (1987, chapter 6) for distinctions among different types of contextuality and Pagonis and Clifton (1994) for a discussion of how the contextuality of Bohm's theory avoids the Gleason and Kochen-Specker no-hidden-variables arguments. Belinfante (1973, 95) also discusses why Gleason's theorem does not apply to Bohm's theory.

68. I use a notation somewhat different from Madelung's in order to have a uniform notation for later developments.

69. The condition of eq. (8.8) on v is equivalent to the requirement $\nabla \times v = 0$, that of irrotational flow (i.e., no vortices).

70. This follows since $\partial v_k / \partial x_j = \partial^2 \phi / \partial x_j \partial x_k = \partial^2 \phi / \partial x_k \partial x_j = \partial v_j / \partial x_k$.

71. Madelung 1926; Jammer 1974, 36. In his hydrodynamical interpretation of a fluid of *charge*, Madelung believed that this energy term (my U) should depend not only upon the local charge density ($R = \sqrt{\rho}$) but also upon the total distribution of charge.

72. de Broglie 1960, 3–6; Jammer 1989, 247–249; Wheaton 1983, 288–292; Brown and Martins 1984.

73. Again, I depart from de Broglie's original *notation*, but not from his *ideas* and *arguments*.

74. Let me make two observations regarding eq. (8.23). Temple (1934, 16) pointed out that, if one uses the relativistic addition formula for velocities

$w = \dfrac{v - u}{1 - u\,v/c^2}$ and demands that de Broglie's "periodic phenomenon" be everywhere simultaneous in the rest frame of the particle (i.e., that this phenomenon propagate *instantaneously* so that $w = \infty$), then it follows that $u\,v = c^2$, which is just eq. (8.23). Also, in classical Hamilton-Jacobi theory (cf. Goldstein 1950, 309) the relation $u = E/mv$ holds between the "pilot" wave velocity u in configuration space and the particle velocity v in real space. With the relativistic relation $E = m\,c^2$ this also reduces to eq. (8.23). De Broglie used relativistic arguments to obtain eq. (8.23) and then employed this relation as a guide to generalizing the Hamilton-Jacobi formalism to obtain a wave equation. It was Schrödinger (1926b, 1926c) who finally made a successful extension of that formalism, using de Broglie's relation $\lambda = h/p$, to obtain his own wave equation.

75. Wheaton (1983, 288–292) discusses at length the paradox between relativity and the quantum condition ($\varepsilon = h\nu$), the resolution of which led de Broglie to this coalescence-of-phase argument.

76. My notation differs slightly from Einstein's own. The details of Einstein's arguments are not wholly clear from this fragment, but the *sense* of his project is fairly evident.

77. These arguments as given here can apply only to a *stationary* state. That is, $i\hbar\partial\psi/\partial t = E\psi$ requires that $\partial R/\partial t = 0$, which guarantees that K is real. Einstein introduced a noneuclidean metric (defined by the kinetic-energy quadratic form) in configuration space as had Schrödinger (1926c) when discussing Hamilton's analogy between mechanics and optics in his basic papers on wave mechanics. In this space, surfaces are defined as the wave fronts (i.e., surfaces of constant phase) of the ψ-wave (Goldstein 1950, section 9.8). Einstein then used standard techniques from differential geometry to find the principal curvatures of these surfaces (cf. Sokolnikoff 1951, sections 63–72, especially section 72; note the correspondences between Sokolnikoff and Einstein: $a_{\alpha\beta} \leftrightarrow g_{\mu\nu}$, $b_{\alpha\beta} \leftrightarrow \psi_{\mu\nu}$).

78. Sokolnikoff 1951, 190–193.

79. See any one of a number of introductory treatments of simple extremum problems for quadratic forms (e.g., Cushing 1975, 113–116). The solutions to this are the principal directions of the surface (and these principal directions are mutually orthogonal). In these locally orthogonal coordinates, the kinetic energy becomes diagonal. Einstein employed unit vectors in these principal directions as a basis in which to express the flow field, assuming that the diagonal terms of eq. (8.35) were equal, component by component, to the corresponding diagonal terms of eq. (8.36). I thank Professor Samir K. Bose and Professor William D. McGlinn for conversations about what Einstein's basic motivations may have been for some of the formal manipulations displayed in his manuscript. They are not, however, to be held responsible for any shortcomings in the reconstruction offered here.

80. Just to make the connection with Madelung's scheme fairly explicit, let me point out that a brief calculation establishes the identity

$$K(\text{Einstein}) = \frac{1}{2m}\left(\frac{1}{m}\nabla S\right)^2 - \frac{\hbar^2}{2m}\frac{\nabla^2 R}{R} = K(\text{Madelung}) - \frac{\hbar^2}{2m}\frac{\nabla^2 R}{R} = K + U.$$

Here I have used eq. (8.8) and $\partial R/\partial t = 0$. Once again, we see the presence of the "quantum potential" $U = -\dfrac{\hbar^2}{2m}\dfrac{\nabla^2 R}{R}$. Of course, Einstein does not point out this connection (since the "quantum potential" U is a later concept). If one takes ψ to be real, then $\nabla S = 0$ and K(Einstein) is just the quantum potential.

81. The root of the difficulty is that $\psi_{\mu\nu}$ is not diagonal in indices referring to different subsystems so that the eigenvalues $\lambda_{(\alpha)}$ of eq. (8.40) will not be just the totality of the eigenvalues that would be obtained by treating each subsystem separately. Because of the (block-diagonal) structure of $g_{\mu\nu}$, however, it remains true that $K = K_1 + K_2$ here. (I thank Sheldon Goldstein for pressing me for a clarification of this point.) Einstein's prescription for the actual flow lines was *not* equivalent to eq. (8.8) (or to eq. [4.2]) and the difficulty Einstein had with his own theory does *not* exist for Bohm's.

82. von Neumann 1955, 308.

83. Selleri 1990, 44ff.

84. This is an abbreviated version of a model given by Selleri (1990, 49ff.), although the basic idea of such a counterexample to von Neumann's crucial axiom goes back to Bell's (1966) original paper on this subject. I do not include a discussion (as Selleri does) of how such a model can handle successive measurements in order to continue to yield the proper statistics. Such an extension is not necessary to make my central point—the existence of a counterexample.

85. Equation (8.54) can also be written as $\hat{\sigma} = \boldsymbol{\sigma} \cdot \hat{n}$ with $\hat{n} = \left(\dfrac{1}{\sqrt{2}}, \dfrac{1}{\sqrt{2}}, 0\right)$.

In the notation of figure 8.1, this corresponds to $a = \hat{n}$.

86. These three values are gotten simply by taking a to lie, successively, along the vertical, x-, and y-directions of figure 8.1. There are, of course, two other dispersion-free subensembles of $\boldsymbol{\lambda}$: those for $-\pi/2 \leq \theta \leq -\pi/4$ and for $\pi/4 \leq \theta \leq \pi/2$.

87. That is, $P(\boldsymbol{\lambda})$ vanishes identically in the lower half plane of figure 8.1.

Chapter Nine

1. Recently David Bohm and Basil Hiley (1993) have given a complete summary of the Bohm causal quantum-theory program. Peter Holland (1993a) has also written a detailed presentation of the modern developments in the de Broglie–Bohm quantum theories, as well as a shorter résumé (1993b) on the subject. I thank Professor Bohm and Dr. Hiley for having sent me a copy of the manuscript of their book prior to publication and Dr. Holland for having provided me with the table of contents and preface of his book, also prior to publication.

2. There was some, but it was very little.

3. Halpern 1952; Bohm 1952b. While this was not the *only* technical response to Bohm, there simply was not much. Takabayasi (1952) objected because of doubts about extensions to spin and relativity and a lack of transformation symmetry. Bohm (1953d) successfully responded to each of these criticisms and also indicated how the necessary extensions could be made. Keller (1953) worried

about the status of probability in Bohm's theory. See Jammer (1974, 288–286) on the early reactions to Bohm's theory.

4. Epstein 1953; Bohm 1953a. Epstein had suggested extending the causal interpretation to other representations besides the position one. Bohm argued that there would be mathematical problems with the Hamilton-Jacobi formulation and with the continuity equation in a momentum representation even for so simple a potential as Coulomb's law.

5. Pauli 1948, 307–309.

6. Bohr 1948, 318.

7. Einstein 1948, 324. Einstein (1948, 322) emphasized that quantum mechanics was incompatible with what he termed the "principle of contiguous action" as employed in field theory.

8. Heisenberg 1948. See section 7.1.1.

9. Reichenbach 1948, 340. See also Rosen (1945) and section 8.4 above.

10. At the end of section 4.1.2 I discussed Einstein's difficulty with the $v = 0$ result for certain stationary states.

11. Born 1971, 192.

12. Deltete and Guy 1990, 673, 677.

13. Born 1971, 170. This attachment to Einstein's letter to Born is just a translation of Einstein's 1948 *Dialectica* article.

14. Born 1971, 170.

15. Born 1971, 158.

16. Fine 1986, 57–58; Stachel 1986, 374–377; Deltete and Guy 1990, 679.

17. Quoted in Fine (1986, 57).

18. Stachel 1986, 374. There we are given the direct quote from a 1954 letter by Einstein to a colleague: "The sore point lies less in the renunciation of causality than in the renunciation of the representation of a reality thought of as independent of observation."

19. Born 1971, 221. Pauli writes that "Einstein's point of departure is 'realistic' rather than 'deterministic'."

20. Quoted in Stachel (1986, 375).

21. George 1953, 13–14; quoted in Stachel (1986, 375–376). The translation used here is taken from Stachel.

22. de Broglie 1949, 126–127.

23. In a book published posthumously (de Broglie, 1990) but the manuscript of which is dated 1950–1951, de Broglie supports and defends in great technical detail the Copenhagen view. For example (p. 148), we find: "Every attempt to attribute physically objective characteristics to entities on the atomic level must be abandoned." What is fascinating about the published version of this book is that the editor (Georges Lochak) includes as notes de Broglie's own criticisms made in the manuscript (beginning within a year or two after the original draft had been written). These show graphically de Broglie's rapid and dramatic reversion to his earlier (circa 1927) views on quantum theory and his rejection of the Copenhagen doctrine. David Bohm's paper was the catalyst for this.

24. Bohm and Hiley 1982, 1003, 1014–1015. Here Bohm stated that he was

unaware of de Broglie's model and of Pauli's (1927) criticism when he wrote his 1952 paper. When de Broglie saw the preprint of Bohm's article, he informed Bohm of this earlier work. In an appendix added in proof, Bohm (1952a) pointed out the relevance of his own treatment of the scattering problem to Pauli's specious criticism.

25. de Broglie 1962, vi.
26. de Broglie 1953a, 221; Einstein and Grommer 1927.
27. de Broglie 1953a, 231.
28. de Broglie 1953a, 237.
29. de Broglie 1953b.
30. de Broglie 1953b.
31. de Broglie 1970, 12.
32. de Broglie 1955a, 41.
33. See, however, the caveat expressed in note 33 of chapter 8 above about the possibility of replacing a configuration-space wave with a set of waves in three-space. I thank Harvey Brown and Pan Kaloyerou for discussions on this point. Bohm's (highly nonlocal) quantum force *can* be pictured as acting in three-dimensional space.
34. Croca et al. 1990, 559; Bohm 1952a, 189–193. In section 10.4.2 I discuss quantum field theory in Bohm's program.
35. Wang, Zou, and Mandel 1991; Zou et al., 1992.
36. David Bohm, private communication (interview of 20 July 1989).
37. This is the spirit of Pauli's (1953) remarks on the pilot-wave theory and of Born's Guthrie lecture (1956, 123–139). In that lecture, Born (1956, 123) states that "Pauli, in a recent letter to me, has used the expression 'styles', styles of thinking, styles not only in art, but also in science" and goes on to argue that a return to the style of Newtonian physics is unlikely.
38. Pauli 1953.
39. Pauli 1953, 33.
40. Pauli 1953, 37–38.
41. Born 1971, 207.
42. Bohm 1953b. See appendix 1.4 to chapter 4.
43. Pauli 1953, 38–39.
44. Laurikainen 1988c, 30.
45. Laurikainen 1988c, 31.
46. Private communication (letter of 11 August 1989).
47. Laurikainen 1988a, 6. See also the discussion in Laurikainen (1987, 214).
48. Laurikainen 1987, 213.
49. Laurikainen 1987, 213.
50. Laurikainen 1987, 214.
51. Laurikainen 1987, 217. The quotation is Laurikainen's view of Pauli's position.
52. Laurikainen 1987, 225.
53. Laurikainen 1988b, 6.
54. Laurikainen 1988b, 13. This is Laurikainen's summary of Pauli's views.

55. Laurikainen 1988c, 28–30.
56. Crain (1988) favors the latter position, based on an examination of Pauli's philosophical papers from the 1950s, such as his 1952 *Dialectica* article. It is clear that Crain disagrees in some ways with Laurikainen.
57. Crain 1988, 4–5.
58. About the only serious and technically competent (even if generally negative) discussion of Bohm's theory from the late 1950s until the early 1980s is that of Belinfante (1973), who concluded (p. 118) "that the generalization of Bohm's theory to a relativistic theory meets complications that have so far not yet satisfactorily been resolved."
59. Jammer 1974, 279. David Bohm confirmed this for me in an interview in London in July of 1989. See also Bohm (1957, 110).
60. Cross 1991.
61. Jammer 1974, 251.
62. Cross 1991.
63. Oddly enough, Blokhintsev's 1965 book on the philosophy of quantum mechanics is in many ways less "philosophical" than his 1944 textbook on quantum mechanics (especially the concluding section 139). See Blokhintsev (1968 [Russian edition, 1965]; 1964 [Russian edition, 1944]). It is also interesting to note that his book on the philosophy of quantum mechanics (1968) does not even *cite* Bohm's work.
64. Cross (1991, 742–743) does, however, seem to border on a stronger claim that Marxism may have been a motivating factor for Bohm.
65. Jammer 1988, 692.
66. Cross 1991, 746–750; Jammer 1974, 290–292.
67. Rabinowich 1949; Oppenheimer 1949.
68. Freistadt 1955; 1956a; 1957b. I thank Edward MacKinnon for having brought to my attention Freistadt's (1957b) review article.
69. Freistadt 1956b; 1957a.
70. Freistadt 1953, 229.
71. Freistadt 1953, 233.
72. Rosenfeld 1953; Freistadt 1957a, 28.
73. Heisenberg 1958, 128.
74. Heisenberg 1958, 129.
75. Heisenberg 1955, 28.
76. Heisenberg 1958, 129–130.
77. Margenau 1954, 7.
78. Margenau 1954, 10.
79. Margenau 1954, 11, 13.
80. Bunge 1956.
81. Bunge 1956, 286.
82. Rosenfeld 1957, 41–42.
83. Körner 1957, 46.
84. Jammer 1974, 288–289; Cross 1991, 747.
85. Bohm 1962.

86. Hanson 1963. He had also written a review article (1959) on the Copenhagen interpretation.
87. Hanson 1963, 31.
88. Hanson 1963, 31 (my emphasis).
89. Hanson 1963, 101.
90. Hanson 1963, 76.
91. Hanson 1963, 105.
92. Hanson 1963, 117.
93. Omnès 1992, 340.
94. Omnès 1992, 339.
95. von Weizsäcker and Görnitz 1991, 320.
96. Bohm's later image as a guru was yet a further put-off for many scientists when it came to considering his theory seriously (e.g., his *Implicate Order* and his interest in the views of J. Krishnamurti).
97. One can, of course, fall back on a distinction between 'locality' and 'separability' to claim that standard quantum mechanics is local, but nonseparable (cf. section 4.4). However, the entanglement of distant systems simply passes for nonlocality at the level of discourse indicated here.
98. Vigier (1979a), for example, has given a physical model of the quantum potential that would have some predictions that differ from those of standard quantum mechanics.
99. Bohm and Vigier 1954.
100. Bohm 1957; 1980; Bohm, Hiley, and Kaloyerou 1987; Vigier 1982; 1989b; Dirac 1951b; 1952b.
101. Dirac 1951a; 1952a; 1954; see especially 1952a, 339.
102. Dirac 1952a, 339.
103. Dirac 1952a, 339.
104. Einstein 1924, 85; Saunders and Brown 1991, 13.
105. Hiley 1991, 219.
106. Bohm and Hiley 1989.
107. Jammer 1974, 418; Nelson 1966; 1967.
108. Nelson 1985.
109. Nelson 1966, 1079.
110. See the discussion of Fürth's work (Fürth 1933) in section 8.4 above and also Beller (1988).
111. Nelson 1966, 1081; 1967, 54.
112. Fényes 1952.
113. Baublitz 1988.
114. Wallstrom 1989. See eq. (9.15b) in appendix 1 to this chapter.
115. Carlen and Loffredo 1989.
116. Lehr and Park 1977; Bohm and Hiley 1989.
117. Nelson 1966, 1084.
118. Nelson 1985, 112–113.
119. Nelson 1966, 1085.
120. Nelson 1985, 112.

121. Nelson 1985, 124. Nelson speaks of a "physically real" theory as a local theory.

122. Nelson 1985, 127.

123. I thank W. V. Quine for this observation (letter of 10 May 1992). My comment here is intended not to recommend this theory because it is a realistic one, but to defuse a possible criticism of the causal program so one can continue to consider the question of its being preferable to standard quantum mechanics for *whatever* other reasons one might have (e.g., improved understanding).

124. de Broglie 1926, 1927a, 1927b.

125. See also the recent work of Dürr, Goldstein, and Zanghi (1990, 1992a, 1992b, 1992c, 1993) and Daumer et al. (1994).

126. Nelson 1966, 1967; Bohm and Hiley 1989.

127. Bohm 1953b.

128. See appendix 1.4 to chapter 4.

129. Bohm and Vigier 1954.

130. Bohm 1952a, 1953b.

131. Valentini 1991a, 1991b. See appendix 2 to this chapter. Both Valentini's work, which I outline in this section, and that of Dürr, Goldstein, and Zanghi, which I outline in the next section, represent extensive research programs. My brief summaries do not do justice to their analyses. My purpose here is to draw attention to this body of work so that the interested reader can pursue it in depth.

132. Valentini 1991a, 1991b.

133. Bohm 1952a, 1953b. See appendix 1.4 to chapter 4 for a summary of Bohm's earlier argument.

134. In general a subsystem does not have a wave function, but it does under the conditions that must obtain for a measurement or an observation (cf. appendix 1.2 to chapter 4 and appendix 2 to this chapter).

135. Here each member of the ensemble has the *same* wave function Ψ, but different initial coordinates X_0.

136. The relevant equations are, respectively, eqs. (9.20), (9.21), and (9.18) of appendix 2.

137. Valentini 1991b.

138. Bohm 1952a; Valentini 1991b.

139. Valentini 1991b, 5.

140. Valentini 1991b, 6. Here Valentini cites Bohm, Hiley, and Kaloyerou (1987).

141. Valentini 1991b, 6.

142. Valentini 1991b, 7.

143. Dürr, Goldstein, and Zanghi 1990, 1992a, 1992b, 1992c, 1993; Daumer et al. 1994.

144. Goldstein 1987.

145. Goldstein 1987, 651.

146. Goldstein 1987, 658–659.

147. Goldstein 1987, 665.

148. Dürr, Goldstein, and Zanghi 1990.

149. Bohmian mechanics may be seen as "Aristotelian" in the sense that the equation for the *velocity* (eq. [9.2]) is the key dynamical element, rather than an equation for the *acceleration* as in Newtonian mechanics.

150. Dürr, Goldstein, and Zanghi 1992a, section 3. Effectively, v defines a velocity flow field. Notice that

$$\nabla \psi = \nabla(Re^{iS/\hbar}) = e^{iS/\hbar} \nabla R + \frac{i}{\hbar} Re^{iS/\hbar}\nabla S$$

so that

$$Im\left(\frac{\nabla \psi}{\psi}\right) = Im\left(\frac{\nabla R}{R} + \frac{i}{\hbar} \nabla S\right) = \frac{1}{\hbar} \nabla S$$

151. Or, equivalently, of eq. (8.8), (8.33), or (9.15b).

152. Dürr, Goldstein, and Zanghi 1992a. They (p. 856) term eq. (9.1) the quantum equilibrium hypothesis.

153. Dürr, Goldstein, and Zanghi 1992a, 856–857. They point out that mixing arguments like Valentini's lack mathematical rigor. Of course, one can argue about just what the proper balance is between mathematical rigor and physical reasonableness in a proof.

154. Dürr, Goldstein, and Zanghi 1992a, 858.

155. Dürr, Goldstein, and Zanghi (1992a, 857) state: "When all is said and done, we shall find that [taking $\wp = |\Psi|^2$ as an initial condition] is an adequate description of the situation *provided the quantum equilibrium hypothesis is interpreted in the appropriate way*" (emphasis in original). Much of their paper is taken up with an adequate formulation and explanation of that hypothesis. The status of an ensemble of identical universes could be problematic for Valentini. One might hope for an argument by which the various "pieces" of the (only actually existing) universe mutually interact with each other to randomize the entire system (as is the case for a classical gas).

156. I thank Professor Goldstein for emphasizing this to me.

157. Dürr, Goldstein, and Zanghi 1992c, 10. Actually, this ψ is what they term the effective wave function (i.e., the ψ_n of eq. [4.55]). Each of the M systems of the ensemble is independent of the others.

158. Dürr, Goldstein, and Zanghi 1992a, 859.

159. Dürr, Goldstein, and Zanghi 1992c, 11. This *in-principle* impossibility obtains, of course, only in quantum equilibrium (which is what these authors assume). They refer to this as absolute uncertainty.

160. Dürr, Goldstein, and Zanghi 1992a, 863. For brevity of discussion, I omit mention of their *conditional wave function* $\Psi(x, Y)$ (1992a, 864) and work instead with eq. (9.3)—a not entirely correct procedure.

161. Here q is the *generic* configuration-space variable and Q is the *actual* configuration of the particles. The notation $q = (x, y)$ and $Q = (X, Y)$ signifies the decomposition of the configuration variables into those (x or X) for the subsystem of interest and into those (y or Y) for the rest of the system.

162. See the discussion at the beginning of appendix 1.4 to chapter 4 for the proof of this equivariance. Here Q_t and Ψ_t are the configuration and wave function into which the initial Q and Ψ_0 evolve dynamically (cf. eqs. [9.4] above).

163. Strictly, I should speak of X_t being in a range dx about $x = X_t$ and write $P(X_t \in dx|X_t) = |\psi_t(x)|^2 dx$, but $P = |\psi|^2$ is easier, given the notation I have used in this and other chapters.

164. As elsewhere, I "renormalize" all wave functions to give unit total probability to P.

165. For a statement of the law of large numbers see, for example, Gnedenko and Khinchin (1962, 96).

166. Dürr, Goldstein, and Zanghì 1992c, 10.

167. The question of actually detecting effects due to fluctuations from quantum equilibrium is discussed at length by Valentini (1992, chapter 8). He considers three possible sources of deviations from this equilibrium condition: (i) finite-ensemble corrections, (ii) residual disequilibrium, and (iii) random fluctuations. The third, even if observed, would not likely be taken seriously (much as a kettle that boils *once* when placed on a block of ice), while the first two depend upon whether or not quantum equilibrium obtained *exactly* for the universe at some time. On this last point, the approaches of Valentini and of Dürr, Goldstein, and Zanghì differ.

168. It is important to appreciate that the quantum-equilibrium distribution is *not* generally a time-independent distribution (unlike, say, the classical Maxwell-Boltzmann distribution of statistical mechanics). Since the wave function for a system is typically time dependent, so will be the quantum-equilibrium distribution for P. Justifying the "Born rule" for samples taken at different times (in our *one* universe) requires a different argument (Dürr, Goldstein, and Zanghì 1992a, 874–882) and I do not go into that here.

169. Dürr, Goldstein, and Zanghì 1992a, 892. This statement is made within the framework of Bohmian mechanics, in which nonlocality is always present.

170. Dürr, Goldstein, and Zanghì 1992a, 899.

171. Dürr, Goldstein, and Zanghì 1992b, 1992c; Daumer et al. 1994.

172. Nelson, 1966, 1081. Bohm and Hiley (1989) question the specific *form* of the equation (essentially my eq. [9.12]) that Nelson assumes for *a*. They point out that a different choice for the form of *a* would not lead to the Schrödinger equation. In their opinion, one needs an *independent* (physical) argument for the form of *a* to avoid an appearance of ad hocness in the derivation. On the other hand, Lehr (1976, 43–56) does claim to justify Nelson's choice, as does Vigier (1979b).

173. Einstein 1926.

174. It should be clear that the effects due to *u*-terms are just those produced by Bohm's quantum potential U. The mathematical manipulations are similar to those that led from eq. (4.20) to eqs. (4.22) and (4.23), or from eq. (8.1) to eqs. (8.3) and (8.4). Once the identifications of eqs. (9.15) have been made, eq. (9.10) guarantees that eq. (9.11) is satisfied, since eq. (9.11) becomes just the gradient of eq. (9.10) (cf. eq. [4.22]) and eq. (9.12) becomes the gradient of eq. (4.23).

175. In eq. (9.23) $d\Sigma = d^{3N}X$ is an element of volume in the N-dimensional *configuration* space.

176. Valentini 1991a; Tolman 1938, 136–142, 165–179; Huang 1963, 68–70. As Valentini (1991a, 8) himself points out, this proof, like the classical one, assumes that the initial distribution is not so peculiar or pathological that mixing cannot occur. For example, mixing would not take place for a highly ordered case in which *all* of the gas molecules in a box are moving with their velocities *exactly* parallel to one of the walls of a cubical container.

177. Just as in the classical case, one computes explicitly $d\overline{H}/dt$ and uses the inequality $(x \ln x - y \ln y - x - y) \geq 0$ for all $x \geq 0$, $y \geq 0$ (cf. Tolman 1938, 169).

Chapter Ten

1. Miller (1984) has discussed the key role that visualization played in the history of quantum mechanics (cf. section 6.2.1 above on intuitiveness). Hendry (1985, 392) sees the loss of visualization as a far more radical shift than does Miller. In any event, it was a major change that did not *have* to take place.

2. Of course, a logically possible, much simpler and more direct (but also farther removed from the *actual* historical background of ideas current at the time) route to such a causal interpretation could proceed via the type of argument given by Dürr, Goldstein, and Zanghi (1992a) as indicated in section 9.4.2, especially the discussion of eq. (9.2).

3. Nelson 1966, 1079.

4. Nelson 1985, 112.

5. I pointed out in section 4.1.2 that in a causal theory the Hilbert-space operator formalism is still readily available. See Daumer et al. (1994) and note 15 in chapter 4.

6. Some modern investigators in this field have themselves done just this (cf. Goldstein 1987 [stochastic mechanics] and then Dürr, Goldstein, and Zanghi 1990 [Bohmian mechanics]).

7. Dorling 1987, 17–18.

8. Dorling 1987, 39.

9. During informal conversation, Anton Zeilinger made just such a remark at a quantum theory conference in Joensuu, Finland, in August of 1990.

10. Howard 1990a, 62. I thank Don Howard for reminding me of the relevance of this essay of his to the question raised here and for calling my attention to Pitowsky's (1989) work. Howard also suggests the intriguing possibility that these concerns, on the part of Einstein and Schrödinger, with distant correlations might well have produced a Bell-type theorem in the 1930s, had not Hitler and the war intervened (letter of 12 August 1992).

11. Howard 1991, 134.

12. Jammer 1989, 190; Howard 1990a, 73. Of course, as Howard himself points out here, these particular distant correlations were susceptible to a common-cause explanation (i.e., no necessary nonlocality here).

13. Pitowsky 1989, 38–39.

14. Stapp 1971; 1985; Peres 1978. See appendix 1 to this chapter for such a derivation. It is true that Bohm's spin version of the EPR thought experiment (Bohm 1951, section 22.16) was necessary for the standard derivations of Bell's theorem. (That is, Bell [1986] showed that the original [1935] EPR experiment and wave functions cannot be used to demonstrate nonlocality. There are, though, other spinless-particles cases that can.) However, by the standards of what was accomplished in the late 1920s and early 1930s in quantum mechanics, such a modification would not have been an implausibly great technical or conceptual development. Electron spin (Uhlenbeck and Goudsmit 1925) was on the scene and the spin matrices (Pauli 1927) were available.

15. See section 4.4.
16. Bohr 1985, 101–103; Jammer 1985, 133.
17. Quoted in Bohr 1985, 102 (emphases in original).
18. The fact that Einstein did *not* actually state that one could, really, transmit information with such correlations does indicate that he was likely aware that this could not be done. That is, an actual example or argument that one could communicate, via these quantum correlations, at a speed faster than that of light would have been much more telling than the claim that, in his *opinion,* there is a "contradiction with the relativity postulate."
19. Bohr 1985, 103.
20. Bohr 1985, 104–105.
21. Bohr 1985, 105–106.
22. A particularly simple proof of such a theorem was given by Shimony (1984, 227–228). This is sketched in appendix 2 to this chapter. The instantaneous, long-range correlations of the quantum potential cannot be used for signaling either (Bohm, Hiley, and Kaloyerou 1987, 345; Bohm and Hiley 1993, section 7.7).
23. Heisenberg 1949, 39.
24. Heisenberg 1949, 39. Also quoted in Eberhard 1989, 50–51.
25. Ben-Menahem 1989, 329.
26. Schrödinger 1935, 811; also quoted in Ben-Menahem 1989, 330 (emphases in original). This English translation is taken from Wheeler and Zurek (1983, 156).
27. See appendix 1 for this proof.
28. See appendix 2 for the details of such a theorem.
29. Shimony 1984, 227.
30. Einstein 1949a, 5.
31. Einstein 1949a, 81.
32. Einstein 1949a, 7.
33. Einstein 1954c, 265–266.
34. Einstein 1954e, 260.
35. Einstein 1949a, 13.
36. Einstein 1949a, 29, 33, 81.
37. Einstein 1949a, 27.
38. Einstein 1949a, 81.
39. Einstein 1949a, 83, 85 (emphasis in original).

40. Fine 1986, 1–2, 101–102.
41. Einstein 1949a, 85; quoted in Fine 1986, 103.
42. Fine 1986, 103 (emphases in original).
43. Fine 1986, 104.
44. Howard 1991.
45. Howard 1991, 131.
46. Howard 1991, 133.
47. In Howard's view, determinism, causality, and conservation laws were grouped together as related concepts for Einstein, as were a separable space-time representation and realism.
48. Howard 1985, 173.
49. Howard 1985, 178.
50. Howard 1985, 179. Let me also point out that Fine (1986, 26–63) does not parse Einstein's separation principle into separability and locality as Howard does.
51. Howard 1985, 186–187; Einstein 1948; 1949a.
52. Although it is not necessary for my present purposes to enter into the details here, I would suggest that, if one wants to take Einstein as claiming separation = separability + locality, then one might argue that for Einstein (1949a, 85) 'separability' is akin to 'reality' (i.e., the existence of an objective reality). However, nothing crucial for my discussion turns on this point.
53. As Howard himself has pointed out, the basic question is whether *properties* of interacting systems or *states* of interacting systems are to be centrally featured in the separation principle (Howard 1985, 182 n. 23). I would characterize Bell, Fine, and myself as coming down on the property side and Howard on the state side.
54. See the passage from Fine (1986, 103) quoted in the text (at note 41) about subsystems S_1 and S_2.
55. I take it that finally Einstein would not have seen locality as a truly a priori concept necessary to do physics.
56. Avramesco 1988, 302–303. His examples include a 1933 lecture in Oxford (Einstein 1954c, 269) and an appendix (Einstein 1956, 165–166) to a late (1954) revision of Einstein's 1921 Stafford lectures. Also, Eberhard (1994) has pointed out that, even in the face of experimental violations of Bell inequalities, one can easily reinstate *local* interactions by allowing superluminal (but *finite*) speeds for the propagation of physical influences.
57. Einstein 1949a, 49.
58. Exhibiting explicit wave solutions with particle-like singularities (Vigier 1989a; Barut 1991) could also have satisfied Schrödinger, who wanted a theory with *waves* as the fundamental physical entities. These developments, which *could* (conceptually and logically) have taken place around 1927, could have overcome the resistance of Einstein and of Schrödinger to supporting a "de Broglie–Bohm" program.
59. Tersoff and Bayer 1983. One then averages over all possible statistical weightings.
60. Tersoff and Bayer 1983, 554.

61. Vigier 1985b.
62. Bohm 1952a, 175. The point is that a particle would spend no measurable time in the neighborhood of such points since $U \to +\infty$ from one side and $U \to -\infty$ from the other.
63. Kyprianidis, Sardelis, and Vigier 1984; Cufaro-Petroni et al., 1984.
64. Kyprianidis, Sardelis, and Vigier 1984; Cufaro-Petroni et al., 1984.
65. Cufaro-Petroni et al., 1984, 4–5. On these "mysterious influences" (Einstein 1925), see also Howard (1990a, 67–69).
66. Brillouin 1927. See also de Broglie and Brillouin 1928, 139.
67. Ψ_0 is given in eq. (2.9).
68. The example I discuss here is essentially that used by Eberhard (1989, 58–68).
69. The calculational details are given in appendix 1 to chapter 2.
70. I introduce the subscripts A and B on the probabilities here to label the single and conditional probabilities that refer to different stations.
71. Equations (10.2) and (10.3) are, respectively, just marginal and conditional probabilities (see section 4.4.1).
72. From note 70 in chapter 2, we see that

$$\Psi_0 \to \Psi_1 = \psi_+(\theta) \otimes \left[-i \sin\left(\frac{\theta - \phi}{2}\right) \psi_+(\phi) + \cos\left(\frac{\theta - \phi}{2}\right) \psi_-(\phi) \right].$$

73. This conditional is just $|\langle \psi_+(\theta) \otimes \psi_+(\phi) | \Psi_1 \rangle|^2$. My notation for the conditional of eq. (10.3) is the same as in the definition of eq. (4.13).
74. $P_{AB}(+ +|\theta, \phi) \equiv P_A(+) P_{BA}(+|\theta, \phi, +)$.
75. $\Psi_0 \to \Psi_2 = \left[-i \sin\left(\frac{\theta - \phi}{2}\right) \psi_+(\theta) - \cos\left(\frac{\theta - \phi}{2}\right) \psi_-(\theta) \right] \otimes \psi_+(\phi)$.
76. $P_{AB}(+ +|\theta, \phi) \equiv P_B(+) P_{AB}(+|\theta, \phi, +)$, as in eq. (4.13).
77. Notice that Ψ_2 is *not* simply the Lorentz-transformed value of Ψ_1.
78. Eberhard 1989, 66.
79. Bohm and Hiley 1993, section 12.6.
80. Bohm and Hiley 1993, section 12.6.
81. Valentini (1992, section 5) discusses this at length.
82. Bell 1987c [77].
83. Maxwell 1985. Stein (1991) has questioned the validity of Maxwell's reasoning. I thank J. B. Kennedy for bringing some of this block-universe literature to my attention.
84. Holland 1993a, section 12.1.
85. Valentini 1992, section 4.1.
86. Bohm and Hiley 1989; 1993, chapter 12; Holland 1993a, section 12.2. Just as for the Pauli equation (cf. section 5.3 and appendix 2 to chapter 5), the "spin" can be handled either in terms of additional intrinsic variables that affect only the motion of the particle or in terms of an additional possessed property of an extended body. Because of the problem of defining an extended body in relativity, the former approach is taken with the Dirac equation.

87. Cushing 1988, 25.
88. I thank Dr. Pan Kaloyerou for stressing this in a letter of 17 October 1991.
89. Bohm, Hiley, and Kaloyerou 1987; Bell 1987b. Valentini (1992) gives a unified treatment of causal quantum field theory for bosons and fermions.
90. Holland 1993a, section 12.4. See eq. (10.32) in appendix 3 for the example of the modification of the scalar wave equation.
91. Bohm, Hiley, and Kaloyerou 1987.
92. Bohm 1952a, 193. Belinfante (1973, 198–209) discusses this procedure of Bohm's.
93. The basic idea is that, just as the quantum potential U produces the quantum effects in particle mechanics (recall the discussion in section 4.1.1), so the super quantum potential (cf. the Q of eq. [10.29] and $\delta Q/\delta \phi$ in eq. [10.32]) produces the quantum effects for fields. Detailed treatments of the measurement process (e.g., for the photoelectric and Compton effects) in causal quantum-field theory have appeared in the literature. The earliest was Bohm's (1952a) paper. More recently, Kaloyerou (Bohm, Hiley, and Kaloyerou 1987; Kaloyerou 1992; 1994) has given careful and complete technical presentations for the scalar and for the electromagnetic fields. Lam and Dewdney (1994a; 1994b) illustrate, both mathematically and graphically, the causal quantum-field theory approach to the scalar field in a cavity and to the interaction of this field with matter (a measurement process).
94. See Bohm, Hiley, and Kaloyerou (1987) for the treatment of the boson case and Holland (1988a; 1988b) for discussion of the fermion case. Potential difficulties for the relativistic field-theory program are considered in Kyprianidis (1985). Holland (1993a, section 10.6) treats the fermion field as a collection of spherical rotators.
95. Valentini 1992.
96. Bohm 1952a. It appears, though, that this equivalence is established there with an *example* or two, rather than as a general *proof*. For more on the observational equivalence between standard quantum field theory and causal quantum field theory, see Kaloyerou (1994) and Lam and Dewdney (1994a). Dürr, Goldstein, and Zanghi (1990, 382) discuss the Lorentz-invariant character of the predictions of Bohm's quantum field theory.
97. Bell 1982 [160].
98. Valentini 1991a, 5.
99. Pitowsky 1991.
100. Valentini 1992.
101. A complete presentation and analysis of Bohm's program (Bohm and Hiley 1993), as well as one of the de Broglie–Bohm causal interpretation (Holland 1993a), has recently appeared in print.
102. Albert (1992) reviews the major contending interpretations and finds difficulties for them all. Some fare less badly than others and Albert levels no killing criticisms against Bohm's theory.
103. On the importance of studying the history of certain types of "dead-end" theories, see also Cushing (1990b, especially sections 10.5 and 10.7).
104. Cushing 1990b.

105. The result of eq. (10.1) has been obtained in appendix 1 to chapter 2.

106. I am aware that I am glossing over several important distinctions and subtle points (cf. Cushing and McMullin [1989] for a discussion of some of these). However, my only purpose here is to outline an argument that most physicists would find (and, in fact, have found) convincing (even though it is not absolutely logically compelling). After all, this type of argument is *at least* on a par with many of the "proofs" marshaled in favor of Copenhagen.

107. For convenience I have taken $r = +1$ to correspond to spin up and $r = -1$ to spin down.

108. I appreciate that eq. (10.16) has been written in the so-called Schmidt representation (which is always possible) and that the $\{\psi_j\}$ and $\{\chi_j\}$ may not be eigenvectors of any particularly interesting or easily measurable hermitian operators. Still, eq. (10.17) does make the necessary point about an entangled state.

109. I am not giving the most general argument possible here, but one that illustrates all of the essential ingredients required for one. See Shimony (1984, 227–228) or Ghirardi, Rimini, and Weber (1980) for more general demonstrations.

110. In this section I give only the briefest outline of how one implements a quantum field theory in the causal interpretation. My purpose is to indicate that there *is* a procedure similar to the canonical approach, not to study this causal field-theory program in depth. For details and actual applications, see Bohm, Hiley, and Kaloyerou (1987), Valentini (1992), Holland (1993a), and Kaloyerou (1994).

111. Bohm (1952a) gave the first causal formulation of quantum electrodynamics. In outline, Bohm and Hiley (1984) and then, in much greater detail, Bohm, Hiley, and Kaloyerou (1987) treated the scalar field (as I present it here). Kaloyerou (1994) has further developed the spin-one case. Kaloyerou (1985, 2) claims that backward motion in time results for a particle interpretation of the Klein-Gordon equation, but that the field interpretation is consistent. Holland (1993a, sections 10.6 and 12.4) and Valentini (1992, section 4) discuss both boson and fermion fields and differ considerably on the latter.

112. I set $\hbar = c = 1$. See Goldstein (1950, chapter 11) for standard details on the classical field case.

113. The Euler-Lagrange equation in this case is just $\frac{\partial \mathcal{L}}{\partial \phi} - \nabla \cdot \left(\frac{\partial \mathcal{L}}{\partial \nabla \phi} \right) - \frac{\partial}{\partial t}\left(\frac{\partial \mathcal{L}}{\partial \dot{\phi}} \right) = 0$. By definition the *functional derivative* $\delta L/\delta \phi$ is $\frac{\delta L}{\delta \phi} \equiv \frac{\partial \mathcal{L}}{\partial \phi} - \nabla \cdot \left(\frac{\partial \mathcal{L}}{\partial \nabla \phi} \right)$.

114. Bohm 1952a, 189–191; Bohm, Hiley, and Kaloyerou 1987, 349–351; Holland 1993a, section 12.4.1; Kaloyerou 1994, section 2. Holland (1993a, section 10.6.1) has a direct and simple demonstration of the equivalence of these two procedures for (second) quantizing a field.

Chapter Eleven

1. As will become clear, Duhem and Quine supported rather different theses.
2. Duhem 1974, 180.
3. Duhem 1974, 190.
4. Duhem 1974, 217.
5. Duhem 1974, 217.
6. Quine 1951, 39.
7. Quine 1951, 39–40.
8. Quine 1951, 40.
9. Quine 1951, 41.
10. Quine 1951, 41.
11. Quine 1951, 43.
12. Laudan 1965, 298.
13. Laudan 1965.
14. Harding 1976, xii.
15. I thank Ernan McMullin for impressing this upon me. (I do *not* thereby mean to imply that Professor McMullin agrees with what I have to say here.)
16. Duhem (1974, 189) points out that in mathematics one can face a strict disjunction between two contradictory theorems. Then he asks (p. 190): "Do two hypotheses in physics ever constitute such a strict dilemma? Shall we ever dare to assert that no other hypothesis is imaginable?" Duhem believed that the necessary exhaustive enumeration was impossible.
17. Of course, the type of underdetermination that is of interest here is *permanent* underdetermination so that I do not always repeat this qualification.
18. Laudan and Leplin 1991.
19. Hanson (1963, 116–118) discussed the claimed equivalence between wave mechanics and matrix mechanics and concluded that "in so far as equivalence between ... physical theories is usually 'with respect to observational consequences' ... , there can never be general proofs of theoretic equivalence" (pp. 117–118). His reason for this (p. 213 n. 1) is that the auxiliary hypotheses employed with any theory "form an ill-defined and constantly changing set" so that the "consequences are not mathematically well defined." This does not, of course, prove that two theories *could not* actually be observationally equivalent. After all, I am concerned with science as it is practiced, not with some argument about what could never happen no matter how science might be practiced.
20. Leplin and Laudan 1993.
21. Laudan and Leplin 1991, 454.
22. Laudan and Leplin 1991, 463.
23. I thank Arthur Fine for pointing out to me this distinction.
24. In the Bohm theory, once quantum equilibrium (i.e., $P = |\psi|^2$) has been reached, we are trapped in a universe in which the Copenhagen and Bohm interpretations are *contingently* equivalent. In principle (i.e., if $P \neq |\psi|^2$) they could differ, but once we reach the equilibrium condition $P = |\psi|^2$ (for the universe), from which there can never be an escape, they are "in principle" forever equivalent.

25. Recall that in section 4.3 and appendix 2 to chapter 4 I discussed what *may* be a difference in predictive ability—but no disagreement—between Bohm and Copenhagen. Even if this were to turn out to be the case (and it is by no means yet clear that it can), no choice between them would necessarily be forced by the evidence.

26. Recall the discussion in section 2.4. I assume, perhaps unreasonably, that a scientific realist believes successful scientific theories to be capable of providing reliable and understandable access to the ontology of the world. If one weakens this demand too much, not much remains, except a *belief* in the existence of an objective reality to which we have little access and whose representation by our theories is nebulous beyond meaningful comprehension. In such a situation, is it worth worrying about whether or not one is a realist? By my criterion, Einstein is clearly a realist and Bohr is not. One can, of course, define 'realism' in other ways, but this is the sense in which I use the term here.

27. This type of move is exemplified, for example, in van Fraassen (1980).

28. McMullin (1984) is representative of this line of argument.

29. Ben-Menahem (1989, 307) claims that "to the end of his life [Schrödinger] continued to entertain the idea that some of the laws of nature are irreducibly probabilistic."

30. Schrödinger 1957, 147 (emphases in the original); also quoted in Ben-Menahem (1989, 308).

31. Schrödinger 1957, xvii–xviii; also quoted in Hanle (1979, 267–268).

32. One could require as well coherence with the rest of science, but then one must be careful, in these other areas of science, to separate out those norms and criteria that have been formed by the Copenhagen worldview.

33. By this I mean that Bohm's program has not been applied to as many areas (e.g., gauge field theories) as has the standard approach. In this book I have discussed mainly the empirical equivalence of the two theories for the nonrelativistic regime. However, as I indicated in section 10.4.2, Valentini (1992) has shown how to extend the pilot-wave theory to massive and spin-1/2 fields and even to gravitation.

34. People may certainly continue to ignore this program if they wish. However, they should at least admit the element of prejudice that is present in that choice when a community of scientists overwhelmingly does this.

35. In some ways there is an interesting parallel between the essential role I have given to contingency in the formation and selection of a scientific theory over its competitors and the central place Gould (1989) assigns to contingency in biological evolution. I mentioned this in the preface too.

36. Forman 1971, 87; Kraft and Kroes 1984, 88. Van Kampen (1991, 274) has returned to this question and has offered a proof of the "theorem": "The ontological determinism à la Laplace cannot be proved or disproved on the basis of observation." He argues that ontological determinism is a physically meaningless concept and that only predictability has observational content.

37. Newton 1952, 369 (my emphasis).

38. Newton 1952, 402 (my emphasis).

39. Laplace 1917, 3–4.

40. Laplace 1886, vi–vii; quoted (in an editorial footnote) in Newton (1934, 677). A nearly identical statement appears in Laplace (1917, 4).

41. Earman (1986) analyzes this at great length. 'Meaningful' here is a caveat about the possibility of even being able to conceive of any way the necessary calculations could be done in the real world.

42. In section 10.4.3 I mentioned the possibility that the conceptual framework of Bohmian mechanics, with its actual trajectories for particles, could provide a useful way to discuss chaos for quantum systems. For more details on this, see Dürr, Goldstein, and Zanghi (1992b).

43. The latter requirement has to do with long-term predictive ability.

44. Moulton 1914, 431–432.

45. Poincaré 1892.

46. 'Phase space' is a term for a mathematical space whose coordinate axes are labeled with the position, say x, and the momentum, p, of a system. A point in this space gives the classical state of the system.

47. Such an "act of faith" need not be irrational, but in this case either of two opposite views seems equally compatible with observations.

48. Ford 1983, 43.

49. This is similar to certain aspects of van Fraassen's (1980) constructive empiricism.

50. The sense of this "dictum" is, it seems to me, a central claim of Quine's *Word and Object* (1960). This particular sentence is my own recollection of a statement made by Quine during a public lecture at Wittenberg University in late April 1992.

REFERENCES

Abragam, A. (1961), *The Principles of Nuclear Magnetism.* Oxford: Clarendon Press.
——— (1988), "Louis Victor Pierre Raymond de Broglie," *Biographical Memoirs of Fellows of the Royal Society* 34: 21–41.
Achinstein, P. (1985), "The Pragmatic Character of Explanation," in Asquith and Kitcher (1985), pp. 275–292.
Aharonov, Y., J. Anandan, and L. Vaidman (1993), "Meaning of the Wave Function," *Physical Review A* 47: 4616–4626.
AHQP (1962–1964), *Archive for the History of Quantum Physics,* interview transcripts on deposit in the Department of History of Science, University of California, Berkeley. [A catalogue of the contents of this archive has been published by Kuhn et al. (1967).]
Albert, D. Z. (1992), *Quantum Mechanics and Experience.* Cambridge, MA: Harvard University Press.
Albert, D. Z., and B. Loewer (1989), "Two No-Collapse Interpretations of Quantum Theory," *Nous* 23: 169–186.
Albert, D. Z., and L. Vaidman (1989), "On a Proposed Postulate of State-Reduction," *Physics Letters A* 139: 1–4.
Albeverio, S., G. Casati, U. Cattaneo, D. Merlini, and R. Moresi (eds.) (1990), *Stochastic Processes, Physics and Geometry.* Singapore: World Scientific Publishing.
Alefeld, B., G. Badurek, and H. Rauch (1981), "Observation of the Neutron Magnetic Resonance Energy Shift," *Zeitschrift für Physik B* 41: 231–235.
Aspect, A., J. Dalibard, and G. Roger (1982), "Experimental Tests of Bell's Inequality Using Time-Varying Analyzers," *Physical Review Letters* 49: 1804–1807.
Asquith, P. D., and R. N. Giere (eds.) (1981), *Proceedings of the 1980 Biennial Meeting of the Philosophy of Science Association,* vol. 2. East Lansing, MI: Philosophy of Science Association.
Asquith, P. D., and P. Kitcher (eds.) (1984), *Proceedings of the 1984 Biennial Meeting of the Philosophy of Science Association,* vol. 1. East Lansing, MI: Philosophy of Science Association.
——— (eds.) (1985), *Proceedings of the 1984 Biennial Meeting of the Philosophy of Science Association,* vol. 2. East Lansing, MI: Philosophy of Science Association.

Avramesco, A. (1988), "The Einstein-Bohr Debate: I. The Background," in van der Merwe, Selleri, and Tarozzi (1988), vol. 1, pp. 299–308.

Bachelard, G. (1934), *Le nouvel esprit scientifique*. Paris: Presses Universitaires de France. [Appeared in translation (A. Goldhammer, 1984) as *The New Scientific Spirit*. Boston: Beacon Press.]

Bacon, F. (1620), *The New Organon*. London. [Appeared in translation (F. H. Anderson, ed., 1960). Indianapolis, IN: Bobbs-Merrill.]

Badurek, G., H. Rauch, and J. Summhammer (1983), "Time-Dependent Superposition of Spinors," *Physical Review Letters* 51: 1015–1018.

Badurek, G., H. Rauch, and D. Tuppinger (1986a), "Neutron Interferometric Double-Resonance Experiment," *Physical Review A* 34: 2600–2608.

——— (1986b), "Polarized Neutron Interferometry," in Greenberger (1986), pp. 133–146.

Badurek, G., H. Rauch, A. Zeilinger, W. Bauspiess, and U. Bonse (1976), "Phase-Shift and Spin-Rotation Phenomena in Neutron Interferometry," *Physical Review D* 14: 1177–1181.

Ballentine, L. E. (1970), "The Statistical Interpretation of Quantum Mechanics," *Reviews of Modern Physics* 42: 358–381. [Also appears in Ballentine (1988).]

——— (ed.) (1988), *Foundations of Quantum Mechanics Since the Bell Inequalities*. College Park, MD: American Association of Physics Teachers.

——— (1990), *Quantum Mechanics*. Englewood Cliffs, NJ: Prentice-Hall.

Barker, P., and C. G. Shugart (eds.) (1981), *After Einstein*. Memphis, TN: Memphis State University Press.

Barut, A. O. (1990a), "Quantum Theory of Single Events: Localized De Broglie Wavelets, Schrödinger Waves, and Classical Trajectories," *Foundations of Physics* 20: 1233–1240.

——— (1990b), "$E = \hbar\omega$," *Physics Letters A* 143: 349–352.

——— (1991), "Quantum Theory of Single Events," in Lahti and Mittelstaedt (1991), pp. 31–46.

Bass, J. (1948), "Lois de probabilité, équations hydrodynamiques et mécanique quantique," *Revue Scientifique* 86: 643–652.

Bates, D. R. (ed.) (1962), *Quantum Theory*, vol. 3. New York: Academic Press.

Baublitz, M. (1988), "Derivation of the Schrödinger Equation from a Stochastic Theory," *Progress of Theoretical Physics* 80: 232–244.

Behm, R. J., N. García, and H. Rohrer (eds.) (1990), *Scanning Tunneling Microscopy and Related Methods*. Dordrecht: Kluwer Academic Publishers.

Belinfante, F. J. (1973), *A Survey of Hidden-Variable Theories*. Oxford: Pergamon Press.

Bell, J. S. (1964), "On the Einstein Podolsky Rosen Paradox," *Physics* 1: 195–200. [Reprinted in Bell (1987a), pp. 14–21.]

——— (1966), "On the Problem of Hidden Variables in Quantum Mechanics," *Reviews of Modern Physics* 38: 447–452. [Reprinted in Bell (1987a), pp. 1–13.]

——— (1975), "On Wave Packet Reduction in the Coleman-Hepp Model," *Helvetica Physica Acta* 48: 93–98. [Reprinted in Bell (1987a), pp. 45–51.]

——— (1980), "De Broglie–Bohm, Delayed-Choice Double-Slit Experiment, and Density Matrix," *International Journal of Quantum Chemistry:* Quantum Chemistry Symposium 14: 155–159. [Reprinted in Bell (1987a), pp. 111–116.]
——— (1981), "Quantum Mechanics for Cosmologists," in Isham, Penrose, and Sciama pp. 611–637. [Reprinted in Bell (1987a), pp. 117–138.]
——— (1982), "On the Impossible Pilot Wave," *Foundations of Physics* 12: 989–999. [Reprinted in Bell (1987a), pp. 159–168.]
——— (1986), "EPR Correlations and EPW Distributions," in Greenberger (1986), pp. 263–266. [Reprinted in Bell (1987a), pp. 196–200.]
——— (1987a), *Speakable and Unspeakable in Quantum Mechanics.* Cambridge: Cambridge University Press. [Page references in the footnotes are made to this collection in brackets for all of Bell's articles that are reprinted in it.]
——— (1987b), "Beables for Quantum Field Theory," in Bell (1987a), pp. 173–180.
——— (1987c), "How to Teach Special Relativity," in Bell (1987a), pp. 67–80.
——— (1990), "Against 'Measurement'," in Miller (1990), pp. 17–31. [Reprinted in *Physics World*, no. 8 (Aug., 1990), 33–40.]
Beller, M. (1983a), "The Genesis of Interpretations of Quantum Physics, 1925–1927." Unpublished Ph. D. dissertation, University of Maryland.
——— (1983b), "Matrix Theory Before Schrödinger," *Isis* 74: 469–491.
——— (1985), "Pascual Jordan's Influence on the Discovery of Heisenberg's Indeterminacy Principle," *Archive for History of Exact Sciences* 33: 337–349.
——— (1988), "Experimental Accuracy, Operationalism, and Limits of Knowledge—1925 to 1935," *Science in Context* 2: 147–162.
——— (1990), "Born's Probabilistic Interpretation: A Case Study of 'Concepts in Flux'," *Studies in History and Philosophy of Science* 21: 563–588.
——— (1992), "The Birth of Bohr's Complementarity: The Context and the Dialogues," *Studies in History and Philosophy of Science* 23: 147–180.
Ben-Menahem, Y. (1989), "Struggling with Causality: Schrödinger's Case," *Studies in History and Philosophy of Science* 20: 307–334.
——— (1990), "Equivalent Descriptions," *British Journal for the Philosophy of Science* 41: 261–279.
Berry, M. (1989), "Quantum Chaology, Not Quantum Chaos," *Physica Scripta* 40: 335–336.
Bialynicki-Birula, I. (1984), "Entropic Uncertainty Relations," *Physics Letters A* 103: 253–254.
Bialynicki-Birula, I., and J. L. Madajczyk (1985), "Entropic Uncertainty Relations for Angular Distributions," *Physics Letters A* 108: 384–386.
Bialynicki-Birula, I., and J. Mycielski (1975), "Uncertainty Relations for Information Entropy in Wave Mechanics," *Communications in Mathematical Physics* 44: 129–132.
Bitsakis, E. I., and C. A. Nicolaides (eds.) (1989), *The Concept of Probability.* Dordrecht: Kluwer Academic Publishers.
Black, T. D., M. M. Nieto, H. S. Pilloff, M. O. Scully, and R. M. Sinclair (eds.) (1992), *Foundations of Quantum Mechanics.* Singapore: World Scientific Publishing.

Bloch, F., and A. Siegert (1940), "Magnetic Resonance for Nonrotating Fields," *Physical Review* 57: 522–527.

Blokhintsev, D. I. (1964), *Quantum Mechanics*. Dordrecht: D. Reidel Publishing.

——— (1968), *The Philosophy of Quantum Mechanics*. Dordrecht: D. Reidel Publishing.

Blum, W., H.-P. Dürr, and H. Rechenberg (eds.) (1989), *Werner Heisenberg: Gesammelte Werke, Series A, Part II*. Berlin: Springer-Verlag.

Bohm, D. (1951), *Quantum Theory*. Englewood Cliffs, NJ: Prentice-Hall.

——— (1952a), "A Suggested Interpretation of the Quantum Theory in Terms of 'Hidden' Variables, I and II," *Physical Review* 85: 166–179, 180–193. [Reprinted in Wheeler and Zurek, pp. 367–396.]

——— (1952b), "Reply to a Criticism of a Causal Re-interpretation of the Quantum Theory," *Physical Review* 87: 389–390.

——— (1953a), "Comments on a Letter Concerning the Causal Interpretation of the Quantum Theory," *Physical Review* 89: 319–320.

——— (1953b), "Proof That Probability Density Approaches $|\psi|^2$ in Causal Interpretation of the Quantum Theory," *Physical Review* 89: 458–466.

——— (1953c), "A Discussion of Certain Remarks by Einstein on Born's Probability Interpretation of the ψ-Function," in Born, pp. 13–19.

——— (1953d), "Comments on an Article of Takabayasi Concerning the Formulation of Quantum Mechanics with Classical Pictures," *Progress of Theoretical Physics* 9: 273–287.

——— (1957), *Causality and Chance in Modern Physics*. Philadelphia: University of Pennsylvania Press.

——— (1962), "Hidden Variables in the Quantum Theory," in Bates, pp. 345–387.

——— (1980), *Wholeness and the Implicate Order*. London: Routledge & Kegan Paul.

——— (1985), "Imagined Worlds: Visions of Theoretical Physics," *Nature* 314: 689–690.

——— (1987), "Hidden Variables and the Implicate Order," in Hiley and Peat, pp. 33–45.

Bohm, D., and B. J. Hiley (1982), "The De Broglie Pilot Wave Theory and the Further Development of New Insights Arising out of It," *Foundations of Physics* 12: 1001–1016.

——— (1984), "Measurement Understood through the Quantum Potential Approach," *Foundations of Physics* 14: 255–274.

——— (1985), "Unbroken Quantum Realism, from Microscopic to Macroscopic Levels," *Physical Review Letters* 55: 2511–2514.

——— (1989), "Non-locality and Locality in the Stochastic Interpretation of Quantum Mechanics," *Physics Reports* 172: 93–122.

——— (1993), *The Undivided Universe: An Ontological Interpretation of Quantum Theory*. London: Routledge, Chapman and Hall.

Bohm, D., B. J. Hiley, and P. N. Kaloyerou (1987), "An Ontological Basis for the Quantum Theory," *Physics Reports* 144: 321–375.

Bohm, D., R. Schiller, and J. Tiomno (1955), "A Causal Interpretation of the Pauli Equation (A)," *Supplemento al Nuovo Cimento* 1: 48–66.

Bohm, D., and J.-P. Vigier (1954), "Model of the Causal Interpretation of Quantum Theory in Terms of a Fluid with Irregular Fluctuations," *Physical Review* 96: 208–216.

Bohr, N. (1913), "On the Constitution of Atoms and Molecules," *Philosophical Magazine* 26: 1–25.

——— (1927), "The Quantum Postulate and the Recent Development of Atomic Theory," *Nature* 121: 580–590. [Also appears in Wheeler and Zurek, pp. 87–126.]

——— (1934), *Atomic Theory and the Description of Nature*, Cambridge, Cambridge University Press.

——— (1935), "Can Quantum-Mechanical Description of Physical Reality be Considered Complete?" *Physical Review* 48, 696–702. [Reprinted in Wheeler and Zurek, pp. 145–151.]

——— (1939), "The Causality Problem in Atomic Physics," in *New Theories in Physics*. Paris: International Institute of Intellectual Cooperation, pp. 11–45.

——— (1948), "On the Notions of Causality and Complementarity," *Dialectica* 2: 312–319.

——— (1949), "Discussion with Einstein on Epistemological Problems in Atomic Physics," in Schilpp, pp. 199–241.

——— (1961), *Atomic Physics and Human Knowledge*. New York: Science Editions.

——— (1985), *Collected Works*, vol. 6. Amsterdam: North-Holland Publishing.

Bonse, U., and H. Rauch (eds.) (1979), *Neutron Interferometry*. Oxford: Clarendon Press.

Boorse, H. A., and L. Motz (1966), *The World of the Atom*, 2 vols. New York: Basic Books.

Bopp, F. (1947), "Quantenmechanische Statistik und Korrelationsrechnung," *Zeitschrift für Naturforschung* 2a: 202–216.

Born, M. (1926a), "Zur Quantenmechanik der Stossvorgänge," *Zeitschrift für Physik* 37: 863–867.

——— (1926b), "Quantenmechanik der Stossvorgänge," *Zeitschrift für Physik* 38: 803–827.

——— (1926c), *Problems of Atomic Dynamics*. Cambridge, MA: Massachusetts Institute of Technology.

——— (1927), "Physical Aspects of Quantum Mechanics," *Nature* 119: 354–357.

——— (1936), *Atomic Physics*. New York: G. E. Stechert.

——— (1949), *Natural Philosophy of Cause and Chance*. Oxford: Clarendon Press.

——— (1951), *The Restless Universe*. New York: Dover Publications.

——— (1953), *Scientific Papers*. New York: Hafner Publishing.

——— (1956), *Physics in My Generation*. London: Pergamon Press.

——— (1971), *The Born-Einstein Letters*. New York: Walker.

Boutroux, É. (1920), *The Contingency of the Laws of Nature*. Chicago: Open Court Publishing.

Brillouin, L. (1927), "Comparaison des différentes statistiques appliquées aux problèmes de quanta," *Annales de Physique* 7: 315–331. [Also appears as "A Comparison of the Different Statistical Methods Applied to Quantum Problems," in de Broglie and Brillouin, pp. 139–151.]

Brown, H., and R. Harré (eds.) (1988), *Philosophical Foundations of Quantum Field Theory*. Oxford: Oxford University Press.

Brown, H. R., and R. de A. Martins (1984), "De Broglie's Relativistic Phase Waves and Wave Groups," *American Journal of Physics* 52: 1130–1140.

Brush, S. G. (1980), "The Chimerical Cat: Philosophy of Quantum Mechanics in Historical Perspective," *Social Studies of Science* 10: 393–447.

Bub, J. (1968), "The Daneri-Loinger-Prosperi Quantum Theory of Measurement," *Il Nuovo Cimento* 57B: 503–520.

——— (1988), "How to Solve the Measurement Problem in Quantum Mechanics," *Foundations of Physics* 18: 701–722.

Bunge, M. (1956), "Survey of the Interpretations of Quantum Mechanics," *American Journal of Physics* 24: 272–286.

Butterfield, J. (1989), "A Space-Time Approach to the Bell Inequality," in Cushing and McMullin, pp. 114–144.

Carlen, E. A., and M. I. Loffredo (1989), "The Correspondence between Stochastic Mechanics and Quantum Mechanics on Multiply Connected Configuration Spaces," *Physics Letters A* 141: 9–13.

Cartwright, N. (1989), *Nature's Capacities and Their Measurement*. Oxford: Clarendon Press.

Cassidy, D. C. (1990), "Werner Karl Heisenberg" in Holmes, vol. 17, pp. 394–403.

——— (1992), *Uncertainty: The Life and Science of Werner Heisenberg*. New York: W. H. Freeman.

Castell, L., M. Drieschner, and C. F. von Weizsäcker (eds.) (1977), *Quantum Theory and the Structures of Time and Space*, vol. 2. Munich: Carl Hanser Verlag.

Clauser, J. F., and A. Shimony (1978), "Bell's Theorem: Experimental Tests and Implications," *Reports on Progress in Physics* 41: 1881–1927.

Colodny, R. G. (1986), *From Quarks to Quasars*. Pittsburgh: University of Pittsburgh Press.

Combourieu, M.-C, and H. Rauch (1992), "The Wave-Particle Dualism in 1992: A Summary," *Foundations of Physics* 22: 1403–1434.

Compton, A. H. (1922), "Secondary Radiations Produced by X-rays, and Some of Their Applications to Physical Problems," *Bulletin of the National Research Council* 4 (part 2, no. 20): 1–56.

——— (1923), "A Quantum Theory of the Scattering of X-rays by Light Elements," *Physical Review* 21: 483–502.

Crain, S. (1988), "Some Remarks on the Philosophy of Wolfgang Pauli." Unpublished University of Notre Dame manuscript.

Croca, J. R., A. Garuccio, V. L. Lepore, and R. N. Moreira (1990), "Quantum-

Optical Predictions for an Experiment on the De Broglie Waves Detection," *Foundations of Physics Letters* 3: 557–564.

Cross, A. (1991), "The Crisis in Physics: Dialectical Materialism and Quantum Theory," *Social Studies of Science* 21: 735–759.

Cufaro-Petroni, N., A. Kyprianidis, Z. Maric, D. Sardelis, and J.-P. Vigier (1984), "Causal Stochastic Interpretation of Fermi-Dirac Statistics in Terms of Distinguishable Non-locally Correlated Particles," *Physics Letters A* 101: 4–6.

Cushing, J. T. (1975), *Applied Analytical Mathematics for Physical Scientists*. New York: John Wiley & Sons.

——— (1982), "Models and Methodologies in Current Theoretical High-Energy Physics," *Synthese* 50: 5–101.

——— (1984a), "The Convergence and Content of Scientific Opinion," in Asquith and Kitcher (1984), pp. 211–223.

——— (1984b), Review of *Scientific Explanation and Atomic Physics*, by Edward M. MacKinnon, *Erkenntnis* 21: 89–100.

——— (1988), "Foundational Problems in and Methodological Lessons from Quantum Field Theory," in Browne and Harré, pp. 25–39.

——— (1989), "A Background Essay," in Cushing and McMullin, pp. 1–24.

——— (1990a), "Is Scientific Methodology Interestingly Atemporal?" *British Journal for the Philosophy of Science* 41: 177–194.

——— (1990b), *Theory Construction and Selection in Modern Physics: The S Matrix*. Cambridge: Cambridge University Press.

——— (1991a), "Quantum Theory and Explanatory Discourse: Endgame for Understanding?" *Philosophy of Science* 58: 337–358.

——— (1991b), "Copenhagen Hegemony: *Need* It Be So?" in Lahti and Mittelstaedt (1991), pp. 89–98.

——— (1992a), "Causal Quantum Theory: Why a Nonstarter?" in Selleri, pp. 37–68.

——— (1992b), "Historical Contingency and Theory Selection in Science" in Hull, Forbes, and Okruhlik, pp. 446–457.

——— (1992c), "What If Bell Had Come *before* 'Copenhagen'?" in van der Merwe, Selleri, and Tarozzi (1992), pp. 125–134.

——— (1993a), "Underdetermination, Conventionalism and Realism: The 'Copenhagen' vs. the Bohm Interpretation of Quantum Mechanics," in French and Kamminga, pp. 261–278.

——— (1993b), "A Bohmian Response to Bohr's Complementarity," in Faye and Folse, pp. 57–75.

Cushing, J. T., and E. McMullin (eds.) (1989), *Philosophical Consequences of Quantum Theory: Reflections on Bell's Theorem*. Notre Dame, IN: University of Notre Dame Press.

Daneri, A., A. Loinger, and G. M. Prosperi (1962), "Quantum Theory of Measurement and Ergodicity Conditions," *Nuclear Physics* 33: 297–319.

——— (1966), "Further Remarks on the Relations between Statistical Mechanics and Quantum Theory of Measurement," *Il Nuovo Cimento* 44B: 119–128.

Darrigol, O. (1992), *From c-Numbers to q-Numbers*. Berkeley, CA: University of California Press.

Daumer, M., D. Dürr, S. Goldstein, and N. Zanghi (1994), "On the Role of Operators in Quantum Theory" (in preparation).

de Broglie, L. (1923a), "Ondes et quanta," *Comptes Rendus des Séances de l'Académie des Sciences* 177: 507–510.

—— (1923b), "Quanta de lumière, diffraction et interférences," *Comptes Rendus des Séances de l'Académie des Sciences* 177: 548–550.

—— (1923c), "Les quanta, la théorie cinétique des gaz et le principe de Fermat," *Comptes Rendus des Séances de l'Académie des Sciences* 177: 630–632.

—— (1925), "Recherches sur la théorie des quanta," *Annales de Physique* 3: 22–128.

—— (1926), "Sur la possibilité de relier phénomènes d'interférence et de diffraction à la théorie des quanta de lumière," *Comptes Rendus des Séances de l'Académie des Sciences* 183: 447–448.

—— (1927a), "La structure atomique de la matière et du rayonnement et la mécanique ondulatoire," *Comptes Rendus des Séances de l'Académie des Sciences* 184: 273–274.

—— (1927b), "Sur le rôle des ondes continues Ψ en mécanique ondulatoire," *Comptes Rendus des Séances de l'Académie des Sciences* 185: 380–382.

—— (1927c), "La mécanique ondulatoire et la structure atomique de la matière et du rayonnement," *Journal de Physique et le Radium* 8: 225–241. [Appears in translation as "The Wave Mechanics and the Atomic Structure of Matter and of Radiation," in de Broglie and Brillouin, pp. 113–138.]

—— (1928a), "Nouvelle dynamique des quanta," in *Electrons et Photons*, pp. 105–132.

—— (1928b), "Discussion générale des idées nouvelles émises," in *Electrons et Photons*, p. 282.

—— (1930), *An Introduction to the Study of Wave Mechanics* (translated from the French by H. T. Flint). London: Methuen.

—— (1936), "Préface" in Meyerson, pp. vii–xiv.

—— (1948), "Sur la complémentarité des idées d'individu et de système," *Dialectica* 2 (3/4): 325–330.

—— (1949), "A General Survey of the Scientific Work of Albert Einstein," in Schilpp, pp. 107–127.

—— (1953a), *The Revolution in Physics*. New York: Noonday Press.

—— (1953b), *La physique quantique restera-t-elle indéterministe?* Paris: Gauthier-Villars.

—— (1955a), "Une interprétation nouvelle de la mécanique ondulatoire est-elle possible?" *Il Nuovo Cimento* 1: 37–50.

—— (1955b), *Physics and Microphysics* (translated from the French by M. Davidson). New York: Pantheon Books.

—— (1960), *Non-linear Wave Mechanics: A Causal Interpretation* (translated from the French by A. J. Knodel and J. C. Miller). Amsterdam: Elsevier Publishing.

—— (1962), *New Perspectives in Physics* (translated from the French by A. J. Pomerans). Edinburgh: Oliver and Boyd.

——— (1965), "The Wave Nature of the Electron" in *Nobel Lectures, Physics, 1922–1941*, pp. 244–256.
——— (1970), "The Reinterpretation of Wave Mechanics," *Foundations of Physics* 1: 5–15.
——— (1973), "The Beginnings of Wave Mechanics," in Price, Chissick, and Ravensdale, pp. 12–18.
——— (1990), *Heisenberg's Uncertainties and the Probabilistic Interpretation of Wave Mechanics*. Dordrecht: Kluwer Academic Publishers.
de Broglie, L., and L. Brillouin (1928), *Selected Papers on Wave Mechanics* (translated from the French by W. M. Deans). London: Blackie & Sons.
Debye, P. (1923), "Zerstreuung von Röntgenstrahlen und Quantentheorie," *Physikalische Zeitschrift* 24: 161–166.
Degen, P. A. (1989), "Einstein's Scientific Orientation as a Search for the God of Spinoza," *Abstracts of the XVIII International Congress of History of Science*. Hamburg-München: ICHS, p. G1 (18).
Deltete, R., and R. Guy (1990), "Einstein's Opposition to the Quantum Theory," *American Journal of Physics* 58: 673–683.
de Regt, H. W. (1993), "Philosophy and the Art of Scientific Discovery." Unpublished Ph.D. dissertation, University of Utrecht.
d'Espagnat, B. (1976), *Conceptual Foundations of Quantum Mechanics*, 2d ed. Reading, MA: W. A. Benjamin.
Deutsch, D. (1983), "Uncertainty in Quantum Measurements," *Physical Review Letters* 50: 631–633.
Dewdney, C., Ph. Gueret, A. Kyprianidis, and J.-P. Vigier (1984), "Testing Wave-Particle Dualism with Time-Dependent Neutron Interferometry," *Physics Letters A* 102: 291–294.
Dewdney, C., P. R. Holland, and A. Kyprianidis (1987), "A Quantum Potential Approach to Spin Superposition in Neutron Interferometry," *Physics Letters A* 121: 105–110.
Dewdney, C., P. R. Holland, A. Kyprianidis, and J.-P. Vigier (1988), "Spin and Non-locality in Quantum Mechanics," *Nature* 336: 536–544.
Dewdney, C., G. Horton, M. M. Lam, Z. Malik, and M. Schmidt (1992), "Wave-Particle Dualism and the Interpretation of Quantum Mechanics," *Foundations of Physics* 22: 1217–1265.
Dieks, D. (1989), "Quantum Mechanics without the Projection Postulate and Its Realistic Interpretation," *Foundations of Physics* 19: 1397–1423.
Dirac, P. A. M. (1951a), "A New Classical Theory of Electrons," *Proceedings of the Royal Society of London* 209: 291–296.
——— (1951b), "Is There an Aether?" *Nature* 168: 906–907.
——— (1952a), "A New Classical Theory of Electrons. II," *Proceedings of the Royal Society of London* 212: 330–339.
——— (1952b), "Comment," *Nature* 169: 702.
——— (1954), "A New Classical Theory of Electrons. III," *Proceedings of the Royal Society of London* 223: 438–445.
——— (1958), *The Principles of Quantum Mechanics*, 4th ed. Oxford, Clarendon Press.

——— (1977), "Recollections of an Exciting Era," in Weiner, pp. 109–146.
Donovan, A., L. Laudan, and R. Laudan (eds.) (1988), *Scrutinizing Science: Empirical Studies of Scientific Change*. Dordrecht: Kluwer Academic Publishers.
Dorling, J. (1987), "Schrödinger's Original Interpretation of the Schrödinger Equation: a Rescue Attempt," in Kilmister, pp. 16–40.
Drabkin, G. M., and R. A. Zhitnikov (1960), "Production of 'Supercold' Polarized Neutrons," *Soviet Physics JETP* 11: 729–730.
Dresden, M. (1987), *H. A. Kramers: Between Tradition and Revolution*. New York: Springer-Verlag.
Duhem, P. (1974), *The Aim and Structure of Physical Theory*. New York: Atheneum.
Dürr, D., S. Goldstein, and N. Zanghi (1990), "On a Realistic Theory for Quantum Physics," in Albeverio et al., pp. 374–391.
——— (1992a), "Quantum Equilibrium and the Origin of Absolute Uncertainty," *Journal of Statistical Physics* 67: 843–907.
——— (1992b), "Quantum Chaos, Classical Randomness, and Bohmian Mechanics," *Journal of Statistical Physics* 68: 259–270.
——— (1992c), "Quantum Mechanics, Randomness, and Deterministic Reality," *Physics Letters A* 172: 6–12.
——— (1993), "A Global Equilibrium as the Foundation of Quantum Randomness," *Foundations of Physics* 23: 721–738.
Earman, J. (1986), *A Primer on Determinism*. Dordrecht: D. Reidel Publishing.
——— (ed.) (1992), *Inference, Explanation, and Other Frustrations*. Berkeley, CA: University of California Press.
Eberhard, P. H. (1989), "The EPR Paradox. Roots and Ramifications" in Schommers, pp. 49–88.
——— (1994), "Restoring Locality with Faster-Than-Light Velocities" in Garuccio and van der Merwe (to be published).
Eder, G., and A. Zeilinger (1976), "Interference Phenomena and Spin Rotation of Neutrons by Magnetic Materials," *Nuovo Cimento B* 34: 76–89.
Edwards, P. (1967), *The Encyclopedia of Philosophy*, 8 vols. New York: Macmillan Publishing.
Einstein, A. (1909a), "Zur gegenwärtigen Stand des Strahlungsproblems," *Physikalische Zeitschrift* 10: 185–193.
——— (1909b), "Über die Entwicklung unserer Anschauungen über das Wesen und die Konstitution der Strahlung," *Physikalische Zeitschrift* 10: 817–826.
——— (1917), "Zur Quantentheorie der Strahlung," *Physikalische Zeitschrift* 18, 121–128. [Also appears as "On the Quantum Theory of Radiation," in van der Waerden (1968), pp. 63–77, and in ter Haar, pp. 167–183.]
——— (1918), "Motive des Forschens," in *Zu Max Plancks sechzigsten Geburtstag Ansprachen, gehalten am 26. April 1918 in der Deutschen Physikalischen Gesellschaft*. Karlsruhe: C. F. Müller, pp. 29–32.
——— (1924), "Über den Äther," *Schweizerische Naturforschende Gesellschaft, Verhandlungen* 105: 85–93. [Also appears in English translation as "On the Aether," in Saunders and Brown, pp. 13–20.]

—— (1925), "Quantentheorie des einatomigen idealen Gases. Zweite Abhandlung," *Sitzungsberichte der Preussischen Akademie der Wissenschaften* 1925: 3–14.

—— (1926), *Investigations on the Theory of the Brownian Movement* (translated by A. D. Cowper). London: Methuen.

—— (1948), "Quanten-Mechanik und Wirklichkeit," *Dialectica* 2: 320–324. [This appears in English translation in Born (1971), pp. 168–173.]

—— (1949a), "Autobiographical Notes," in Schilpp, pp. 2–95.

—— (1949b), "Remarks on the Essays Appearing in the Collective Volume," in Schilpp, pp. 663–688.

—— (1953), "Elementare Überlegungen zur Interpretation der Grundlagen der Quanten-Mechanik" in Born (1971), pp. 33–40.

—— (1954a), *Ideas and Opinions*. New York: Dell Publishing.

—— (1954b), "Principles of Research," in Einstein (1954a), pp. 219–222.

—— (1954c), "On the Method of Theoretical Physics," in Einstein (1954a), pp. 263–270.

—— (1954d), "What Is the Theory of Relativity?" in Einstein (1954a), pp. 222–227.

—— (1954e), "Maxwell's Influence on the Evolution of the Idea of Physical Reality," in Einstein (1954a), pp. 259–263.

—— (1956), *The Meaning of Relativity*, Princeton, Princeton University Press.

Einstein, A., and P. Ehrenfest (1922), "Quantentheoretische Bemerkungen zum Experiment von Stern und Gerlach," *Zeitschrift für Physik* 11: 31–34.

Einstein, A., and J. Grommer (1927), "Allgemeine Relativitätstheorie und Bewegungsgesetz," *Sitzungsberichte der Preussischen Akademie der Wissenschaften* 1927: 2–13, 235–245.

Einstein, A., B. Podolsky, and N. Rosen (1935), "Can Quantum-Mechanical Description of Physical Reality Be Considered Complete?" *Physical Review* 47: 777–780. [Reprinted in Wheeler and Zurek, pp. 138–141.]

Eisenstaedt, J., and A. J. Kox (eds.) (1992), *Studies in the History of General Relativity*. Boston: Birkhäuser.

Electrons et Photons, Rapports et Discussions du Cinquième Conseil de Physique (1928). Paris: Gauthier-Villars.

Enz, C. P., and K. v. Meyenn (eds.) (1988), *Wolfgang Pauli: Das Gewissen der Physik*. Braunschweig: Friedr. Vieweg & Sohn.

Epstein, S. T. (1953), "The Causal Interpretation of Quantum Mechanics," *Physical Review* 89: 319.

Favrholdt, D. (1992), *Niels Bohr's Philosophical Background*. Copenhagen: Munksgaard.

Faye, J. (1991), *Niels Bohr: His Heritage and Legacy*. Dordrecht: Kluwer Academic Publishers.

Faye, J., and H. J. Folse (eds.) (1993), *Niels Bohr and Contemporary Philosophy*. Dordrecht: Kluwer Academic Publishers.

Fényes, I. (1952), "Eine wahrscheinlichkeitstheoretische Begründung und Interpretation der Quantenmechanik," *Zeitschrift für Physik* 132: 81–106.

Fermi, E. (1926), "Zur Wellenmechanik des Stossvorganges," *Zeitschrift für Physik* 40: 399–402.
Fertig, H. A. (1990), "Traversal-Time Distribution and the Uncertainty Principle in Quantum Mechanics," *Physical Review Letters* 65: 2321–2324.
––––––– (1993), "Path Decomposition and the Traversal-Time Distribution in Quantum Tunneling," *Physical Review B* 47: 1346–1358.
Feuer, L. S. (1974), *Einstein and the Generations of Science*. New York: Basic Books.
Feyerabend, P. (1975), *Against Method*. London: Verso.
––––––– (1989), "Realism and the Historicity of Knowledge," *The Journal of Philosophy* 86 (8): 393–406.
Feynman, R. P. (1965), "Probability and Uncertainty—the Quantum Mechanical View of Nature," in *The Character of Physical Law*. Cambridge, MA: MIT Press, pp. 127–148.
Fine, A. (1981a), "Correlations and Physical Locality," in Asquith and Giere, pp. 535–562.
––––––– (1981b), "Einstein's Critique of Quantum Theory: The Roots and Significance of EPR," in Barker and Shugart, pp. 147–158.
––––––– (1982), "Hidden Variables, Joint Probability, and the Bell Inequalities," *Physical Review Letters* 48: 291–295.
––––––– (1986), *The Shaky Game: Einstein, Realism and the Quantum Theory*. Chicago: University of Chicago Press.
––––––– (1989), "Do Correlations Need to Be Explained?" in Cushing and McMullin, pp. 175–194.
Folse, H. J. (1985), *The Philosophy of Niels Bohr*. Amsterdam: North-Holland Publishing.
––––––– (1989), "Bohr on Bell," in Cushing and McMullin, pp. 254–271.
––––––– (1993), "The Bohr-Einstein Debate and the Philosophers' Debate over Realism versus Anti-realism," in *Proceedings of the Beijing Conference on the Philosophy of Science*. Dordrecht: Kluwer Academic Publishers (to be published).
Ford, J. (1983), "How Random Is a Coin Toss?" *Physics Today* 36 (4): 40–47.
Forman, P. (1967), "The Environment and Practice of Atomic Physics in Weimar Germany: A Study in the History of Science." Unpublished Ph. D. dissertation, University of California at Berkeley.
––––––– (1971), "Weimar Culture, Causality, and Quantum Theory, 1918–1927: Adaptation by German Physicists and Mathematicians to a Hostile Intellectual Environment," *Historical Studies in the Physical Sciences* 3: 1–115.
––––––– (1979), "The Reception of an Acausal Quantum Mechanics in Germany and Britain," in Mauskopf, pp. 11–50.
––––––– (1984), "*Kausalität, Anschaulichkeit*, and *Individualität*, or How Cultural Values Prescribed the Character and Lessons Ascribed to Quantum Mechanics," in Stehr and Meja, pp. 333–347.
Frank, P. (1947), *Einstein: His Life and Times*. New York: Alfred Knopf.
Freistadt, H. (1953), "The Crisis in Physics," *Science and Society* 17: 211–237.

——— (1955), "Classical Field Theory in the Hamilton-Jacobi Formalism," *Physical Review* 97: 1158–1161.
——— (1956a), "Quantized Field Theory in the Hamilton-Jacobi Formalism," *Physical Review* 102: 274–278.
——— (1956b), "Dialectical Materialism: A Friendly Interpretation," *Philosophy of Science* 23: 97–110.
——— (1957a), "Dialectical Materialism: A Further Discussion," *Philosophy of Science* 24: 25–40.
——— (1957b), "The Causal Formulation of Quantum Mechanics of Particles (The Theory of de Broglie, Bohm and Takabayasi)," *Supplemento al Nuovo Cimento* 5: 1–70.
French, A. P., and P. J. Kennedy (eds.) (1985), *Niels Bohr: A Centenary Volume*. Cambridge, MA: Harvard University Press.
French, S., and H. Kamminga (eds.) (1993), *Correspondence, Invariance and Heuristics*. Dordrecht: Kluwer Academic Publishers.
Friedman, M. (1974), "Explanation and Scientific Understanding," *Journal of Philosophy* 71: 5–19.
Fürth, R. (1933), "Über einige Beziehungen zwischen klassischer Statistik und Quantenmechanik," *Zeitschrift für Physik* 81: 143–162.
Galison, P. (1987), *How Experiments End*. Chicago: University of Chicago Press.
Garuccio, A., and A. van der Merwe (eds.) (1994), *Waves and Particles in Light and Matter*. New York: Plenum Publishing (to be published).
Gell-Mann, M. (1981), "Questions for the Future," in Mulvey, pp. 169–186.
George, A. (ed.) (1953), *Louis de Broglie, Physicien et Penseur*, Paris, Éditions Albin Michel.
Gershenson, D. E., and D. A. Greenberg (eds.) (1964), *The Natural Philosopher*, vol. 3. New York: Blaisdell Publishing.
Ghirardi, G. C., and A. Rimini (1990), "Old and New Ideas in the Theory of Quantum Measurement," in Miller (1990), pp. 167–191.
Ghirardi, G. C., A. Rimini, and T. Weber (1980), "A General Argument against Superluminal Transmission through the Quantum-Mechanical Measurement Process," *Lettere al Nuovo Cimento* 27: 293–298.
——— (1986), "Unified Dynamics for Microscopic and Macroscopic Systems," *Physical Review D* 34: 470–491.
Ghose, P., and D. Home (1992), "Wave-Particle Duality of Single-Photon States," *Foundations of Physics* 22: 1435–1447.
Ghose, P., D. Home, and G. S. Agarwal (1991), "An Experiment to Throw More Light on Light," *Physics Letters A* 153: 403–406.
——— (1992), "An 'Experiment to Throw More Light on Light': Implications," *Physics Letters A* 168: 95–99.
Gillispie, C. C. (1960), *The Edge of Objectivity*. Princeton, NJ: Princeton University Press.
——— (ed.) (1970–1978), *Dictionary of Scientific Biography*, vols. 1–16. New York: Charles Scribner's Sons.

Gnedenko, B. V. and Khinchin, A. Ya. (1962), *An Elementary Introduction to the Theory of Probability.* New York: Dover Publications.

Goldstein, H. (1950), *Classical Mechanics.* Reading, MA: Addison-Wesley.

Goldstein, S. (1987), "Stochastic Mechanics and Quantum Theory," *Journal of Statistical Physics* 47: 645–667.

Gould, S. J. (1989), *Wonderful Life: The Burgess Shale and the Nature of History.* New York: W. W. Norton.

Greenberger, D. M. (1983), "The Neutron Interferometer as a Device for Illustrating the Strange Behavior of Quantum Systems," *Reviews of Modern Physics* 55: 875–906.

——— (ed.) (1986), *New Techniques and Ideas in Quantum Measurement Theory.* New York: New York Academy of Sciences.

Güntherodt, H.-J., and R. Wiesendanger (eds.) (1993), *Scanning Tunneling Microscopy III.* Berlin: Springer-Verlag.

Halpern, Q. (1952), "A Proposed Re-interpretation of Quantum Mechanics," *Physical Review* 87: 389.

Hanle, P. A. (1979), "Indeterminacy before Heisenberg: The Case of Franz Exner and Erwin Schrödinger," *Historical Studies in the Physical Sciences* 10: 225–269.

Hanson, N. R. (1959), "Copenhagen Interpretation of Quantum Theory," *American Journal of Physics* 27: 1–15.

——— (1963), *The Concept of the Positron.* Cambridge: Cambridge University Press.

Harding, S. G. (ed.) (1976), *Can Theories Be Refuted?* Dordrecht: D. Reidel Publishing.

Hauge, E. H., and J. A. Støvneng (1989), "Tunneling Times: A Critical Review," *Reviews of Modern Physics* 61: 917–936.

Hauschildt, D. (1990), "On the Impossibility of Observing Particle Paths and Interference Simultaneously," in Mizerski et al., pp. 308–333.

Healey, R. (1989), *The Philosophy of Quantum Mechanics: An Interactive Interpretation.* Cambridge: Cambridge University Press.

Heilbron, J. L. (1985), Review of *The Historical Development of Quantum Theory,* by Jagdish Mehra and Helmut Rechenberg. *Isis* 76: 388–393.

——— (1988), "The Earliest Missionaries of the Copenhagen Spirit," in Ullmann-Margalit, pp. 201–233.

Heisenberg, W. (1925), "Über quantentheoretische Umdeutung kinematischer und mechanischer Beziehungen," *Zeitschrift für Physik* 33: 879–893. [Also appears as "Quantum-Theoretical Re-interpretation of Kinematic and Mechanical Relations," in van der Waerden (1968), pp. 261–276. Page references are to this English translation.]

——— (1926), "Schwankungserscheinungen und Quantenmechanik," *Zeitschrift für Physik* 40: 501–506.

——— (1927), "Über den anschaulichen Inhalt der quantentheoretischen Kinematik und Mechanik," *Zeitschrift für Physik* 43: 172–198. [Also appears as "The Physical Content of Quantum Kinematics and Mechanics" in Wheeler and Zurek, pp. 62–84.]

——— (1948), "Der Begriff 'Abgeschlossene Theorie' in der modernen Naturwissenschaft," *Dialectica* 3: 331–336.

——— (1949), *The Physical Principles of the Quantum Theory*. New York: Dover Publications.

——— (1955), "The Development of the Interpretation of the Quantum Theory," in Pauli (1955), pp. 12–29.

——— (1958), *Physics and Philosophy*. New York: Harper & Row.

——— (1971), *Physics and Beyond*. New York: Harper & Row.

——— (1976), "The Nature of Elementary Particles," *Physics Today* 29 (3): 32–39.

Heitler, W. (1961), "Erwin Schrödinger," *Biographical Memoirs of Fellows of the Royal Society* 7: 221–228.

Hendry, J. (1980), "Weimar Culture and Quantum Causality," *History of Science* 18: 155–180.

——— (1984), *The Creation of Quantum Mechanics and the Bohr-Pauli Dialogue*. Dordrecht: D. Reidel Publishing.

——— (1985), "The History of Complementarity: Niels Bohr and the Problem of Visualization," *Rivista di Storia della Scienza* 3: 391–407.

——— (1993), Review of *Niels Bohr*, by Jan Faye. *Isis* 84: 169.

Hepp, K. (1972), "Quantum Theory of Measurement and Macroscopic Observables," *Helvetica Physica Acta* 45: 237–248.

Hiley, B. J. (1991), "Vacuum or Holomovement," in Saunders and Brown, pp. 217–249.

Hiley, B. J., and D. F. Peat (eds.) (1987), *Quantum Implications*. London: Routledge & Kegan Paul.

Holland, P. R. (1988a), "Causal Interpretation of Fermi Fields," *Physics Letters A* 128: 9–18.

——— (1988b), "Causal Interpretation of a System of Two Spin-1/2 Particles," *Physics Reports* 169: 293–327.

——— (1993a), *The Quantum Theory of Motion*. Cambridge: Cambridge University Press.

——— (1993b), "The De Broglie–Bohm Theory of Motion and Quantum Field Theory," *Physics Reports* 224: 95–150.

Holmes, F. L. (ed.) (1990), *Dictionary of Scientific Biography*, vols. 17 and 18. New York: Charles Scribner's Sons.

Home, D. (1986), "Interview with David Bohm," *Science Today* 20 (11): 25–27, 48–49.

——— (1992), "Optical Tunnelling of Single Photon States: Wave-Particle Complementarity Revisited." Talk given at the 4th International Symposium on the Foundations of Quantum Mechanics in the Light of New Technology, Tokyo, 23–27 August 1992.

Home, D., and M. A. B. Whitaker (1992), "Ensemble Interpretations of Quantum Mechanics. A Modern Perspective," *Physics Reports* 210: 223–317.

Honig, W. M., D. W. Kraft, and E. Panarella (eds.) (1987), *Quantum Uncertainties*. New York: Plenum Publishing.

Honner, J. (1987), *The Description of Nature*. Oxford: Oxford University Press.

Howard, D. (1985), "Einstein on Locality and Separability," *Studies in History and Philosophy of Science* 16: 171–201.

——— (1989), "Holism, Separability, and the Metaphysical Implications of the Bell Experiments," in Cushing and McMullin, pp. 224–253.

——— (1990a), "'*Nicht Sein Kann Was Nicht Sein Darf,*' or the Prehistory of EPR, 1909–1935: Einstein's Early Worries about the Quantum Mechanics of Composite Systems," in Miller (1990), pp. 61–111.

——— (1990b), "Einstein and Duhem," *Synthese* 83: 363–384.

——— (1991), Review of *The Shaky Game,* by Arthur Fine. *Synthese* 86: 123–141.

——— (1992), "Einstein and *Eindeutigkeit*: A Neglected Theme in the Philosophical Background to General Relativity," in Eisenstaedt and Kox, pp. 154–243.

——— (1993), "A Peek behind the Veil of Maya: The Historical Background of the Conception of Space as a Ground for the Individuation of Physical Systems." Paper delivered to the Center for Philosophy of Science Annual Lecture Series at the University of Pittsburgh on 12 February 1993 (to be published).

Huang, K. (1963), *Statistical Mechanics.* New York: John Wiley & Sons.

Hull, D., M. Forbes, and K. Okruhlik (eds.) (1992), *Proceedings of the 1992 Biennial Meeting of the Philosophy of Science Association,* vol. 1. East Lansing, MI: Philosophy of Science Association.

Isham, C., R. Penrose, and D. Sciama (eds.) (1981), *Quantum Gravity 2,* Oxford, Oxford University Press.

Jammer, M. (1966), *The Conceptual Development of Quantum Mechanics.* New York: McGraw-Hill.

——— (1974), *The Philosophy of Quantum Mechanics.* New York: John Wiley & Sons.

——— (1985), "The EPR Problem in Its Historical Development," in Lahti and Mittelstaedt (1985), pp. 129–149.

——— (1988), "David Bohm and His Work—On the Occasion of His Seventieth Birthday," *Foundations of Physics* 18: 691–699.

——— (1989), *The Conceptual Development of Quantum Mechanics,* 2d ed. New York: Tomash.

Jankovic, V. (1991), "Matrix Mechanics and the Meaning of Concepts: Operationalism or Not?" Unpublished, University of Notre Dame.

Jarrett, J. P. (1984), "On the Physical Significance of the Locality Conditions in the Bell Arguments," *Nous* 18: 569–589.

——— (1989), "Bell's Theorem: A Guide to the Implications," in Cushing and McMullin, pp. 60–79.

Jones, M. R., and R. K. Clifton (1993), "Against Experimental Metaphysics," in *Midwest Studies in Philosophy* 18. Notre Dame, IN: University of Notre Dame Press, pp. 295–316.

Jordan, P. (1927), "Philosophical Foundations of Quantum Theory," *Nature* 119: 566–569.

——— (1936a), *Die Physik des 20. Jahrhunderts.* Braunschweig: Friedr. Viewig & Sohn.

——— (1936b), *Anschauliche Quantentheorie*. Berlin: Verlag von Julius Springer.
——— (1944), *Physics of the 20th Century*. New York: Philosophical Library.
Kafatos, M. (ed.) (1989), *Bell's Theorem, Quantum Theory and Conceptions of the Universe*. Dordrecht: Kluwer Academic Press.
Kaiser, D. (1994), "Bringing the Human Actors Back on Stage: The Personal Context of the Einstein-Bohr Debate," *British Journal for the History of Science* (to be published).
Kalckar, J. (1985), "General Introduction to Volumes 6 and 7," in Bohr (1985), pp. xvii–xxvi.
Kaloyerou, P. N. (1985), "Investigation of the Quantum Potential in the Relativistic Domain." Unpublished Ph. D. dissertation, University of London.
——— (1992), "The Causal Interpretation of the Electromagnetic Field: The EPR Experiment," in van der Merwe, Selleri, and Tarozzi (1992), pp. 315–337.
——— (1994), "The Causal Interpretation of the Electromagnetic Field," *Physics Reports* (to be published).
Kamefuchi, S., H. Ezawa, Y. Murayama, M. Namiki, S. Nomura, Y. Ohnuki, and T. Yojima (eds.) (1984), *Proceedings of the International Symposium on the Foundations of Quantum Mechanics*. Tokyo: Physical Society of Japan.
Kant, I. (1952), *The Critique of Pure Reason in The Western World* vol. 42. Chicago: Encyclopaedia Britannica, pp. 1–250.
Keller, J. B. (1953), "Bohm's Interpretation of the Quantum Theory in Terms of 'Hidden' Variables," *Physical Review* 89: 1040–1041.
Kennard, E. H. (1928), "On the Quantum Mechanics of a System of Particles," *Physical Review* 31: 876–890.
Kilmister, C. W. (ed.) (1987), *Schrödinger: Centenary Celebration of a Polymath*. Cambridge: Cambridge University Press.
Kirsten, C., and H. -J. Treder (1979), *Albert Einstein in Berlin 1913–1933*. Berlin: Akademie-Verlag.
Klein, M. J. (1964), "Einstein and the Wave-Particle Duality," in Gershenson and Greenberg, pp. 1–49.
Kobayashi, S., H. Ezawa, Y. Murayama, and S. Nomura (eds.) (1990), *Proceedings of the 3rd International Symposium on the Foundations of Quantum Mechanics in the Light of New Technology*. Tokyo: Physical Society of Japan.
Kochen, S. (1985), "A New Interpretation of Quantum Mechanics," in Lahti and Mittelstaedt (1985), pp. 151–169.
Koopman, B. O. (1931), "Hamiltonian Systems and Transformations in Hilbert Space," *Proceedings of the National Academy of Sciences of the United States of America* 17: 315–318.
Körner, S. (ed.) (1957), *Observation and Interpretation*. London: Butterworths Scientific Publications.
Kraft, P., and P. Kroes (1984), "Adaptation of Scientific Knowledge to an Intellectual Environment. Paul Forman's 'Weimar Culture, Causality, and Quantum Theory, 1918–1927': Analysis and Criticism," *Centaurus* 27: 76–99.
Kragh, H. (1982), "Erwin Schrödinger and the Wave Equation: The Crucial Phase," *Centaurus* 26: 154–197.

——— (1990), *Dirac, A Scientific Biography*. New York: Cambridge University Press.
Krieger, M. H. (1992), *Doing Physics*. Bloomington, IN: Indiana University Press.
Krips, H. (1987), *The Metaphysics of Quantum Theory*. Oxford: Oxford University Press.
Krüger, L., L. J. Daston, and M. Heidelberger (eds.) (1987), *The Probabilistic Revolution*, vol. 1. Cambridge, MA: MIT Press.
Kubli, F. (1970), "Louis de Broglie und die Entdeckung der Materiewellen," *Archive for History of Exact Sciences* 7: 26–68.
Kuhn, T. S., J. L. Heilbron, P. Forman, and L. Allen (1967), *Sources for History of Quantum Physics*. Philadelphia: American Philosophical Society.
Kyprianidis, A. (1985), "Particle Trajectories in Relativistic Quantum Mechanics," *Physics Letters A* 111: 111–116.
Kyprianidis, A., D. Sardelis, and J. -P. Vigier (1984), "Causal Non-local Character of Quantum Statistics," *Physics Letters A* 100: 228–230.
Lahti, P., and P. Mittelstaedt (eds.) (1985), *Symposium on the Foundations of Modern Physics*. Singapore: World Scientific Publishing.
——— (eds.) (1987), *Symposium on the Foundations of Modern Physics 1987*. Singapore: World Scientific Publishing.
——— (eds.) (1991), *Symposium on the Foundations of Modern Physics 1990*. Singapore: World Scientific Publishing.
Lam, M. M., and C. Dewdney (1994a), "The Bohm Approach to Cavity Quantum Scalar Field Dynamics. Part I: The Free Field," *Foundations of Physics* (to be published).
——— (1994b), "The Bohm Approach to Cavity Quantum Scalar Field Dynamics. Part II: The Interaction of the Field with Matter," *Foundations of Physics* (to be published).
Laplace, P. -S. (1886), *Théorie Analytique des Probabilités*, in *Oeuvres Complètes de Laplace*, Vol. 7. Paris: Gauthier-Villars.
——— (1917), *A Philosophical Essay on Probabilities*. New York: John Wiley & Sons.
Laudan, L. (1965), "Grünbaum on 'The Duhemian Argument,'" *Philosophy of Science* 32: 295–299.
——— (1981), *Science and Hypothesis*. Dordrecht: D. Reidel Publishing.
——— (1984), *Science and Values*. Berkeley: University of California Press.
Laudan, L., and J. Leplin (1991), "Empirical Equivalence and Underdetermination," *Journal of Philosophy* 88 (9): 449–472.
Laurikainen, K. V. (1987), "Wolfgang Pauli's Conception of Reality," in Lahti and Mittelstaedt (1987), pp. 209–228.
——— (1988a), "Can the Goal of Science Be an Objective Reality?" Lecture at the Finnish Academy of Sciences on 14 November 1988.
——— (1988b), "Quantum Physics and Philosophy." Preprint HU-TFT-88-36 of the Research Institute for Theoretical Physics of the University of Helsinki.
——— (1988c), *Beyond the Atom: The Philosophical Thought of Wolfgang Pauli*. Berlin: Springer-Verlag.

Leavens, C. R. (1990a), "Transmission, Reflection and Dwell Times within Bohm's Causal Interpretation of Quantum Mechanics," *Solid State Communications* 74: 923–928.

——— (1990b), "Traversal Times for Rectangular Barriers within Bohm's Causal Interpretation of Quantum Mechanics," *Solid State Communications* 76: 253–261.

——— (1991), "The Quantum Traversal Times of Nassar and of Jonson," *Solid State Communications* 77: 571–574.

——— (1993a), "On the Olkhovsky-Recami Approach to the 'Tunneling Time Problem,'" *Solid State Communications* 85: 115–119.

——— (1993b), "Application of the Quantum Clock of Salecker and Wigner to the 'Tunneling Time Problem,'" *Solid State Communications* 86: 781–788.

——— (1993c), "Arrival Time Distributions," *Physics Letters A* 178: 27–32.

Leavens, C. R., and G. C. Aers (1990), "Tunneling Times for One-Dimensional Barriers," in Behm, García, and Rohrer, pp. 59–76.

——— (1991), "The Time-modulated Barrier Approach to Traversal Times from the Bohm Trajectory Point of View," *Solid State Communications* 78: 1015–1023.

——— (1993), "Bohm Trajectories and the Tunneling Time Problem," in Güntherodt and Wiesendanger, pp. 105–140.

Lehr, W. (1976), "The Stochastic Approach to Quantum Mechanics." Unpublished Ph. D. dissertation, Washington State University.

Lehr, W. J., and J. L. Park (1977), "A Stochastic Derivation of the Klein-Gordon Equation," *Journal of Mathematical Physics* 18: 1235–1240.

Leplin, J. (ed.) (1984), *Scientific Realism*. Berkeley, CA: University of California Press.

Leplin, J., and L. Laudan (1993), "Determination Undeterred," *Analysis* 53 (1): 8–16.

MacKinnon, E. (1980), "The Rise and Fall of the Schrödinger Interpretation," in Suppes, pp. 1–57.

——— (1982), *Scientific Explanation and Atomic Physics*. Chicago: University of Chicago Press.

Macrae, N. (1992), *John von Neumann*. New York: Pantheon.

Madelung, E. (1926), "Quantentheorie in hydrodynamischer Form," *Zeitschrift für Physik* 40: 322–326.

Margenau, H. (1954), "Advantages and Disadvantages of Various Interpretations of the Quantum Theory," *Physics Today* 7 (10): 6–13.

Mauskopf, S. H. (ed.) (1979), *The Reception of Unconventional Science, AAAS Selected Symposium 25*. Boulder, CO: Westview Press.

Maxwell, J. C. (1890), *The Scientific Papers of James Clerk Maxwell*, 2 vols. (W. D. Niven, ed.). Cambridge: Cambridge University Press.

Maxwell, N. (1985), "Are Probabilism and Special Relativity Incompatible?" *Philosophy of Science* 52: 23–43.

McMullin, E. (1978), *Newton on Matter and Activity*. Notre Dame, IN: University of Notre Dame Press.

——— (1984), "A Case for Scientific Realism," in Leplin, pp. 8–40.
——— (ed.) (1988), *Construction and Constraint: The Shaping of Scientific Rationality.* Notre Dame, IN: University of Notre Dame Press.
——— (1989), "The Explanation of Distant Action: Historical Notes," in Cushing and McMullin, pp. 272–302.
Mehra, J. (ed.) (1973), *The Physicist's Conception of Nature.* Dordrecht: D. Reidel Publishing.
——— (1975), *The Solvay Conferences on Physics.* Dordrecht: D. Reidel Publishing.
——— (1987), "Niels Bohr's Discussion with Albert Einstein, Werner Heisenberg, and Erwin Schrödinger: The Origins of the Principles of Uncertainty and Complementarity," in Lahti and Mittelstaedt (1987), pp. 19–64.
Mehra, J., and H. Rechenberg (1982), *The Historical Development of Quantum Theory,* vol. 1, part 2. New York: Springer-Verlag.
Mendelsohn, E., P. Weingart, and R. Whitley (eds.) (1977), *The Social Production of Scientific Knowledge.* Dordrecht: D. Reidel Publishing.
Messiah, A. (1965), *Quantum Mechanics,* vol. 1. Amsterdam: North-Holland Publishing.
Métadier, J. (1931), "Sur l'équation générale du mouvement brownien," *Comptes Rendus des Séances de l'Académie des Sciences* 193: 1173–1176.
Meyerson, É. (1908), *Identité et realité.* Paris: Libraries Félix Alcan et Guillaumin Réunies. [Appeared in translation (K. Loewenberg, 1930) as *Identity and Reality.* London: George Allen & Unwin Ltd. Page references are to the English edition.]
——— (1936), *Essais.* Paris: Librarie Philosophique J. Vrier.
Miller, A. I. (1984), *Imagery in Scientific Thought.* Boston: Birkhäuser.
——— (ed.) (1990), *Sixty-Two Years of Uncertainty.* New York: Plenum Press.
Mizerski, J., A. Posiewnik, J. Pykacz, and M. Zukowski (eds.) (1990), *Problems in Quantum Physics II; Gdansk '89.* Singapore: World Scientific Publishing.
Moritz, R. E. (1914), *Memorabilia Mathematica.* New York: Macmillan.
Mott, N. F., and H. S. W. Massey (1933), *The Theory of Atomic Collisions.* Oxford: Clarendon Press.
Moulton, F. R. (1906), *An Introduction to Astronomy.* New York: Macmillan.
——— (1914), *An Introduction to Celestial Mechanics.* New York: Macmillan.
Muga, J. G., S. Brouard, and R. Sala (1992), "Transmission and Reflection Tunneling Times," *Physics Letters A* 167: 24–28.
Mugur-Schächter, M. (1964), *Étude du Caractère Complet de la Théorie Quantique.* Paris: Gauthier-Villars.
Mulvey, J. H. (ed.) (1981), *The Nature of Matter.* Oxford: Oxford University Press.
Murdoch, D. (1987), *Niels Bohr's Philosophy of Physics.* Cambridge: Cambridge University Press.
Nelson, E. (1966), "Derivation of the Schrödinger Equation from Newtonian Mechanics," *Physical Review* 150: 1079–1085.
——— (1967), *Dynamical Theories of Brownian Motion.* Princeton, NJ: Princeton University Press.

——— (1985), *Quantum Fluctuations*. Princeton, NJ: Princeton University Press.
Newton, I. (1934), *Mathematical Principles of Natural Philosophy*. Berkeley, CA: University of California Press.
——— (1952), *Optics*. New York: Dover Publications.
Nobel Lectures, Physics, 1942–1962 (1964). Amsterdam: Elsevier Publishing.
Nobel Lectures, Physics, 1922–1941 (1965). Amsterdam: Elsevier Publishing.
Olkhovsky, V. S., and E. Recami (1992), "Recent Developments in the Time Analysis of Tunneling Processes," *Physics Reports* 214: 339–356.
Omnès, R. (1992), "Consistent Interpretations of Quantum Mechanics," *Reviews of Modern Physics* 64: 339–382.
Oppenheimer, J. R. (1949), "A Letter to Senator McMahon," *Bulletin of the Atomic Scientists* 5: 163, 178.
The Oxford English Dictionary, 2d ed. (1989). Oxford: Clarendon Press.
Pagonis, C., and R. Clifton (1994), "Unremarkable Contextualism: Dispositions in the Bohm Theory," *Foundations of Physics* (to be published).
Pais, A. (1991), *Niels Bohr's Times, In Physics, Philosophy, and Polity*. Oxford: Clarendon Press.
Partovi, M. H. (1983), "Entropic Formulation of Uncertainty for Quantum Measurements," *Physical Review Letters* 50: 1883–1885.
Pauli, W. (1921), "Relativitätstheorie," *Enzyklopädie der mathematischen Wissenschaften*, Vol. 19. Leipzig: B. G. Teubner, pp. 539–775. [Also appears in translation (by G. Field, 1981) as *Theory of Relativity*. New York: Dover Publications. Page references are to the Dover edition.]
——— (1926), "Über das Wasserstoffspektrum vom Standpunkt der neuen Quantenmechanik," *Zeitschrift für Physik* 36: 336–363. [Also appears as "On the Hydrogen Spectrum from the Standpoint of the New Quantum Mechanics," in van der Waerden (1968), pp. 387–415.]
——— (1927), "Zur Quantenmechanik des magnetischen Elektrons," *Zeitschrift für Physik* 43: 601–623.
——— (1928), "Discussion générale des idées nouvelles émises," in *Electrons et Photons*, pp. 280–282.
——— (1933), "Die allgemeinen Prinzipien der Wellenmechanik," in *Handbuch der Physik*, 2d ed., vol. 24, pt. 1, pp. 83–272. Berlin: Verlag von Julius Springer.
——— (1948), "Editorial on the Concept of Complementarity," *Dialectica* 2 (3/4): 307–311.
——— (1952), "Phänomen und physikalische Realität," *Dialectica* 11: 36–48.
——— (1953), "Remarques sur le problème des paramètres cachés dans la mécanique quantique et sur la théorie de l'onde pilote," in George, pp. 33–42.
——— (ed.) (1955), *Niels Bohr and the Development of Physics*. New York: Pergamon.
——— (1964), "Exclusion Principle and Quantum Mechanics," in *Nobel Lectures, Physics, 1942–1962*, pp. 27–43. [Also appears in Boorse and Motz, pp. 970–984.]
——— (1980), *General Principles of Quantum Mechanics*. Berlin: Springer-Verlag.

——— (1985), *Wissenschaftlicher Briefwechsel mit Bohr, Einstein, Heisenberg, u.a., Band II: 1919–1929*. New York: Springer-Verlag.

Pearle, P. (1990), "Toward a Relativistic Theory of Statevector Reduction," in Miller (1990), pp. 193–214.

Peierls, R. E. (1959), "Wolfgang Ernst Pauli," *Biographical Memoirs of Fellows of the Royal Society* 5: 175–192.

Penrose, R., and C. J. Isham (eds.) (1986), *Quantum Concepts in Space and Time*. Oxford: Oxford University Press.

Peres, A. (1978), "Unperformed Experiments Have No Results," *American Journal of Physics* 4: 745–747.

Petersen, A. (1963), "The Philosophy of Niels Bohr," *Bulletin of the Atomic Scientists* 19 (7): 8–14. [Reprinted in French and Kennedy, pp. 299–310.]

Pickering, A. (1984), *Constructing Quarks: A Sociological History of Particle Physics*. Chicago: University of Chicago Press.

Pinch, T. J. (1976), "Hidden Variables, Impossibility Proofs, and Paradoxes: A Sociological Study of Non-relativistic Quantum Mechanics." Unpublished M.S. dissertation, University of Manchester.

——— (1977), "What Does a Proof Do If It Does Not Prove?" in Mendelsohn, Weingart, and Whitley, pp. 171–215.

——— (1979), "The Hidden-Variables Controversy in Quantum Mechanics," *Physics Education* 14: 48–52.

Pitowsky, I. (1989), "From George Boole to John Bell—The Origins of Bell's Inequality," in Kafatos, pp. 37–49.

——— (1991), "Bohm's Quantum Potentials and Quantum Gravity," *Foundations of Physics* 21: 343–352.

Poincaré, H. (1892), *Les Méthodes Nouvelles de la Mécanique Céleste*, vol. 1. Paris: Gauthier-Villars.

Post, H. R. (1971), "Correspondence, Invariance and Heuristics," *Studies in History and Philosophy of Science* 2: 213–255.

Price, W. C., S. S. Chissick, and T. Ravensdale (eds.) (1973), *Wave Mechanics: The First Fifty Years*. London: Butterworths.

Quine, W. V. (1951), "Two Dogmas of Empiricism," *Philosophical Review* 60: 20–43.

——— (1960), *Word and Object*. Cambridge, MA: MIT Press.

Rabi, I. I. (1937), "Space Quantization in a Gyrating Magnetic Field," *Physical Review* 51: 652–654.

Rabinowich, E. (1949), "The 'Cleansing' of AEC Fellowships," *Bulletin of the Atomic Scientists* 5: 161–162.

Raman, V. V., and P. Forman (1969), "Why Was It Schrödinger Who Developed de Broglie's Ideas?" *Historical Studies in the Physical Sciences* 1: 291–314.

The Random House Dictionary of the English Language (1966). New York: Random House.

Rauch, H., and M. Suda (1974), "Intensitätsberechnung für ein Neutronen-Interferometer," *Physica Status Solidi (A)* 25: 495–505.

Rauch, H., W. Triemer, and U. Bonse (1974), "Test of a Single Crystal Neutron Interferometer," *Physics Letters A* 47: 369–371.

Redhead, M. L. G. (1987), *Incompleteness, Nonlocality, and Realism*. Oxford: Clarendon Press.
Reichenbach, H. (1948), "The Principle of Anomaly in Quantum Mechanics," *Dialectica* 2: 337–350.
Robertson, P. (1979), *The Early Years: The Niels Bohr Institute 1921–1930*. Copenhagen: Akademisk Forlag.
Rohrlich, F. (1986), "Reality and Quantum Mechanics," in Greenberger (1986), pp. 373–381.
——— (1987), *From Paradox to Reality*. Cambridge: Cambridge University Press.
Rorty, R. (1988), "Is Natural Science a Natural Kind?" in McMullin, pp. 49–74.
Rosen, N. (1945), "On Waves and Particles," *Journal of the Elisha Mitchell Scientific Society* 61: 67–73.
Rosenfeld, L. (1953), "Strife about Complementarity," *Science Progress* 41: 393–410.
——— (1957), "Misunderstandings about the Foundations of Quantum Theory," in Körner, pp. 41–45.
——— (1961), "Foundations of Quantum Theory and Complementarity," *Nature* 190: 384–388.
——— (1970), "Niels Henrich David Bohr," in Gillispie (1970–1978), Vol. 2, pp. 239–254.
Salmon, W. C. (1984), *Scientific Explanation and the Causal Structure of the World*. Princeton, NJ: Princeton University Press.
——— (1985), "Scientific Explanation: Three Basic Conceptions," in Asquith and Kitcher, pp. 293–305.
Saunders, S., and H. R. Brown (eds.) (1991), *The Philosophy of Vacuum*. Oxford: Clarendon Press.
Schilpp, P. A. (ed.) (1949), *Albert Einstein: Philosophy-Scientist*. La Salle, IL: Open Court.
Schommers, W. (ed.) (1989), *Quantum Theory and Pictures of Reality*. Berlin: Springer-Verlag.
Schrödinger, E. (1926a), "Zur Einsteinschen Gastheorie," *Physikalische Zeitschrift* 27: 95–101.
——— (1926b), "Quantisierung als Eigenwertproblem I," *Annalen der Physik* 79: 361–376. [Also appears as "Quantisation as a Problem of Proper Values (Part I)," in Schrödinger (1928), pp. 1–12.]
——— (1926c), "Quantisierung als Eigenwertproblem II," *Annalen der Physik* 79: 489–527. [Also appears as "Quantisation as a Problem of Proper Values (Part II)" in Schrödinger (1928), pp. 13–40.]
——— (1926d), "Über das Verhältnis der Heisenberg-Born-Jordanschen Quantenmechanik zu der meinen," *Annalen der Physik* 79: 734–756. [Also appears as "On the Relation between the Quantum Mechanics of Heisenberg, Born, and Jordan, and That of Schrödinger," in Schrödinger (1928), pp. 45–61.]
——— (1927), "Über den Comptoneffekt," *Annalen der Physik* 82: 257–264.
——— (1928), *Collected Papers on Wave Mechanics*. London: Blackie & Sons.

[This English translation was reissued by Chelsea Publishing (New York) in 1978.]

——— (1931), "Über die Umkehrung der Naturgesetze," *Sitzungsberichte der Preussischen Akademie der Wissenschaften* 1931: 144–153.

——— (1932), "Sur la théorie relativiste de l'électron et l'interprétation de la mécanique quantique," *Annales de l'Institut Henri Poincaré* 2: 267–310.

——— (1935), "Die gegenwärtige Situation in der Quantenmechanik," *Die Naturwissenschaften* 23: 807–812, 824–828, 844–849. [Also appears in translation as "The Present Situation in Quantum Mechanics," in Wheeler and Zurek, pp. 152–167.]

——— (1957), *Science, Theory and Man*. New York: Dover Publications.

Schweber, S. S. (1986), "Feynman and the Visualization of Space-Time Processes," *Reviews of Modern Physics* 58: 449–508.

Scully, M. O., and H. Walther (1989), "Quantum Optical Test of Observation and Complementarity in Quantum Mechanics," *Physical Review A* 39: 5229–5236.

Scully, M. O., B.-G. Englert, and H. Walther (1991), "Quantum Optical Tests of Complementarity," *Nature* 351: 111–116.

Selleri, F. (1990), *Quantum Paradoxes and Physical Reality*. Dordrecht: Kluwer Academic Publishers.

——— (ed.) (1992), *Wave-Particle Duality*, London, Plenum Publishing.

Shapin, S., and S. Schaffer (1985), *Leviathan and the Air-Pump*. Princeton, NJ: Princeton University Press.

Shimony, A. (1963), "Role of the Observer in Quantum Theory," *American Journal of Physics* 31: 755–773.

——— (1984), "Controllable and Uncontrollable Non-locality," in Kamefuchi et al., pp. 225–230.

——— (1986), "Events and Processes in the Quantum World," in Penrose and Isham, pp. 182–203.

——— (1993), *Search for a Naturalistic World View*, vol. 2. Cambridge: Cambridge University Press.

Sinha, S., and R. D. Sorkin (1991), "A Sum-over-Histories Account of an EPR(B) Experiment," *Foundations of Physics Letters* 4: 303–335.

Slater, J. C. (1973), "The Development of Quantum Mechanics in the Period 1924–1926," in Price, Chissick, and Ravensdale, pp. 19–25.

Sokolnikoff, I. S. (1951), *Tensor Analysis*. New York: John Wiley & Sons.

Sokolovsky, D. and J. N. L. Connor (1993), "Quantum Interference and Determination of the Traversal Time," *Physical Review A* 47: 4677–4680.

Sopka, K. R. (1988), *Quantum Physics in America: The Years through 1935*. New York: Tomash Publishers.

Sprinkle, H. C. (1933), *Concerning the Philosophical Defensibility of a Limited Indeterminism*. Scottdale, PA: Mennonite Press.

Squires, E. J. (1993), "A Local Hidden-Variable Theory That, FAPP, Agrees with Quantum Theory," *Physics Letters A* 178: 22–26.

Stachel, J. (1986), "Einstein and the Quantum: Fifty Years of Struggle," in Colodny, pp. 349–385.

Stapp, H. P. (1971), "S-Matrix Interpretation of Quantum Mechanics," *Physical Review D* 3: 1303–1320.

——— (1972), "The Copenhagen Interpretation," *American Journal of Physics* 40: 1098–1116. [Reprinted in Ballentine (1988).]

——— (1985), "Bell's Theorem and the Foundations of Quantum Physics," *American Journal of Physics* 53: 306–317.

Stehr, N., and V. Meja (eds.) (1984), *Society and Knowledge*. New Brunswick, NJ: Transaction Books.

Stein, H. (1991), "On Relativity Theory and Openness of the Future," *Philosophy of Science* 58: 147–167.

Sudarshan, E. C. G. (1992), "Measurement Theory," in Black et al., pp. 148–156.

Summhammer, J., G. Badurek, H. Rauch, U. Kischko, and A. Zeilinger (1983), "Direct Observation of Fermion Spin Superposition by Neutron Interferometry," *Physical Review A* 27: 2523–2532.

Suppes, P. (ed.) (1980), *Studies in the Foundations of Quantum Mechanics*. East Lansing, MI: Philosophy of Science Association.

Takabayasi, T. (1952), "On the Formulation of Quantum Mechanics Associated with Classical Pictures," *Progress of Theoretical Physics* 8: 143–182.

Teller, P. (1986), "Relational Holism and Quantum Mechanics," *British Journal for the Philosophy of Science* 37: 71–81.

——— (1989), "Relativity, Relational Holism, and the Bell Inequalities," in Cushing and McMullin, pp. 208–223.

Temple, G. (1934), *An Introduction to Quantum Theory*. New York: D. Van Nostrand.

——— (1951), *The General Principles of Quantum Theory*. London: Methuen.

ter Haar, D. (ed.) (1967), *The Old Quantum Theory*. London: Pergamon Press.

Tersoff, J., and D. Bayer (1983), "Quantum Statistics for Distinguishable Particles," *Physical Review Letters* 50: 553–554.

Tolman, R. C. (1938), *The Principles of Statistical Mechanics*. Oxford: Oxford University Press.

Toulmin, S. (1961), *Foresight and Understanding*. London: Hutchinson.

Uhlenbeck, G. E., and S. Goudsmit (1925), "Ersetzung der Hypothese vom unmechanischen Zwang durch eine Forderung bezüglich des inneren Verhaltens jedes einzelnen Elektrons," *Die Naturwissenschaften* 13: 953–954.

Ullmann-Margalit, E. (ed.) (1988), *Science in Reflection*. Dordrecht: Kluwer Academic Press.

Unnerstall, T. (1990), "Comment on the Rauch-Vigier Experiments on Neutron Interferometry," *Physics Letters A* 151: 263–268.

Valentini, A. (1991a), "Signal-Locality, Uncertainty, and the Subquantum H-Theorem. I," *Physics Letters A* 156: 5–11.

——— (1991b), "Signal-Locality, Uncertainty, and the Subquantum H-Theorem. II," *Physics Letters A* 158: 1–8.

——— (1992), "On the Pilot-Wave Theory of Classical, Quantum and Subquantum Physics." Unpublished Ph.D. dissertation, ISAS—International School for Advanced Studies, Trieste, Italy. [To be published in 1994 by Springer-Verlag

(Berlin) as *On the Pilot-Wave Theory of Classical, Quantum and Subquantum Physics*.]
van der Merwe, A., F. Selleri, and G. Tarozzi (eds.) (1988), *Microphysical Reality and Quantum Formalism*, 2 vols. Dordrecht: Kluwer Academic Publishers.
——— (eds.) (1992), *Bell's Theorem and the Foundations of Modern Physics*. Singapore: World Scientific Publishing.
van der Waerden, B. L. (ed.) (1968), *Sources of Quantum Mechanics*. New York: Dover Publications.
——— (1973), "From Matrix Mechanics and Wave Mechanics to Unified Quantum Mechanics," in Mehra (1973), pp. 276–293.
van Fraassen, B. C. (1980), *The Scientific Image*. Oxford: Clarendon Press.
——— (1985), "EPR: When Is a Correlation Not a Mystery?" in Lahti and Mittelstaedt (1985), 113–128.
——— (1989), *Laws and Symmetry*, Oxford, Clarendon Press.
van Hove, L. (1958), "Von Neumann's Contribution to Quantum Theory," *Bulletin of the American Mathematical Society* 64: 95–99.
van Kampen, N. G. (1991), "Determinism and Predictability," *Synthese* 89: 273–281.
van Lunteren, F. H. (1991), "Framing Hypotheses: Conceptions of Gravity in the 18th and 19th Centuries." Unpublished Ph.D. dissertation, University of Utrecht.
Vigier, J.-P. (1979a), "Superluminal Propagation of the Quantum Potential in the Causal Interpretation of Quantum Mechanics," *Lettere al Nuovo Cimento* 24: 258–264.
——— (1979b), "Model of Quantum Statistics in Terms of a Fluid with Irregular Stochastic Fluctuations Propagating at the Velocity of Light: A Derivation of Nelson's Equations," *Lettere al Nuovo Cimento* 24: 265–272.
——— (1982), "Non-locality, Causality and Aether in Quantum Mechanics," *Astronomische Nachrichten* 303: 55–80.
——— (1985a), "Causal Non-local Interpretation of Neutron Interferometry Experiments, EPR Correlations and Quantum Statistics," in Lahti and Mittelstaedt (1985), pp. 653–675.
——— (1985b), "Causal Stochastic Interpretation of Quantum Statistics," *Pramāna* 25: 397–418.
——— (1986), "Trajectories, Spin, and Energy Conservation in Time-Dependent Neutron Interferometry," in Greenberger (1986), pp. 503–511.
——— (1987), "Theoretical Implications of Time-Dependent Double Resonance Neutron Interferometry" in Honig, Kraft, and Panarella, pp. 1–18.
——— (1988), "New Theoretical Implications of Neutron Interferometric Double Resonance Experiments," *Physica B* 151: 386–392.
——— (1989a), "Particular Solutions of a Non-linear Schrödinger Equation Carrying Particle-like Singularities Represent Possible Models of de Broglie's Double Solution Theory," *Physics Letters A* 135: 99–105.
——— (1989b), "Comments on the 'Uncontrollable' Character of Non-locality," in Bitsakis and Nicolaides, pp. 133–140.

——— (1990), "Real Physical Paths in Quantum Mechanics—Equivalence of the Einstein–de Broglie and Feynman Points of Views [sic] on Quantum Particle Behavior," in Kobayashi et al., pp. 140–152.

——— (1991), "Explicit Mathematical Construction of Relativistic Nonlinear De Broglie Waves Described by Three-Dimensional (Wave and Electromagnetic) Solitons 'Piloted' (Controlled) by Corresponding Solutions of Associated Linear Klein-Gordon and Schrödinger Equations," *Foundations of Physics* 21: 125–148.

von Neumann, J. (1927a), "Mathematische Begründung der Quantenmechanik," *Nachrichten von der Gesellschaft der Wissenschaften zu Göttingen*. 1927: 1–57.

——— (1927b), "Wahrscheinlichkeitstheoretischer Aufbau der Quantenmechanik," *Nachrichten von der Gesellschaft der Wissenschaften zu Göttingen*. 1927: 245–272.

——— (1927c), "Thermodynamik quantenmechanischer Gesamtheiten," *Nachrichten von der Gesellschaft der Wissenschaften zu Göttingen*. 1927: 273–291.

——— (1932), *Mathematische Grundlagen der Quantenmechanik*. Berlin: Springer-Verlag.

——— (1955), *Mathematical Foundations of Quantum Mechanics*. Princeton, NJ: Princeton University Press.

von Weizsäcker, C. F. (1977), "Heisenberg's Conception of Physics," in Castell, Drieschner, and von Weizsäcker, pp. 9–19.

——— (1987), "Heisenberg's Philosophy" in Lahti and Mittelstaedt (1987), pp. 277–293.

——— (1989), "Structure and Properties of Nuclei (1932–1935), An Annotation," in Blum, Dürr, and Rechenberg, pp. 183–196.

von Weizsäcker, C. F., and Th. Görnitz (1991), "Quantum-Realistic Interpretation," *Foundations of Physics* 21: 311–321.

Wallstrom, T. C. (1989), "On the Derivation of the Schrödinger Equation from Stochastic Mechanics," *Foundations of Physics Letters* 2: 113–126.

Wang, L. J., and X. Y. Zou, and L. Mandel (1991), "Experimental Test of the De Broglie Guided-Wave Theory for Photons," *Physical Review Letters* 66: 1111–1114.

Watkins, J. (1984), *Science and Skepticism*. Princeton: Princeton University Press.

Webster's Third New International Dictionary of the English Language (1986). Springfield, MA: Merriam-Webster.

Weiner, C. (ed.) (1977), *History of Twentieth Century Physics*. New York: Academic Press.

Weinfurter, H., G. Badurek, H. Rauch, and D. Schwahn (1988), "Inelastic Action of a Gradient Radio-Frequency Neutron Spin Flipper," *Zeitschrift für Physik B* 72: 195–201.

Weizel, W. (1953), "Ableitung der Quantentheorie aus einem klassischer, kausal determinierten Modell," *Zeitschrift für Physik* 134, 264–285.

Wessels, L. (1975), "Schrœdinger's Interpretations of Wave Mechanics." Unpublished Ph. D. dissertation, Indiana University.

——— (1979), "Schrödinger's Route to Wave Mechanics," *Studies in History and Philosophy of Science* 10: 311–340.

——— (1980), "The Intellectual Sources of Schrödinger's Interpretations" in Suppes, pp. 59–76.

——— (1981), "What Was Born's Statistical Interpretation?" in Asquith and Giere (1981), pp. 187–200.

Wheaton, B. R. (1983), *The Tiger and the Shark: Empirical Roots of Wave-Particle Dualism*. Cambridge: Cambridge University Press.

Wheeler, J. A., and W. H. Zurek (eds.) (1983), *Quantum Theory and Measurement*. Princeton: Princeton University Press.

Whewell, W. (1857), *History of the Inductive Sciences*, 3 vols. London: John W. Parker and Son. [Reprinted in 1967. London: Frank Cass & Co., Ltd.]

Wigner, E. P. (1976), "Interpretation of Quantum Mechanics," in Wheeler and Zurek, pp. 260–314.

Wise, M. N. (1987), "How Do Sums Count? On the Cultural Origins of Statistical Causality," in Krüger, Daston, and Heidelberger, pp. 395–425.

Zeh, H. D. (1993), "There Are No Quantum Jumps, Nor Are There Particles!" *Physics Letters A* 172: 189–192.

Zou, X. Y., T. Grayson, L. J. Wang, and L. Mandel (1992), "Can an 'Empty' De Broglie Pilot Wave Induce Coherence?" *Physical Review Letters* 68, 3667–3669.

AUTHOR INDEX

This index contains only citation references, listed by author. Proper names appearing in the text, or those actually discussed as such in the notes, appear in the subject index that follows. Note numbers are given in parentheses. Multiple note references for a given page are given in the order in which they appear on that page.

Abragam, A., 238 (15), 241 (53), 245 (67)
Achinstein, P., 220 (9)
Aers, G. C., 232 (57, 60)
Agarwal, G. S., 226 (67)
Aharonov, Y., 229 (13)
Albert, D. Z., 217 (2), 226 (69), 227 (76, 77), 229 (11), 230 (17), 232 (53), 238 (4), 267 (102)
Alefeld, B., 238 (15)
Anandan, J., 229 (13)
Aspect, A., 221 (25)
Avramesco, A., 265 (56)

Bachelard, G., 220 (11), 223 (63–65)
Bacon, F., 218 (11)
Badurek, G., 238 (11), 239 (17, 19–21, 25), 241 (53), 242 (60)
Ballentine, L. E., 224 (18, 19), 227 (78), 232 (56)
Barut, A. O., 251 (23), 265 (58)
Bass, J., 252 (43)
Baublitz, M., 259 (113)
Bauspiess, W., 238 (15)
Bayer, D., 265 (59, 60)

Belinfante, F. J., 237 (110), 253 (55, 67), 258 (58), 267 (92)
Bell, J. S., 221 (25, 33), 222 (52), 227 (80, 81), 230 (25), 235 (90, 91), 239 (27, 28), 253 (61, 66), 255 (84), 264 (14), 266 (82), 267 (89, 97)
Beller, M., 225 (44–46, 51), 246 (87, 92), 247 (119), 248 (128, 134, 137, 8), 249 (16, 18, 21), 250 (3, 5, 6), 252 (42), 259 (110)
Ben-Menahem, Y., 218 (22), 246 (92), 248 (134), 264 (25, 26), 270 (29, 30)
Berry, M., 227 (85)
Bialynicki-Birula, I., 232 (48)
Bloch, F., 238 (15), 242 (57)
Blokhintsev, D. I., 258 (63)
Bohm, D., 217 (8, 10), 218 (19), 220 (22), 222 (53), 223 (66, 67, 68, 3, 5), 226 (70), 228 (89, 2, 3, 6, 7), 229 (10, 12, 14), 230 (16, 21, 22, 26, 27), 231 (30, 33, 39, 40, 42, 44–47), 232 (51, 53), 235 (92), 236 (98), 237 (106, 107, 111), 238 (5–7), 239 (27), 240 (45–47), 243

Bohm, D. *(cont.)*
(65, 67, 69–72), 250 (31), 255 (1, 3), 256 (4, 24), 257 (34, 36, 42), 258 (59, 85), 259 (99, 100, 106, 116), 260 (126, 127, 129, 130, 133, 138, 140), 262 (172), 264 (14, 22), 266 (62, 79, 80, 86), 267 (89, 91–94, 96, 101), 268 (110, 111, 114)

Bohr, N., 223 (2), 224 (20–25), 225 (41, 56–60), 226 (63), 246 (90, 93–97), 247 (98), 250 (32), 252 (49, 50), 256 (6), 264 (16, 17, 19–21)

Bonse, U., 238 (15)

Bopp, F., 252 (43)

Born, M., 224 (35, 36, 37), 225 (38, 39), 247 (118), 250 (3, 4), 251 (26, 28), 252 (51), 256 (11, 13–15, 19), 257 (37, 41)

Boutroux, É., 243 (4)

Brillouin, L., 251 (21), 266 (66)

Brouard, S., 232 (57)

Brown, H. R., 253 (72), 259 (104)

Brush, S. G., 243 (1), 244 (30, 31)

Bub, J., 227 (78, 79)

Bunge, M., 258 (80, 81)

Butterfield, J., 221 (27)

Carlen, E. A., 259 (116)

Cartwright, N., 221 (27)

Cassidy, D. C., 247 (106), 249 (16, 19)

Clauser, J. F., 221 (25)

Clifton, R., 234 (83), 253 (67)

Combourieu, M.-C., 239 (22)

Compton, A. H., 245 (55)

Connor, J. N. L., 232 (58)

Crain, S. D., 258 (56, 57)

Croca, J. R., 257 (34)

Cross, A., 258 (60, 62, 64, 66, 84)

Cufaro-Petroni, N., 266 (63–65)

Cushing, J. T., 217 (11, 2), 218 (5, 8, 9), 219 (25, 2, 3), 220 (20, 21), 221 (35), 222 (52, 55), 225 (43), 230 (25), 236 (99), 240 (50), 245 (34, 50), 254 (79), 267 (87, 103, 104), 268 (106)

Dalibard, J., 221 (25)

Daneri, A., 227 (77, 78)

Darrigol, O., 247 (106), 249 (25)

Daumer, M., 229 (15), 260 (125, 143), 262 (171), 263 (5)

de Broglie, L., 222 (47), 245 (59, 61, 69), 248 (2), 249 (28, 29), 250 (31, 33, 35–37, 41–43, 45–47, 55), 251 (11–16, 19–21, 23, 24), 252 (45), 256 (22, 23), 257 (25–32), 260 (124), 266 (66)

Debye, P., 245 (54)

Degen, P. A., 245 (35)

Deltete, R., 256 (12, 16)

d'Espagnat, B., 224 (13), 227 (78), 253 (57)

de Regt, H. W., 244 (7, 8)

Deutsch, D., 232 (48)

Dewdney, C., 226 (68), 237 (124), 239 (19, 23, 29, 32), 240 (37), 243 (65, 69, 73), 267 (93, 96)

Dieks, D., 226 (69)

Dirac, P. A. M., 225 (40), 247 (124), 259 (100–103)

Donovan, A., 218 (8, 9)

Dorling, J., 263 (7, 8)

Drabkin, G. M., 238 (15)

Dresden, M., 247 (112)

Duhem, P., 219 (24, 7), 269 (2–5, 16)

Dürr, D., 229 (15), 260 (125, 131, 143, 148), 261 (150, 152–155, 157–160), 262 (166, 168–171), 263 (2, 5, 6), 267 (96), 271 (42)

Earman, J., 219 (4), 271 (41)

Eberhard, P. H., 222 (53), 264 (24), 265 (56), 266 (68, 78)

Eder, G., 238 (15), 242 (59)

Edwards, P., 222 (44), 245 (37, 63)

Ehrenfest, P., 245 (56)

Einstein, A., 218 (23, 24), 220 (14,

22), 224 (12), 230 (21), 245 (37, 51–53, 56, 58), 256 (7), 257 (26), 259 (104), 262 (173), 264 (30–39), 265 (41, 51, 52, 56, 57), 266 (65)
Englert, B.-G., 226 (65)
Enz, C. P., 250 (44, 55)
Epstein, S. T., 256 (4)

Favrholdt, D., 243 (7), 244 (8, 29), 245 (40, 43)
Faye, J., 223 (62), 225 (52), 243 (2, 7), 244 (7, 29), 245 (38, 43–47)
Fényes, I., 252 (43), 259 (112)
Fermi, E., 250 (38)
Fertig, H. A., 232 (57)
Feuer, L. S., 243 (6), 244 (29), 245 (60, 64), 247 (122)
Feyerabend, P., 217 (9), 248 (1)
Feynman, R. P., 223 (1), 238 (14)
Fine, A., 221 (27), 222 (39), 234 (81), 251 (25, 30), 256 (16, 17), 265 (40–43, 50, 54)
Folse, H. J., 223 (3, 6, 7, 9), 224 (10, 11), 225 (47–49), 226 (62), 244 (7)
Ford, J., 271 (48)
Forman, P., 218 (9), 244 (10, 14–18, 20–27), 247 (126), 248 (3), 270 (36)
Frank, P., 245 (36)
Freistadt, H., 258 (68–72)
Friedman, M., 219 (5)
Fürth, R., 252 (38, 39, 41), 259 (110)

Galison, P., 218 (13)
Garuccio, A., 257 (34)
Gell-Mann, M., 223 (1)
George, A., 256 (21)
Ghirardi, G. C., 221 (34), 227 (76), 234 (83), 268 (109)
Ghose, P., 226 (66, 67)
Gillispie, C. C., 217 (5)
Gnedenko, B. V., 262 (165)
Goldstein, H., 227 (83), 228 (4, 7), 242 (63–64), 243 (65), 254 (74, 77), 268 (112)

Goldstein, S., 229 (15), 260 (125, 131, 143–148), 261 (150, 152–155, 157–160), 262 (166, 168–171), 263 (2, 5, 6), 267 (96), 271 (42)
Görnitz T., 259 (95)
Goudsmit, S., 264 (14)
Gould, S. J., 217 (6), 270 (35)
Grayson, T., 257 (35)
Greenberger, D. M., 239 (16)
Grommer, J., 257 (26)
Guéret, Ph., 239 (19)
Guy, R., 256 (12, 16)

Halpern, O., 255 (3)
Hanle, P. A., 270 (31)
Hanson, N. R., 259 (86–92), 269 (19)
Harding, S. G., 218 (21), 269 (14)
Hauge, E. H., 232 (57, 59), 237 (115, 117, 120–122)
Hauschildt, D., 226 (65)
Healey, R., 226 (69), 238 (4)
Heilbron, J. L., 248 (131–133, 135, 136), 249 (20), 250 (50)
Heisenberg, W., 219 (26), 224 (23, 27–34), 225 (54), 244 (20), 247 (117, 120, 121), 248 (7, 10, 11), 249 (12–14), 250 (50, 6), 256 (8), 258 (73–76), 264 (23, 24)
Heitler, W., 246 (71)
Hendry, J., 244 (33), 245 (43), 246 (86, 92), 247 (101, 105, 106, 108, 109, 111), 248 (129, 130), 249 (25), 263 (1)
Hepp, K., 227 (79)
Hiley, B. J., 217 (10), 222 (53), 223 (67, 68), 230 (27), 231 (33, 40, 42, 44), 232 (53), 239 (27), 240 (45–47), 250 (31), 255 (1), 256 (24), 259 (100, 105, 106, 116), 260 (126, 140), 262 (172), 264 (22), 266 (79, 80, 86), 267 (89, 91, 93, 94, 101), 268 (110, 111, 114)
Holland, P. R., 217 (10), 231 (43), 232 (57), 239 (23, 27, 29, 30, 32),

Author Index

Holland, P. R. *(cont.)*
240 (37, 43, 44, 46), 243 (65, 69, 72, 73), 255 (1), 266 (84, 86), 267 (90, 94, 101), 268 (110, 111, 114)
Home, D., 224 (18), 226 (64, 66, 67), 232 (54)
Honner, J., 225 (53)
Horton, G., 226 (68), 237 (124)
Howard, D., 219 (24), 222 (43), 235 (84), 245 (69), 248 (8), 251 (30), 263 (10–12), 265 (44–46, 48–51, 53), 266 (65)
Huang, K., 220 (20), 263 (176)

Jammer, M., 219 (2), 232 (56), 243 (2, 6), 244 (7, 8, 11–13), 245 (39, 57, 68), 246 (70, 80, 82, 83, 89), 249 (26), 250 (31, 55, 1, 2, 7, 8), 251 (19, 29), 252 (37), 253 (56, 63, 71, 72), 256 (3), 258 (59, 61, 65, 84), 259 (107), 263 (12), 264 (16)
Jankovic, V., 247 (104)
Jarrett, J. P., 221 (26, 28), 234 (76, 83), 235 (84)
Jones, M. R., 234 (83)
Jordan, P., 250 (5), 252 (47, 48)

Kaiser, D., 244 (28)
Kalckar, J., 248 (127)
Kaloyerou, P. N., 223 (68), 232 (53), 240 (36), 259 (100), 260 (140), 264 (22), 267 (89, 91, 93, 94, 96), 268 (110, 111, 114)
Kant, I., 222 (45)
Keller, J. B., 229 (12), 255 (3)
Kennard, E. H., 251 (31, 33), 252 (34)
Khinchin, A. Ya., 262 (165)
Kischko, U., 239 (15)
Kirsten, C., 251 (25)
Klein, M. J., 245 (51, 52), 246 (72, 78, 79)
Kochen, S., 226 (69)
Koopman, B. O., 229 (15)

Körner, S., 258 (83)
Kraft, D., 244 (33), 270 (36)
Kragh, H., 246 (81), 247 (123, 125)
Krieger, M. H., 222 (45)
Krips, H., 238 (4)
Kroes, P., 244 (33), 270 (36)
Kronig, R. 249 (23), 250 (48)
Kubli, F., 245 (66)
Kyprianidis, A., 239 (19, 23, 29, 32), 240 (37), 243 (65, 69, 73), 266 (63–65), 267 (94)

Lahti, P., 226 (69)
Lam, M. M., 226 (68), 237 (124), 267 (93, 96)
Laplace, P.-S., 271 (39, 40)
Laudan, L., 217 (4, 3), 218 (8, 9), 269 (12, 13, 18, 20–22)
Laudan. R., 218 (8, 9)
Laurikainen, K. V., 223 (8), 257 (44, 45, 47–54), 258 (55)
Leavens, C. R., 232 (57, 60), 237 (123–125)
Lehr, W. J., 259 (116), 262 (172)
Leplin, J., 217 (4), 269 (18, 20–22)
Lepore, V. L., 257 (34)
Loewer, B., 238 (4)
Loffredo, M. I., 259 (115)
Loinger, A., 227 (77, 78)

MacKinnon, E., 219 (2), 222 (56), 246 (74, 77), 252 (33)
Macrae, N., 253 (65)
Madajczyk, J. L., 232 (48)
Madelung, E., 250 (7, 8), 253 (71)
Malik, Z., 226 (68), 237 (124)
Mandel, L., 257 (35)
Margenau, H., 223 (69), 258 (77–79)
Maric, Z., 266 (63–65)
Martins, R. de A., 253 (72)
Massey, H. S. W., 236 (99)
Maxwell, J. C., 222 (51)
Maxwell, N., 266 (83)
McMullin E., 222 (41, 48, 49, 52), 230 (25), 268 (106), 270 (28)

Mehra, J., 245 (65), 246 (76, 88), 248 (4)
Messiah, A., 224 (13), 227 (86, 88), 236 (99, 100), 237 (114)
Métadier, J., 252 (37)
Meyerson, É., 220 (11), 222 (46), 223 (57, 59–61), 245 (62)
Miller, A. I., 223 (58), 263 (1)
Mittelstaedt, P., 226 (69)
Moreira, R. N., 257 (34)
Moritz, R. E., 217 (5)
Mott, N. F., 236 (99)
Moulton, F. R., 217 (5), 271 (44)
Muga, J. G., 232 (57)
Mugur-Schächter, M., 253 (66)
Murdoch, D., 244 (7), 245 (48)
Mycielski, J., 232 (48)

Nelson, E., 230 (17), 252 (40, 43), 259 (107–109, 111, 117–120), 260 (121, 122, 126), 262 (172), 263 (3, 4)
Newton, I., 270 (37, 38), 271 (40)

Olkhovsky, V. S., 232 (57)
Omnès, R., 227 (78), 259 (93, 94)
Oppenheimer, J. R., 258 (67)

Pagonis, C., 253 (67)
Pais, A., 226 (71), 245 (41, 42)
Park, J. L., 259 (116)
Partovi, M. H., 232 (48)
Pauli, W., 225 (40, 55), 232 (56), 247 (102, 103, 109), 249 (17, 30), 252 (46, 52), 256 (5), 257 (37–40, 43), 264 (14)
Pearle, P., 221 (34)
Peierls, R. E., 246 (91), 247 (107, 110)
Peres, A., 264 (14)
Petersen, A., 245 (40)
Pickering, A., 218 (9,15), 219 (25)
Pinch, T. J., 218 (9), 252 (44), 253 (64)

Pitowsky, I., 263 (10, 13), 267 (99)
Podolsky, B., 220 (22)
Poincaré, H., 271 (45)
Post, H. R., 218 (16, 18–20)
Prosperi, G. M., 227 (77, 78)

Quine, W. V., 269 (6–11), 271 (50)

Rabi, I. I., 238 (15)
Rabinowich, E., 258 (67)
Raman, V. V., 248 (3)
Rauch, H., 238 (11, 15), 239 (17, 19–22, 25), 241 (53), 242 (60)
Recami, E., 232 (57)
Redhead, M. L. G., 253 (67)
Rechenberg, H., 245 (65), 248 (4)
Reichenbach, H., 256 (9)
Rimini, A., 221 (34), 227 (76), 234 (83), 268 (109)
Robertson, P., 249 (24)
Roger, G., 221 (25)
Rohrlich, F., 221 (31)
Rorty, R., 217 (4)
Rosen, N., 220 (22), 252 (35, 36), 256 (9)
Rosenfeld, L., 238 (2, 3), 247 (100), 258 (72, 82)

Sala, R., 232 (57)
Salmon, W. C., 219 (6), 220 (12)
Sardelis, D., 266 (63–65)
Saunders, S., 259 (104)
Schaffer, S., 218 (9)
Schiller, R., 243 (65, 67, 69–72)
Schmidt, M., 226 (68), 237 (124)
Schrödinger, E., 227 (82), 246 (72, 78–80, 82, 84), 249 (15), 252 (33, 37, 38), 254 (74, 77), 264 (26), 270 (30, 31)
Schwahn, D., 239 (15)
Schweber, S. S., 222 (54)
Scully, M. O, 226 (65)
Selleri, F., 255 (83, 84)
Shapin, S., 218 (9)

Shimony, A., 221 (25, 28, 31), 234 (75, 76, 83), 264 (22, 29), 268 (109)
Siegert, A., 238 (15), 242 (57)
Sinha, S., 228 (1)
Slater, J. C., 247 (113–116), 250 (34)
Sokolnikoff, I. S., 254 (77, 78)
Sokolovski, D., 232 (58)
Sopka, K. R., 250 (51–54)
Sorkin, R. D., 228 (1)
Sprinkle, H. C., 243 (2)
Squires, E. J., 222 (53)
Stachel, J., 256 (16, 18, 20, 21)
Stapp, H. P., 221 (32), 224 (16, 17, 26), 264 (14)
Stein, H., 266 (83)
Støvneng, J. A., 232 (57, 59), 237 (115, 117, 120–122)
Suda, M., 238 (15)
Sudarshan, E. C. G., 229 (15)
Summhammer, J., 238 (15)

Takabayasi, T., 255 (3)
Teller, P., 221 (36), 222 (37), 235 (86)
Temple, G., 250 (9), 251 (10), 253 (74)
Tersoff, J., 265 (59, 60)
Tiomno, J., 243 (65, 67, 69–72)
Tolman, R. C., 263 (176, 177)
Toulmin, S., 220 (10)
Treder, H.-J., 251 (25)
Treimer, W., 238 (15)
Tuppinger, D., 238 (11), 239 (15, 17, 19–21, 25), 241 (53), 242 (60)

Uhlenbeck, G. E., 264 (14)
Unnerstall, T., 239 (24)

Vaidman, L., 229 (13), 230 (17)
Valentini, A., 217 (10), 234 (77, 78), 240 (44), 260 (131, 132, 137–142), 262 (167), 263 (176), 266 (81, 85), 267 (89, 95, 98, 100), 268 (110, 111), 270 (33)
van der Waerden, B. L., 245 (58), 249 (15)
van Fraassen, B. C., 217 (7), 222 (39), 270 (27), 271 (49)
van Hove, L., 252 (53)
van Kampen, N. G., 270 (36)
van Lunteren, F. H., 222 (42, 50)
Vigier, J.-P., 222 (53), 223 (5), 229 (15), 239 (19, 23, 29, 32), 240 (33, 35, 37), 243 (65, 69, 73, 74), 250 (49), 251 (23), 259 (98–100), 260 (129), 262 (172), 265 (58), 266 (61, 63–65)
von Meyenn, K., 250 (44, 55)
von Neumann, J., 224 (13), 225 (40), 237 (107), 250 (37), 252 (44), 253 (54, 56, 58–60), 255 (82)
von Weizsäcker, C. F., 219 (2), 248 (6), 259 (95)

Wallstrom, T. C., 259 (114)
Walther, H., 226 (65)
Wang, L. J., 257 (35)
Watkins, J., 218 (7)
Weber, T., 221 (34), 227 (76), 234 (83), 268 (109)
Weinfurter, H., 239 (15)
Weizel, W., 252 (43)
Wessels, L., 246 (73, 75, 78, 81, 82), 250 (3), 252 (33)
Wheaton, B. R., 253 (72), 254 (75)
Wheeler, J. A., 226 (71), 264 (26)
Whewell, W., 218 (12), 220 (13, 17)
Whitaker, M. A. B., 224 (18)
Wigner, E. P., 224 (13)
Wise, M. N., 244 (32), 245 (49)

Zanghi, N., 229 (15), 260 (125, 131, 143, 148), 261 (150, 152–155, 157–160), 262 (166, 168–171), 263 (2, 5, 6), 267 (96), 271 (42)
Zeh, H. D., 227 (78)
Zeilinger, A., 238 (15), 242 (59)
Zhitnikov, R. A., 238 (15)
Zou, X. Y., 257 (35)
Zurek, W. H., 226 (71), 264 (26)

SUBJECT INDEX

Note numbers are given in parentheses.

absolute-Ψ approach, 27, 28
absolute space, 18
absolute uncertainty, 169, 261 (159)
acausality, 98, 99
accommodation, 99
action at a distance, 13, 18, 47, 222 (42); field concept and, 220 (18); versus contact action, 19, 222 (50)
Adkins, M., 234 (75)
Aharonov-Bohm effect, 166
Anschaulichkeit, 99, 246 (84)
anticoincidence, 34
antirealism, 204. *See also* realism
Aristotle, 18, 19

Bachelard, G., 21
Bacon, F. 3, 180
Bacon-Descartes ideal, 2
Ballentine, L., 28
Balmer formula, 106, 117
beauty, 7, 96, 206
Bell, J., 16, 19, 60; on Bohm's theory, 192; locality, 56; on Lorentz versus Einstein on relativity, 189–190; on measurement, 37–38; nonlocality, 59; on observational equivalence, 53
Bell's theorem, 16, 56, 161, 247 (99); counterfactual scenario for, 176–179, 185, 186; derivation of, 193–195

Beller, M., xv, 32, 111, 248 (8), 249 (21)
Bergson, H., 104
Bernoulli, D., 220 (19)
Besso, M., 147
Bloch-Siegert effect, 242 (57)
block universe, 190, 266 (83)
Blokhintsev, D., 152
Bode's law, 10
Bohm, D., 6, 144, 154, 215, 255 (1), 258 (59); on alternative interpretations, 77–78; on contextuality, 53; on de Broglie's pilot wave, 217 (8); implicate order, 159; on irreversibility, 51–52; mixing argument, 70–72; on nonlocality, 22; and Pauli's objection, 120, 126, 257 (24); on probability, 44; quantum mechanics textbook, 146; and the uncertainty relation, 53; on understandability, 223 (66)
Bohm interpretation, xi, xiii, xiv, 255 (1); classical limit of, 52; decoherence and, 227 (78); Dirac equation in, 190; effective collapse in, 164; EPRB correlations in, 83–84; field theory in, 191, 196–198, 267 (93–95); formalism of, 42–44, 61–63; Gleason/Kochen-Specker and, 253 (67); in momentum space, 256 (4); Lorentz-invariance in, 188–191, 258 (58), 267 (96); Marxism and,

308 Subject Index

Bohm interpretation *(cont.)*
  152–153; measurement in, 47–52, 63–68, 230 (23), 231 (34), 240 (48); mixing in, 70–72, 162–165, 171–172; nonlocality in, 47, 59–60; nonlocality/nonseparability in, 235 (85–87); Pauli equation in, 91–95, 190; photons in, 191; probability in, 255 (3); quantum statistics in, 187; realism and, 204–205; reception of, 76, 144–158; relativity in, 188–191, 255 (3); spin in, 84, 190, 255 (3), 266 (86); status of position in, 43, 45, 47, 66, 230 (23), 231 (34), 240 (48); tunneling times in, 54–55, 72–75; uncertainty relations and, 52–53, 58, 68–70; and understanding, 12, 21; versus Bohm theory, 224 (15); von Neumann's theorem and, 134. *See also* Bohmian mechanics; neutron interferometry; quantum potential; trajectory
Bohm-Vigier model, 158–159, 162
Bohmian mechanics, 165–169, 175; as Aristotelian, 261 (149). *See also* Bohm interpretation
Bohr, N.: and causality, 25, 108, 145; Como lecture by, 28, 32, 33, 131; and complementarity, 26, 32–34, 100, 145; on contextuality, 225 (41); and free will, 102; on the future of physics, 117; and intelligibility, 21; on interpretation, 28–29; versus Einstein, 24–26; on von Neumann's theorem, 132. *See also* Høffding
Bohr Institute, 117, 119, 249 (24)
Bohr-Kramers-Slater theory, 109
Boltzmann, L., 220 (19)
Bopp, F., 154
Born, M.: on Bohm's theory, 150; and complementarity, 31; "defection" by, 116; on Einstein's hidden variables, 128; on interpretation,
  31–32; MIT lectures by, 122; statistical interpretation of, 115, 124; on von Neumann's theorem, 132
Born's law, 168, 262 (168)
Bose, S. K., 254 (79)
Bose-Einstein statistics, 186, 187
boson, 191
Bothe, W., 128, 139, 176
Boutroux, É., 96
Bowman, G., xv
Boyle's law, 12
Bragg diffraction, 78
Brillouin, L., 104; on indistinguishability, 187–188
Brown, H. R., 257 (33)
Brownian motion, 52, 130, 160, 169, 170, 174
Bunge, M., 154

Capelin, S., xv
Cassidy, D., 251 (25)
catastrophe of helium, 247 (106)
causal explanation, 11, 13, 19
causality, 2; Bohr on, 29, 223 (62); and energy conservation, 104, 265 (47); and explanation, 25; metaphysical, 96, 97; and space-time description, 244 (19); and understanding, 20–22, 96; in Weimar era, 98, 99. *See also* Bohr; determinism; Einstein; Pauli; statistical causality
causal interpretation. *See* Bohm interpretation
cause-effect, 11
Cayley-Klein parameters, 243 (65)
chaos, 231 (37); in Bohm's theory, 51, 157, 192, 271 (42); in classical mechanics, 209–214
Clarke, S., 19
classical limit, 39–40, 52, 227 (87)
Clifton, R., 252 (44)
coherence, 3, 7, 96, 206
collapse, 31, 39, 177; in Bohm's the-

ory, 68. *See also* projection postulate; reduction
common cause, 11, 16, 58, 221 (27), 234 (81)
Como lecture, 28, 131, 224 (23), 249 (21); and complementarity, 32, 33
complementarity, 25, 28, 30, 77, 112; and Bohm's interpretation, 46; and Høffding, 97, 102; wave-particle duality as, 32–34, 225 (53, 61), 226 (62, 65). *See also* Bohr; Born; de Broglie
completeness, 29, 30, 31, 76, 119; and hidden variables, 133; Jarrett, 58, 59
Compton, A., 103
Compton scattering, 176
conditional probability. *See* probability
conditional wave function, 261 (160)
configuration space, 49, 120, 124, 149; versus physical space, 246 (72), 252 (33), 257 (33)
conservation laws, 21, 28, 104; determinism, causality and, 265 (47)
constructive theory, 11, 12
contact action, 19
contextual factors, 11
contextuality, 46, 47, 53, 134
contingency, xiii, 96, 270 (35); historical, xi, xii, 7, 96, 174, 192, 193, 203–207; in understanding, 18–22
continuity, 107, 125; versus discontinuity, 249 (21)
continuity equation, 135
controllable nonlocality. *See* nonlocality
convective derivative, 236 (94)
conventionalism, 101, 181. *See also* Einstein
convergence. *See* science
Copenhagen interpretation, xi, xiii, 6, 24–32; and realism, 205; versus Copenhagen theory, 224 (15); versus formalism, 10

correlations. *See* EPRB correlations
cosmology, 192. *See also* quantum equilibrium
Courant, R., 106
Crain, S. D., 258 (56)
Crowe, M., 217 (5)
cultural milieu, 96–97, 99
current velocity, 160, 169
covering-law model, 10

de Broglie, L., 105, 106, 217 (8); on Bohm's theory, 148–149; and the Copenhagen interpretation, 121, 126; on complementarity, 19; double solution/pilot wave, 126–128; on guidance condition, 136–138; nonlinear equation, 127; and Pauli's objection, 120, 121, 250 (43); Poincaré, influence by, 96; and a realistic interpretation, 104; reconversion to pilot wave, 148–149, 256 (23, 24); and von Neumann's theorem, 131; wave-particle duality, 113. *See also* double solution; pilot wave; Solvay congress
de Broglie, M., 105
de-Broglie-Bohm theory, 149, 186, 265 (58)
Debye, P., 103
decoherence, 227 (78)
deductive-nomological model, 10
Dennison, D., 121
Descartes, R., 19, 97
determinism, xiii; Bohr on, 29; Born on, 31; and causality, 25; Einstein on, 118, 175, 183; in equations, 213; Forman on, 98; God and belief in, 208, 210; Laplace, 208–209; in nature, 213; in Newtonian mechanics, 208–210; ontological, 270 (36); versus indeterminism, 193, 207–214. *See also* Bohr; Born; Einstein; Schrödinger
deterministic chaos. *See* chaos
Dickson, W., xv

Dirac, P., 111; ether, 158, 159; on interpretation, 118; and measurement, 29, 178
Dirac equation, 160, 190, 266 (86)
discontinuity, 97, 107. *See also* continuity
dispersion, 140
dispersion-free states, 133, 134, 140, 142. *See also* von Neumann's theorem
double-slit experiment. *See* neutron interferometry
double solution, 106, 118, 126–127, 148, 149, 251 (23)
Duhem, P., 199, 200, 201
Duhem-Quine thesis, 199, 201. *See also* underdetermination
Dürr, D., 165, 166, 168, 169
dwell time, 54, 72, 74, 75

Eddington, A., 109
effective wave function, 167, 261 (157)
Eger, M., 243 (6)
Ehrenfest, P., 103
Ehrenfest's theorem, 39
eigenvalue, 27
eikonal approximation, 39, 40
Einstein, A.: and action at a distance, 22; on Bohm's theory, 46, 146–148, 182; and Brownian motion, 170, 174; on causality, 104, 118, 182, 183, 256 (18); on collapse and relativity, 177; on completeness, 119, 145; on constructive versus principle theories, 11–12; on conventionalism, 181; on Copenhagen interpretation, 100; on de Broglie's pilot wave, 118–119, 129; and determinism, 183, 265 (47); on energy fluctuations, 103; and entanglement, 176; general relativity, 13, 18; ghost field, 245 (69); hidden-variables theory, 119, 128–129, 139–140, 251 (25), 254 (77, 79, 80); and locality, 179, 184–186, 265 (55); and local realism, 185; and nonlocality, 175, 179–186, 251 (29, 30), 264 (18); and nonseparability, 176; on objective reality, 100, 180, 185, 256 (18); and operationalism, 114; on positivism, 29, 110–111; separability, 184; separation principle, 184, 265 (52, 53); and space-time, 183, 265 (47); on statistical interpretation, 28, 99, 182; on theoretical physics, 101, 180–181; on underdetermination, 7–8, 218 (24); versus Bohr, 24–26. *See also* Solvay congress
Empedocles, 18
empirical adequacy, 9, 10, 11, 15, 96, 174
empirical equivalence, 5, 203, 204, 217 (4), 269 (19). *See also* observational equivalence
ensemble, 28, 37, 133, 134; of universes, 167, 168. *See also* dispersion-free states
entanglement, 58, 59, 175, 251 (27); Einstein on, 176, 185; Schrödinger on, 178–179
epistemology, xiv, 25, 30, 102, 212; naturalized, 1–5. *See also* ontology
EPR (Einstein-Podolsky-Rosen) paper, 25, 33, 108, 179, 264 (14)
EPRB (Einstein-Podolsky-Rosen-Bohm) correlations, 11, 12, 17, 20, 22–23, 222 (39)
EPRB experiment, 14–16, 56, 161, 188, 234 (81); in Bohm's theory, 83–84, 240 (36); and nonseparability, 32
equilibrium distribution. *See* quantum equilibrium
equivariance, 168
ether, 11, 12, 13, 19, 222 (53); Dirac on, 158–159
Eudoxus, 19
Euler angles, 83, 93, 240 (46)

Euler-Lagrange equation, 196, 268 (113)
evidential criteria, 7, 203
exclusion principle. *See* Pauli
Exner, F., 98
expectation value, 27
explanation, 2, 3; versus understanding, 10–16. *See also* common cause; understanding
explanatory discourse, 9, 18

factorizability, 57; common-cause explanation and, 221 (27), 234 (81)
Fermi, E., 120, 122
Fermi-Dirac statistics, 187
fermion, 191
fertility, 3, 47, 96, 206
Feynman, R., 20, 192, 223 (1)
Fierz, M., 150, 154, 257 (46)
Fine, A., xv, 183, 251 (25), 269 (23); on Einstein separability/locality, 184
first-signal principle, 58, 177. *See also* no-signaling theorem
Fock, V., 152
formal explanation, 9, 10. *See also* explanation
formalism, 9, 10, 15, 25, 219 (1); of quantum mechanics, 26–27
Forman, P., 97, 98, 100, 152
Forman thesis, 97–100
foundationism, 2, 4
Frank, P., 110
Freistadt, H., 152, 153
functional derivative, 268 (113)
Fürth, R., 130

Geiger, H., 176
Gell-Mann, M., 24
general correspondence principle, 6
general relativity. *See* relativity
geometrical optics, 39
ghost field, 245 (69)
Gibbs, J., 220 (19)
Gleason's theorem, 253 (67)

goals. *See* science
God: determinism and, 208, 210; Pauli and, 151
Gödel, K., 134
Goldstein, S., xv, 165, 166, 169, 255 (81), 261 (156)
gravity, 13, 18, 47
Grommer, J., 148
guidance condition: in Bohm's theory, 42, 66, 166; in de Broglie's theory, 119, 120, 136–138; in the Pauli equation, 93; in stochastic mechanics, 160

Hamiltonian, 27
Hamilton-Jacobi equation, 92, 152; and Bohm's theory, 228 (7), 242 (62), 256 (4); in field theory, 198; importance for de Broglie, 104, 126, 254 (74); and Schrödinger, 106, 246 (81)
Hamilton's equations, 92
Hanson, N., 155; on alternative interpretations, 155–156
Heilbron, J., 112
Heisenberg, W.: on alternative interpretations, 153; on complementarity, 33; and discontinuities, 107; on epistemology versus ontology, 30, 224 (30); on the interpretation of a formalism, 8, 29–30, 111, 114–118, 248 (6, 8); on a materialistic ontology, 153; matrix mechanics, 98; and measurement, 29, 178; on no signaling, 178; and observables, 110; and the old quantum theory, 109; and *potentia,* 17; and reality, 224 (34); uncertainty principle, 32, 99, 224 (23); and von Neumann's theorem, 131. *See also* matrix mechanics
Heitler, W., 105
heuristic guidelines, 5–7
hidden variables, 134; Bohm on, 77–78; in Bohm's theory, 58, 59,

hidden variables *(cont.)*
144; in de Broglie's theory, 127; and the statistical interpretation, 28. *See also* Bell's theorem; completeness; dispersion-free states; Einstein; von Neumann
Hilbert, D., 106, 118, 134
Hilbert space, 27
Hiley, B., 159, 255 (1)
historical contingency. *See* contingency
historicism, 2
Høffding, H., 97, 100, 101, 102, 243 (7), 244 (29), 245 (43)
holism, 17, 46
Holland, P., xv, 248 (1), 255 (1)
Home, D., 232 (54)
Honner, J., 225 (53)
Howard, D., xv, 184, 218–219 (24), 248 (8), 263 (10); on complementarity, 225 (61); on Einstein separability/locality, 184–185; locality, 57, 59, 60; separability, 58, 59, 60
H-theorem: Boltzmann, 172; quantum, 171–173, 263 (176)
Huygens, C., 19
hypothetico-deductive method, 3

ideals of natural order, 11
implicate order, 159, 259 (96)
impulsive approximation, 64, 236 (98), 240 (41)
incompleteness, 39. *See also* completeness
indeterminism, xiii, 17, 24, 32, 96, 100; versus determinism, 193, 207–214. *See also* Schrödinger
individuality, 98, 99
individuation and space, 18–19, 222 (43)
instrumentalism, 2, 11
integrable system, 212
intelligibility, 19, 47

interferometry. *See* neutron interferometry
interpretation: of a formalism, 11, 15; of quantum mechanics, 8, 27–32; physical, 9; of a scientific theory, 9; versus theory, 9. *See also* Bohm interpretation; causal interpretation; Copenhagen interpretation; statistical interpretation
intuitiveness, 98
irrationality, 96, 98, 151
irreversibility, 51

Jammer, M., 97, 98, 152
Jarrett, J., 57, 58; completeness, 58; locality, 57, 58, 59
joint probability. *See* probability
Jordan, P., 107, 124, 131
Jung, C., 151

Kaloyerou, P., 225 (61), 257 (33), 267 (88)
Kant, I., 18, 96, 102, 181, 222 (56)
Kennard, E., 129, 251 (33)
Kennedy, J. B., xv, 217 (3), 222 (43), 266 (83)
Kepler, J., 15
Kepler's laws, 12, 13, 220 (15)
Kierkegaard, S., 97, 102, 244 (29)
kinetic theory, 12, 14
Kirchhoff, G., 22
Klein-Gordon equation, 190
Kochen-Specker proof, 253 (67)
Krishnamurti, J., 259 (96)
Kronig, R., 117, 121
Krueger, L., xv
Kuhn, T., 114, 115, 116, 248 (9)

Lagrange, J., xii, 211
Landau, L., 152
Landé, A., 117
Langevin, P., 120
Laplace, P.-S., 208, 209, 211
Larmor precession, 85, 86, 88, 89
Laudan, L., 202, 203

Laudan-Leplin thesis, 202–203. *See also* empirical equivalence; observational equivalence; underdetermination
Leavens, C., 230 (20)
law of large numbers, 168, 262 (165)
Leibniz, G., 19
Lenzo, A., xv
Leplin, J., 202, 203
Lindsay, R., 121
locality, 2, 16, 175, 235 (85–87); Bell, 56; Howard, 57, 59, 60; and individuation, 18–19; Jarrett, 57, 58, 59; in Nelson's theory, 161; and understanding, 20, 22; versus separability, 57–59, 184–185, 259 (97). *See also* separability
local realism, 18–21, 185, 221 (29)
Lochak, G., 256 (23)
logical positivism. *See* positivism
logicism, 2
Lorentz, H., 11, 120
Lorentz covariance, 165; wave function and, 188–189. *See also* Bohm interpretation; quantum field theory

Mach, E., 101, 185
MacKinnon, E., 258 (68)
Madelung, E., 124, 125, 126, 127, 128, 135–136, 186; fluid, 158
many-body system, 62
Margenau, H., 154
marginal. *See* probability
materialism, 150, 153
matrix mechanics, 28, 98, 106, 107–112; as an abstract formalism, 114; and the challenge of wave mechanics, 113–118; equivalence with wave mechanics, 116, 249 (15); lack of success, 117
Maxwell, J. C., 19, 22, 220 (19)
Maxwell-Boltzmann distribution, 163, 187, 262 (168)
McCarthy, J., 152

McGlinn, W. D., 254 (79)
McMullin, E., xv, 269 (15)
measurement: as discovery, 47–52, 63–68; Nelson on, 161. *See also* Bohm interpretation; measurement problem
measurement problem, 17, 30, 34–38, 40–41, 227 (77–79). *See also* Bohm interpretation
metalevels, 1, 3, 4
methodological rules, 1, 2
Meyerson, É., 19, 20, 21, 104, 223 (59)
minimum mutilation, 7
mixed state. *See* state
mixing. *See* quantum equilibrium
Mugur-Schächter, M., 253 (66)
Muller, F., 222 (42)

Napoléon, xii
Nash, N., xv
naturalness, 206
Nelson, E., 124, 130, 165, 174; stochastic mechanics, 159–162, 169–170
Nernst, W., 98
nesting property, 172
neutron interferometry, 34, 78–82; beam characteristics of, 80; in Bohm's interpretation, 82–84; self-interference in, 80. *See also* polarization; spin flipping
Newcomb, S., 211
Newton, I., xii, 18, 19, 22, 208, 210
Newton's laws, 12, 13, 39, 44, 166; determinism and, 208–210. *See also* gravity
nonevidential criteria, 7–8
nonlocality, 18–20, 22, 235 (85–87); Bell, 59; benign, 21, 186; controllable, 57, 59; and the pilot wave, 129; uncontrollable, 58, 59, 179; various types of, 56–60, 176. *See also* Bohm interpretation; Einstein; locality; no-signaling theorem

nonseparability, 16, 32, 47, 183, 235 (85–87). *See also* separability
norms. *See* science
no-signaling theorem, 58, 59, 233 (75), 268 (109); in Bohm's theory, 163, 165; in a counterfactual scenario, 179, 183, 185, 186; Heisenberg on, 178; proof of, 195–196. *See also* quantum equilibrium

objective reality, 31, 52, 77, 180, 185. *See also* Einstein
observable, 27
observational equivalence, xiv, 5, 203; between the Bohm and Copenhagen theories, xi, 5, 53–55, 267 (96). *See also* empirical equivalence
one and the many, 18
ontic blurring, 16
ontology, 9, 32, 174, 203; in Bohm's theory, 52, 83, 165, 197; in the Copenhagen theory, 17, 26, 204; versus epistemology, 25, 30, 212. *See also* epistemology
operationalism, 28, 31, 114, 247 (104)
operator, 27, 44
Oppenheimer, J., 152, 157
orbit, 30
osmotic velocity, 160, 169
Ostwald, W., 185
outcome independence, 58, 59

parameter independence, 57, 59
Parmenides, 18
particularism, 17
path integral, 192
Pauli, W.: Balmer formula, 106, 117; and Bohm's theory, 149–151; *Dialectica* article, 145, 258 (56); on Einstein's realism/determinism, 256 (19); and electron spin, 117; exclusion principle, 109; and free will, 151; and God, 151; matrix/wave mechanics equivalence, 249 (15); objection to pilot wave, 119–120, 122–123, 128, 129, 149, 250 (55); and the old quantum theory, 109; on operationalism, 108–109; and Sommerfeld, 107; and statistical causality, 26, 151; on von Neumann's theorem, 131, 145; on wave-particle duality, 33
Pauli equation, 82, 83, 91–95, 190, 243 (65), 266 (86)
Pauli-Fierz correspondence, 150
Peres, A., 176
perturbation theory, 211
phase shifting, 79, 80, 90
phase space, 271 (46)
phase times, 74, 237 (117)
phase wave, 128, 137, 138
picturability, 125
pilot wave, 106, 118, 119, 126–128, 148, 149, 254 (74)
Planck, M., 7
Planck's law, 103
Plato, 18
Poincaré, H., 96, 97, 101, 211, 223 (58)
Poisson, S., 211
polarization, 78, 80–83, 87–91, 94
Polkinghorne, J., xv
positivism, 3, 104; and the Copenhagen interpretation, 26, 28–31, 110, 153. *See also* Bohr; Born; Heisenberg; operationalism; Pauli
Post, H., 5, 6
potentia, 17
pragmatic virtue, 21
predictive accuracy, 3
primitives, 18
principle of causality, 20, 21, 31, 223 (59)
principle of lawfulness, 20, 21
principle theory, 11, 12, 183
probability: conditional, 56, 189; density, 44; joint, 56, 188; marginal, 56, 188
professional dominance, 114, 117

Subject Index 315

projection postulate, 27, 31, 49. *See also* collapse; reduction
psychological factors, 100
Ptolemy, C., 15
pure state. *See* state

quantum chaos. *See* chaos
quantum chromodynamics, 228 (5)
quantum correlations. *See* EPRB correlations
quantum equilibrium, 229 (14), 262 (168), 269 (24); in Bohm-Vigier theory, 158; and cosmology, 262 (167); Lorentz invariance and, 189; and signal locality, 58, 59, 163, 165; via mixing, 70–72, 162–164, 261 (153, 155), 267 (96)
quantum field theory, 20. *See also* Bohm interpretation
quantum mechanics, 9, 12, 20, 24. *See also* Bohm interpretation; Copenhagen interpretation; formalism; interpretation; measurement problem
quantum particle, 16
quantum postulate, 25, 28, 32
quantum potential, 43, 47, 61, 175; in Madelung's theory, 126, 136; Rosen on, 130. *See also* super quantum potential
quantum realism, 17
quantum statistics, 186–188
quantum torque, 83, 94
Quine, W., 200, 201, 215, 260 (123), 271 (50)

randomness, 100
rational reconstruction, 207
Rauch, H., 79
realism, 104, 175; local, 18–20; understanding and, 219 (8). *See also* Bohm interpretation; scientific realism
Rechenberg, H., 248 (9)

reduction, 11, 17, 36, 51. *See also* collapse; projection postulate
reflection time, 54, 72, 75
refutation. *See* scientific theory
regulative principles, 11
Reichenbach, H., 98, 145, 146
relational holism, 17, 222 (38)
relativity, 24, 58; in Bohm's theory, 188–191; general, 13, 18, 39; special, 39. *See also* locality; nonlocality
Renn, J., 251 (25)
Renouvier, C., 96
rigid rotator, 239 (30)
Rohrlich, F., xv
Rosen, N., 112, 124, 129–130
Rosenfeld, L., 76, 77, 153, 154
Rosenkranz, Z., 251 (25)

scattering process, 49, 64
Schilpp, P. A., 108
Schmidt representation, 268 (108)
Schottky, W., 98
Schrödinger, E., 98, 105, 106; on determinism/indeterminism, 205–206; on entanglement, 178–179; and de Broglie's pilot wave, 118; and the diffusion equation, 130; and probabilistic laws, 270 (29); and a realistic interpretation, 124; and the wave concept, 105, 118. *See also* wave mechanics
Schrödinger cat paradox, 38–39, 227 (82)
Schrödinger equation, 27, 40, 61, 86; linearity of, 36, 41, 226 (74); for the super wave function, 197
Schulmann, R., 251 (25)
Schwarz inequality, 253 (62)
Schwinger, J., 20
science: convergence of, 4–5; goals of, 1–3, 10; simple model of, 3
scientific practice. *See* science
scientific realism, xi, 5, 16, 204–206, 270 (26); and understanding, 11,

scientific realism *(cont.)*
  219 (8). *See also* underdetermination
scientific theory, 9; refutation of, 7.
  See also formalism; interpretation
second quantization, 197
separability, 17, 59, 60, 235 (85–87);
  Howard, 58, 59, 60; and reality,
  265 (52); signaling and, 234 (83);
  versus locality, 57–59, 184–185,
  259 (97), 265 (52). *See also* locality; nonseparability
separation principle. *See* Einstein
Shimony, A., 57, 59, 179, 221 (31)
signal locality. *See* no-signaling
  theorem
signaling, 84; superluminal, 58. *See also* no-signaling theorem
simple model. *See* science
simplicity, 3, 7, 47
Slater, J., 109, 110, 121, 247 (115)
sociological factors, 2, 100, 112, 114
Solvay congress, 118–119, 130, 131,
  217 (8); and de Broglie, 106, 113,
  128; and Einstein, 177. *See also*
  Pauli
Sommerfeld, A., 98, 105, 107
space quantization, 109
space-time, 24, 28, 29, 31, 183; realism and, 265 (47). *See also* absolute space
spin: in Bohm's theory, 82–84, 190,
  238 (9), 240 (46), 266 (86); flipping, 79–80, 85–91
Spinoza, B., 101
spontaneous reduction, 17
Stapp, H., 27, 29, 176
state: in Bohm's theory, 59, 66, 166;
  entangled, 235 (88); pure versus
  mixed, 37, 87, 226 (73); separable,
  235 (88); vector, 27, 31
stationary-phase approximation, 236
  (100)
statistical causality, 26, 77, 151
statistical determinism, 26

statistical interpretation, 28, 31, 224
  (37). *See also* Born
statistical mechanics, 12, 14
Stern-Gerlach apparatus, 48, 82, 83,
  84, 230 (23); experiment, 103, 109
stochastic mechanics, 159, 162, 174,
  175. *See also* Nelson
sum over histories, 228 (1)
superposition, 38, 87, 90, 226 (74)
super quantum potential, 191, 267
  (93)
super wave function, 197
support, 66, 67, 167, 231 (31)

Tamm, I., 152
Terletskii, J., 152
Teller, P., xv
Temple, G., 125, 250 (9)
theory choice, 5–8
theory construction and selection, xii,
  xiv, 1–8
time, 54, 232 (56)
Tomonaga, S.-I., 20
trajectory, 25, 26, 31, 43, 47, 52
transcendental aesthetic, 18
transformation symmetry, 45, 77,
  175, 255 (3)
transmission time, 55, 72, 75
tunneling, 34; times, 54–55, 72–75,
  232 (58). *See also* Bohm interpretation
typicality, 168

uncertainty principle. *See* uncertainty
  relation
uncertainty relation, 29, 34, 99, 117;
  Bohr on, 25, 28; and information,
  231 (48); and no signaling, 163,
  165. *See also* Bohm interpretation;
  Heisenberg
uncontrollable nonlocality. *See* nonlocality
underdetermination, xi, xii, 7, 199–
  204, 207; realism and, 204–206.
  See also scientific realism

understandability, 21, 77, 174
understanding, 9–16, 20, 47, 78; and causal models, 20–22; realism and, 219 (8); of quantum phenomena, 16–17
unification, 11
uniqueness, 8. *See also* underdetermination
Urey, H., 121

vacuum, 159
Valentini, A., xv, 163, 164, 166, 167, 192
van Hove, L., 132
van Lunteren, F., 222 (50)
velocity potential, 135. *See also* guidance condition
Vigier, J.-P., 148, 152, 154, 158
vis a tergo, 19
visualization, 28, 29, 104, 263 (1)
Voltaire, 148
von Mises, R., 98
von Neumann, J., 119, 134
von Neumann's theorem, 131–134, 140–141, 145, 252 (50), 253 (66); counterexample to, 141–143. *See also* Jordan

wave mechanics, 32, 103–107; equivalence with matrix mechanics, 116, 249 (15); versus matrix mechanics, 113–118
wave-particle duality. *See* complementarity
Weimar milieu, 97, 99
weltanschauung, 101
Wessels, L., 244 (9), 252 (33)
Weyl, H., 98
Whewell, W., 3, 11, 13
"which-path" information, 34
Wien, W., 105
Wigner, E., 134
WKB (Wentzel-Kramers-Brillouin) approximation, 40, 228 (7)

Zanghi, N., 165, 166, 168, 169
Zeeman effect, 109
Zeilinger, A., 263 (9)
Zeno, 18